MILK and MILK PRODUCTS

Technology, chemistry and microbiology

Alan H. Varnam
Consultant Microbiologist
Southern Biological
Reading
UK

and

Jane P. Sutherland
Head of Food and Beverage Microbiology Section
AFRC Institute of Food Research
Reading
UK

An Aspen Publication®
Aspen Publishers, Inc.
Gaithersburg, Maryland
2001

The author has made every effort to ensure the accuracy of the information herein. However, appropriate information sources should be consulted, especially for new or unfamiliar procedures. It is the responsibility of every practitioner to evaluate the appropriateness of a particular opinion in the context of actual clinical situations and whit due considerations to new developments. The author, editors, and the publisher cannot be held responsible for any typographical or other errors found in this book.

Aspen Publishers, Inc., is not affiliated with the American Society of Parenteral and Enteral Nutrition.

Copyright © 1994, 2001 by Alan H. Varnam and Jane P. Sutherland

Originally published: New York: Chapman & Hall, 1994
Includes bibliographical references and index.
(Formerly published by Chapman & Hall, ISBN 0-412-45730-X) ISBN 0-8342-1955-7

Aspen Publishers, Inc., grants permission for photocopying for limited personal or internal use. This consent does not extend to other kinds of copying, such as copying for general distribution, for advertising or promotional purposes, for creating new collective works, or for resale. For information, address Aspen Publishers, Inc., Permissions Department, 200 Orchard Ridge Drive, Suite 200, Gaithersburg, Maryland 20878.

Orders: (800) 638-8437
Customer Service: (800) 234-1660

About Aspen Publishers • For more than 40 years, Aspen has been a leading professional publisher in a variety of disciplines. Aspen's vast information resources are available in both print and electronic formats. We are committed to providing the highest quality information available in the most appropriate format for our customers. Visit Aspen's Internet site for more information resources, directories, articles, and a searchable version of Aspen's full catalog, including the most recent publications: **www.aspenpublishers.com**
Aspen Publishers, Inc. • The hallmark of quality in publishing
Member of the worldwide Wolters Kluwer group

Editorial Services: Ruth Bloom
Library of Congress Catalog Card Number: 93-074881
ISBN: 0-8342-1955-7
Printed in the United States of America
2 3 4 5

MILK and MILK PRODUCTS

Contents

Preface		*vii*
A note on using the book		*ix*
Acknowledgements		*x*
1	**Introduction**	**1**
	1.1 The nature of milk	1
	1.2 Milk production as an activity of man	1
	1.3 Biosynthesis of milk	6
	1.4 The composition of milk	8
	1.5 The flavour and sensory properties of milk	27
	1.6 Potential hazardous substances in milk	28
	1.7 The microbiology of milk at farm level	33
2	**Liquid milk and liquid milk products**	**42**
	2.1 Introduction	42
	2.2 Technology	46
	2.3 Chemistry	76
	2.4 Microbiology	88
3	**Concentrated and dried milk products**	**103**
	3.1 Introduction	103
	3.2 Technology	104
	3.3 Chemistry	138
	3.4 Microbiology	149
4	**Dairy protein products**	**159**
	4.1 Introduction	159
	4.2 Technology	162
	4.3 Chemistry	179
	4.4 Microbiology	180

Contents

5 Cream and cream-based products — **183**
- 5.1 Introduction — 183
- 5.2 Technology — 183
- 5.3 Chemistry — 211
- 5.4 Microbiology — 216

6 Butter, margarine and spreads — **224**
- 6.1 Introduction — 224
- 6.2 Technology — 227
- 6.3 Chemistry — 257
- 6.4 Microbiology — 268

7 Cheese — **275**
- 7.1 Introduction — 275
- 7.2 Technology — 276
- 7.3 Chemistry — 321
- 7.4 Microbiology — 332

8 Fermented milks — **346**
- 8.1 Introduction — 346
- 8.2 Technology — 347
- 8.3 Chemistry — 370
- 8.4 Microbiology — 380

9 Ice cream and related products — **387**
- 9.1 Introduction — 387
- 9.2 Technology — 391
- 9.3 Chemistry — 420
- 9.4 Microbiology — 426

Bibliography — *432*

Index — *436*

Preface

Milk has been an important food for man since the domestication of cattle and the adoption of a pastoralist agriculture. It is also the most versatile of the animal-derived food commodities and is a component of the diet in many physical forms. In addition to milk itself, a rural technology evolved which permitted the manufacture of cheese, fermented milks, cream and butter. At a later date, successive advances in technology were exploited in the manufacture of ice cream, concentrated and dried milks and, at a later date, of ultra-heat-treated dairy products, new dairy desserts and new functional products. At the same time, however, dairy products have been increasingly perceived as unhealthy foods and a number of high quality dairy substitutes, or analogues, have been developed which have made significant inroads into the total dairy food market. Paradoxically, perhaps, the technology which, on the one hand, presents a threat to the dairy industry through making possible high quality substitutes offers, on the other hand, an opportunity to exploit new uses for milk and its components and to develop entirely new dairy products. Further, the development of products such as low fat dairy spreads has tended to blur the distinction between the dairy industry and its imitators and further broadened the range of knowledge required of dairy scientists and technologists.

One of the most striking features of the traditional dairy industry is the manner in which technology, chemistry and, subsequently, microbiology were integrated to allow the manufacture of high quality and safe products. It is considered that a similar integration of disciplines is required both to enable the dairy industry to meet the many future challenges, and also to permit the student of Food Science, Food Technology and related subject areas to gain a true knowledge of the nature of dairy products.

In writing this book, we have been very conscious of the requirements, not only of undergraduate and equivalent students, but of the new graduate entering industry and facing new and potentially frightening situations. To this end, the book is structured to meet the requirements both of the student, with a basic knowledge of chemistry, biochemistry and microbiology and of persons working in the dairy industry. The basic approach is to discuss the manufacturing process in the context of technology and its related chemistry and microbiology, followed by a more fundamental appraisal of the underlying science. The dairy industry is defined in a broad context and information is included on imitation products and analogues.

A number of innovations have been adopted in the presentation of the book. Information boxes and * points are used to place the text in a wider scientific and commercial context, and exercises are included in most chapters to encourage the reader to apply the knowledge gained from the book to unfamiliar situations. It is also our firm belief that the control of food manufacturing processes should be considered as an integral part of the technology and for this reason control points, based on the HACCP system, are included where appropriate.

A note on using the book

EXERCISES

Exercises are not intended to be treated like an examination question. Indeed in many cases there is no single correct, or incorrect, answer. The main intention is to encourage the reader in making the transition from an acquirer of knowledge to a user. In many cases the exercises are based on 'real' situations and many alternative solutions are possible. In some cases provision of a full solution will require reference to more specialist texts and 'starting points' are recommended.

CONTROL POINTS

In most chapters control points, derived from HACCP analysis, are included for the **main** processing stages. These are linked to process flow-diagrams. Points designated as CCP 1 are those which ensure elimination of a hazard; those designated CCP 2 are either points at which a hazard can be controlled but not eliminated *or* points which must be controlled to ensure satisfactory product quality. The inclusion of control points is intended to encourage a way of thinking in which control is an integral part of technology. If required, readers may use the control points included to further develop the HACCP approach for themselves.

Acknowledgements

The Authors wish to thank all who gave assistance in the writing of this book. Special appreciation is due to:

Debbie and Phil Andrews for providing, respectively, hand drawn and computer-generated illustrations. Dr G. Ellen, NIZO, The Netherlands and Dr A.I. Alvarez de Felipe, University of Leon, Spain for providing electron micrographs.

Those manufacturers of food processing equipment, food ingredients and laboratory equipment, who willingly provided information concerning the 'state-of-the art' (APV Pasilac Anhydro A/S, Copenhagen, Denmark; Cerestar SA/NV, Brussels, Belgium; Charm Sciences Inc., Malden, Massachusetts USA; Gadan UK Ltd, Wem, UK; Niro Atomizer A/S, Soeborg, Denmark; Oxford Instruments Ltd, Abingdon, UK; R.P. Texel Ltd, Stockport, UK; G.E.A. Wiegand GmbH, Ettlingen, Germany).

The libraries of the AFRC Institute of Food Research, Reading Laboratory and the University of Reading for their assistance in obtaining information.

Our colleagues in Reading and elsewhere for their help and interest during the preparation of the book.

1

INTRODUCTION

OBJECTIVES

After reading this chapter you should understand
- The nature of milk
- The importance of milk as an agricultural commodity
- The biosynthesis of milk
- The chemical constituents of milk and their relation to processing
- The nutritional importance of milk constituents
- Milk flavour
- Undesirable substances in milk
- The microbiology of milk at farm level

1.1 THE NATURE OF MILK

For young mammals, including human infants, milk is the first food ingested and, in most cases, it continues to be the sole constituent of the diet for a considerable period of time. Milk is a complex biological fluid, the composition and physical characteristics of which vary from species to species, reflecting the dietary needs of the young mammal. The major constituent of milk is water, but according to species milk contains varying quantities of lipids, proteins and carbohydrates which are synthesized within the mammary gland. Smaller quantities of minerals and other fat-soluble and water-soluble components derived directly from blood plasma, specific blood proteins and intermediates of mammary synthesis are also present (Table 1.1).

1.2 MILK PRODUCTION AS AN ACTIVITY OF MAN

Domestication of animals such as the cow and the availability of milk surplus to that required to feed the young, meant that animal milk became part of the adult human diet. Many animals are exploited to produce milk for human consumption; cows, goats, sheep, buffaloes,

Table 1.1 Average composition of cow's milk

Component	Percentage	Percentage of solids
Lactose	4.8	37.5
Fat	3.7	28.9
Protein	3.4	26.6
Non-protein nitrogen	0.19	1.5
Ash	0.7	5.5

camels and mares all forming the basis of commercial milk production in various parts of the world. In general, the dominant milk-producing animal in a region reflects the geographic and climatic conditions. Goats, for example, can be successfully farmed in mountainous regions with poor grazing, which would be quite unsuitable for cattle. In many parts of the world the cow is of overwhelming importance in milk production and in some countries, including the UK, milk of species other than the cow is not legally defined as 'milk'.

The discussion in this book refers to cows' milk unless other types are mentioned by name.

BOX 1.1 The friendly cow

The so-called 'cow culture' has come under severe attack by some environmentalists in recent years. The cow is considered to be an inefficient means of food production and to require land which could otherwise be used for direct production of human food. The cow has also been accused of being a major agent of global warming through production of rumen gases. With respect to food production, the anti-cow argument fails to take account of the ability of the rumen to synthesize nutrients from fibrous and cellulosic plant materials and from simple nitrogen sources such as urea, none of which play any direct role in human nutrition. Further it is simplistic to suggest that substitution of milk analogues made from soya would vitalize farming in developing countries. Quite apart from the disruptive effect on traditional agricultural patterns and the energy cost of transporting and processing soya, it must be recognized that farmers in the third world do not grow rich by supplying cash crops to western markets.

1

INTRODUCTION

OBJECTIVES

After reading this chapter you should understand
- The nature of milk
- The importance of milk as an agricultural commodity
- The biosynthesis of milk
- The chemical constituents of milk and their relation to processing
- The nutritional importance of milk constituents
- Milk flavour
- Undesirable substances in milk
- The microbiology of milk at farm level

1.1 THE NATURE OF MILK

For young mammals, including human infants, milk is the first food ingested and, in most cases, it continues to be the sole constituent of the diet for a considerable period of time. Milk is a complex biological fluid, the composition and physical characteristics of which vary from species to species, reflecting the dietary needs of the young mammal. The major constituent of milk is water, but according to species milk contains varying quantities of lipids, proteins and carbohydrates which are synthesized within the mammary gland. Smaller quantities of minerals and other fat-soluble and water-soluble components derived directly from blood plasma, specific blood proteins and intermediates of mammary synthesis are also present (Table 1.1).

1.2 MILK PRODUCTION AS AN ACTIVITY OF MAN

Domestication of animals such as the cow and the availability of milk surplus to that required to feed the young, meant that animal milk became part of the adult human diet. Many animals are exploited to produce milk for human consumption; cows, goats, sheep, buffaloes,

Table 1.1 Average composition of cow's milk

Component	Percentage	Percentage of solids
Lactose	4.8	37.5
Fat	3.7	28.9
Protein	3.4	26.6
Non-protein nitrogen	0.19	1.5
Ash	0.7	5.5

camels and mares all forming the basis of commercial milk production in various parts of the world. In general, the dominant milk-producing animal in a region reflects the geographic and climatic conditions. Goats, for example, can be successfully farmed in mountainous regions with poor grazing, which would be quite unsuitable for cattle. In many parts of the world the cow is of overwhelming importance in milk production and in some countries, including the UK, milk of species other than the cow is not legally defined as 'milk'.

The discussion in this book refers to cows' milk unless other types are mentioned by name.

BOX 1.1 **The friendly cow**

The so-called 'cow culture' has come under severe attack by some environmentalists in recent years. The cow is considered to be an inefficient means of food production and to require land which could otherwise be used for direct production of human food. The cow has also been accused of being a major agent of global warming through production of rumen gases. With respect to food production, the anti-cow argument fails to take account of the ability of the rumen to synthesize nutrients from fibrous and cellulosic plant materials and from simple nitrogen sources such as urea, none of which play any direct role in human nutrition. Further it is simplistic to suggest that substitution of milk analogues made from soya would vitalize farming in developing countries. Quite apart from the disruptive effect on traditional agricultural patterns and the energy cost of transporting and processing soya, it must be recognized that farmers in the third world do not grow rich by supplying cash crops to western markets.

1.2.1 The economic importance of milk production

The relative economic importance of milk production to a national or regional economy is largely dictated by the suitability of the area for grass production. Other factors can also be of importance including the extent of government intervention through subsidies or other economic mechanisms and the availability of export markets. In the UK, for example, milk is probably the most important single agricultural commodity and in the years up to 1991 milk sales accounted for more than 20% of total farm sales. The importance of milk in the UK agricultural economy has been attributed to the temperate climate, heavy rainfall and suitability for grass production as well as favourable governmental attitudes including heavy subsidies and guaranteed payments.

In recent years, however, the dairy industry in the UK and other member countries of the European Economic Community (EEC) have been affected by attempts to balance milk production with milk consumption. This has involved, since 1984, a complex system of quotas for each producer. The overall effect has been to reduce milk production within the EEC, but the economic consequences for milk-producing regions have been severe and have led to many farmers abandoning totally dairy farming, or adopting collaborative working practices with neighbours. In the UK the number of dairy cows, which had risen over the years 1965 to 1983, fell over the 5 years from 1984 to 1989 from 3 328 000 to 2 868 000. This was reflected in a loss of confidence amongst dairy farmers and a significant fall in land prices. Dairy farming tends to be concentrated in high-rainfall regions of the west and economic consequences extended beyond the producers. The closure of creameries following reduced milk production, for example, can cause considerable hardship in areas where little alternative employment is available.

BOX 1.2 **Three acres and a cow**

The milk quota system is often seen as an example of bureaucratic interference. For many years, however, the dairy industry and its structure have been the subject of political and economic controversy. During agitation for land reform during 1885, for example, it was envisaged that farm labourers could attain self-sufficiency if resources were distributed as three acres and a cow per family. The slogan 'Three acres and a cow' thus became an important part of land reform propaganda.

Introduction

The availability of export markets is of particular importance to countries such as Eire and New Zealand which have a large surplus of milk and which have developed large processing industries. A similar situation exists within the internal economy of the US where states such as Michigan are exporters of milk and dairy products to other parts of the country. Exports of dairy produce from New Zealand and Australia have been affected by the creation of the EEC and it has been necessary to develop alternative markets in the Far East and Middle East.

1.2.2 The structure of the dairy industry

The structure of the dairy industry varies widely from country to country. At one extreme the producer is selling surplus milk from one, or a small number of animals, while at the other extreme a single herd may comprise several hundred animals. In the developed world there has been a number of long-term trends. The number of producers have tended to fall, while at the same time the average herd size has increased. At the same time the average milk yield per cow has steadily increased. The increase in milk yield has resulted from improvements in cattle breeding, including herd improvement through artificial insemination schemes, improvements in nutrition and general herd management, including reduction in the incidence of mastitis. More recently the use of bovine somatotrophin (BST) as a means of increasing milk yield has been a cause of considerable controversy in a number of countries.

Increase in herd size and yield has been paralleled by increasing automation of milk production. Hand-milking has almost disappeared from commercial agriculture and machine-milking increasingly involves highly automated parlour systems. In some cases electronic 'tags' allow individual cows to be automatically identified as the parlour is approached, allowing automated feeding and recording of yield.

---

* Mastitis is inflammation of the udder. The usual cause is bacterial infection, although other factors including milking practices and type of housing may predispose animals to infection. Mastitis may be present in a clinical form, where visible changes may be seen in the milk or, more commonly, in sub-clinical form. Milk yield is reduced by mastitis while severe cases ('August bag') often result in permanent loss of milk production in the affected part of the udder and even in death. A large number of micro-organisms can cause mastitis, some of which are also human pathogens (see page 35–6). Mastitis due to specific mastitis pathogens, which are spread from udder to udder, may be controlled by disinfecting teats, maintaining the hygienic status of milking machines and by prophylactic and therapeutic intra-mammary administration of antibiotics to non-lactating cattle (dry cow therapy). In such circumstances mastitis due to organisms derived from the environment becomes of much greater significance.

Marketing systems for milk also vary widely. In less developed countries, all of the milk may be sold directly to the public, but in the major milk producing countries most milk is sold from the farm on a wholesale basis. The commercial relationships between producer and processor differ from country to country. In countries such as Eire and Australia many of the large-scale processors are owned by the farmers on a co-operative basis, while in the US farmers agree individual contracts with processors. Until recently, wholesale sales in the UK were made through the Milk Marketing Boards, but legislative changes will enable farmers to deal directly with processors.

A common feature to milk marketing arrangements in many countries is the principle of payment according to milk compositional quality. In the past, fat has been considered to be, commercially, the most important component of milk and most quality payment schemes have been based on fat content. More recently other components, especially protein, have increased in relative value and this is reflected in the introduction of protein and/or lactose into payment schemes. The use of compositional quality in determining price paid to the farmer may be illustrated by reference to the situation pertaining in England and Wales during July 1990 in which the producers' price was determined by multiplying the levels of fat, protein and lactose by assigned values for each constituent (Table 1.2).

The producers' price is, however, subject to adjustment. A seasonal price differential is applied to encourage a more even milk supply. This varies from −14.5% of the basic price for milk produced in May, to +30.3% for that produced in August. Milk produced in July has a seasonal differential of +21.6% and thus a basic producers' price of 17.187 pence per litre (ppl; Table 1.2) is increased to 20.899 ppl.

For many years, since the recognition of milk as a vehicle of human

Table 1.2 Calculation of milk prices on the basis of compositional quality, England and Wales, July 1990

Composition	(%)[1]		Assigned value (ppl per 1%)	Producer price (ppl)
Fat	3.95	x	2.141	8.456
Protein	3.21	x	2.254	7.235
Lactose	4.55	x	0.329	1.496
				17.187

[1]Assumed values for example only.

disease, there have been continuing efforts both to eradicate milkborne zoonotic diseases, such as tuberculosis and brucellosis, and to raise the overall standard of milk hygiene. Specific programmes were employed to eradicate zoonotic diseases by elimination of the causative organisms from the national herd (see page 38–9), but improvement of the overall standard of milk hygiene largely involved the education of milk producers, supported by laboratory testing to determine the microbiological status of the milk at farm level. In recent years, a number of countries have introduced a weighting into milk payment schemes according to the microbiological status of the milk. The underlying rationale is that all producers should be capable of meeting high standards as determined by total viable count (total bacterial count; TBC). The payment schemes tend, therefore, to be punitive, with only a small, or no, additional payment for achieving the highest hygienic category, but with increasingly severe financial penalties being applied with increasingly high bacterial counts. The schemes vary in detail (Table 1.3), more rigorous standards being applied where the general standards of hygiene are already high.

1.3 BIOSYNTHESIS OF MILK

Milk is the product of the mammary gland, that of the cow being illustrated in Figure 1.1. Milk originates in the secretory tissue and collects in a series of ducts, which increase in size as the milk moves

Table 1.3 Adjustments to milk payments according to microbiological status

England and Wales Milk Marketing Board		
Grade	Criteria[1]	Adjustment
A	$\not> 2 \times 10^4$	+0.23
B	$>2 \times 10^4$ but $<1 \times 10^5$	0
C_1	$>1 \times 10^5$, but no deduction in previous 6 months	-1.50[2]
C_2	$>1 \times 10^5$, and c_1 deduction in previous 6 months	-6.00[2]
C_3	$>1 \times 10^5$ and C_2 or C_3 deduction has been applied	-10.0
Aberdeen and District Milk Marketing Board		
	Average total bacterial count/ml	Adjustment
	$\leq 4.5 \times 10^4$	0
	4.6×10^4 to 9×10^4	-0.243
	9.1×10^4 to 1.3×10^5	-1.215
	$>1.3 \times 10^5$	-2.43

[1] Based on the average of at least two valid determinations of total bacterial count per millilitre in the previous month.

[2] Provided at least two results exceeded 1×10^5, otherwise milk qualifies for Grade B.

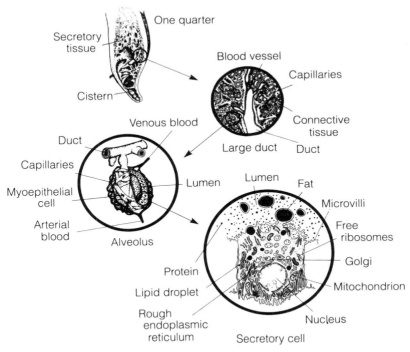

Figure 1.1 The bovine mammary gland. Redrawn with permission from Swaisgood, H.E. 1985. In *Food Chemistry*, 2nd edn (ed. Fennema, O.R.). Marcel Dekker, New York.

towards the teat. The alveolus may be considered as the smallest complete unit of milk production and is approximately spherical in shape with a central storage lumen surrounded by a single layer of secretory epithelial cells. Secretory cells are orientated so that the apical end, which has a unique membrane, is positioned adjacent to the lumen, while the basal end is separated by a basement membrane from blood and lymph. Metabolites enter the secretory cell from the bloodstream *via* the basement membrane and are utilized in milk synthesis by the endoplasmic reticulum, energy being supplied by mitochondria. The endoplasmic reticulum appears as a series of tubes, the cisternae, which empty into the Golgi apparatus. The Golgi apparatus is transformed into Golgi vesicles which transport the aqueous phase milk components to the apical plasma membrane. The Golgi vesicles then merge with the apical plasma membrane, fuse to become part of the membrane and discharge the aqueous phase into the lumen. In this process the inside of the vesicle membrane becomes the outside of the cell plasma membrane.

The lipid phase is also synthesized in the endoplasmic reticulum and collects as droplets on the cytoplasmic side of the membrane. The droplets move to the apical plasma membrane and pass into the lumen by pinocytosis. In this process the surface acquires a coating of plasma membrane which has important future consequences for the manufacturing properties of milk.

Synthesis is completed in the alveolar lumen where lactose is synthesized and proteins glycosylated and phosphorylated. Casein micelles appear both in the Golgi vesicles and in the lumen.

The secretory epithelial cells are surrounded by a layer of myoepithelial cells and blood capillaries. When the circulating pituitary hormone, oxytocin, is bound to the myoepithelial cells the alveolus contracts and the milk is expelled from the lumen into the duct system ('let down').

1.4 THE COMPOSITION OF MILK

1.4.1 The milk proteins

Milk proteins are of two distinct types, whey proteins (serum proteins) and caseins. Caseins constitute over 80% of the total protein of milk, although the relative proportion of whey proteins to casein varies according to the stage of lactation. Milk produced in the first few days after calving and towards the end of lactation is of substantially higher whey protein content than that produced in mid-lactation. This increase is accompanied by elevated levels of blood serum proteins.

(a) The caseins

The caseins of milk may be subdivided into five main classes, α_{s1}-, α_{s2}-, β-, γ- and κ-caseins (Table 1.4). Of these all except γ-casein are mammary gland gene products, γ-casein resulting from post-translational proteolyis of β-casein. This may be due either to native milk proteinases, primarily plasmin, or to the proteolytic activity of bacteria. The relative proportions of α-, β- and γ-caseins are subject to genetic variation within individual herds and there can be significant differences in casein composition of milk from different cows. The casein composition of bulk milk, however, varies very little at any single stage in lactation.

The caseins are globular proteins and have an amino acid content similar

Table 1.4 The caseins of milk

Fraction	Molecular weight[1]	Phosphoserine residues
Alpha$_{s1}$	23 000	7–9
Alpha$_{s2}$	25 000	10–13
Beta	24 000	5
Gamma	11 600–20 500	0 or 1
Kappa	1980	1[2]

[1] Molecular weight of monomer.
[2] Only carbohydrate containing casein.

to that of other types, although cysteine is present, in small quantities, in only α_{s2}- and κ-casein. An unusual feature of caseins is the post-ribosomal modification which results in the phosphorylation of the hydroxyl groups of serine. The phosphoserine residues are involved in providing caseins with their unique properties.

Alpha$_{s1}$-casein contains seven to nine phosphoserine residues per mole, α_{s2}, 10 to 13 and β-, five. Phosphoserine residues are concentrated in clusters and are responsible for the existence of hydrophilic areas of strong negative charge. The molecules also contain blocks of hydrophobic residues. Beta-casein contains the most hydrophobic component and forms aggregates with the N-terminal hydrophilic parts exposed to solvent and hydrophobic parts in the interior.

Alpha$_s$-caseins are sensitive to calcium due to the presence of phosphate groups and precipitate in the presence of Ca^{2+} ions at a pH value of 7.0. The polypeptide chains contain 8.5% proline which restricts the extent of α-helix formation.

Kappa-casein differs from α- and β-casein in having only one phosphoserine group and in containing a charged oligosaccharide moiety. The κ-casein molecule appears to consist of a relatively stable, single disulphide bonded structure within which are both α-helical and β-sheet regions. The chymosin-sensitive Phe_{105}–Met_{106} bond is thought to protrude from the molecular surface. One third of the κ-casein molecule is represented by the strongly ionic C-terminal section, which contains the three oligosaccharide residues. The remainder of the molecule is highly hydrophobic and corresponds to the *para*-κ-casein formed after hydrolysis of the Phe–Met bond (see Chapter 7, page 322–3).

The ampiphilic nature of caseins and their phosphorylation facilitate interactions with each other and with calcium phosphate to form highly hydrated spherical complexes known as micelles. These are of varying size, the diameter varying from 30 to 300 nm (mean diameter *ca.* 120 nm, molecular weight 10^8). The protein content of micelles is 92%, composed of α_{s1}-, α_{s2}-, β- and κ-caseins in an average ratio of 3:1:3:1. The remaining 8% is composed of inorganic constituents, primarily colloidal calcium phosphate. This is believed to be in the form of amorphous tertiary calcium phosphate distributed throughout the micelle and the presence of calcium ions is an absolute essential for micelle formation. Transformation to more stable forms, such as hydroxyapatite, is prevented by the presence of other ions, especially magnesium. Phosphate also appears to be essential in the formation of temperature-stable micelles.

A number of models, often contradictory, have been proposed for the casein micelle. The most satisfactory envisages the casein micelle consisting of an aggregate of almost spherical sub-micelles, which in turn consist of more limited aggregates of casein molecules (Figure 1.2). The calcium phosphate and α_s- and β-caseins are linked by the involvement of the phosphoserine residues in the structure of the calcium phosphate. Kappa-casein is localized on, or very close to, the surface of the casein micelle. The hydrophobic part of the κ-casein molecule is bound to the core of the micelle, while the hydrophilic macropeptide forms a layer of highly hydrated 'hairs', which project into the aqueous phase. Kappa-casein hairs are responsible for the steric stabilization of casein micelles.

Components of the casein micelle appear to be in equilibrium with the

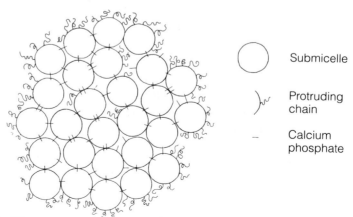

Figure 1.2 The casein micelle: schematic diagram.

aqueous phase, and while freshly secreted milk contains only small quantities of soluble casein, some casein together with colloidal calcium phosphate dissociates from the micelle during storage at *ca.* 0°C. Beta-casein is primarily involved and as much as 40% can enter solution. Relatively little α_s- or κ-casein dissociates and for this reason it has been postulated that these two components have the major structural role. Dissociation is reversible, but it is not known whether, or not, the native structure is re-attained.

(b) Whey proteins

Whey proteins comprise two gene products, β-lactoglobulins and α-lactalbumins, proteose-peptones (partly derived from hydrolysis of β-casein) and small quantities of the blood-derived proteins, serum albumin and immunoglobulins (Table 1.5).

In structure, whey proteins are typical compact globular proteins, with a relatively uniform sequence distribution of nonpolar, polar and charged residues. The proteins undergo intramolecular folding as a result of the formation of disulphide bonds between cysteinyl residues, which buries most of the hydrophobic residues in the interior of the molecule. For this reason whey proteins do not aggregate strongly, or interact with other proteins, in the native state.

The major whey protein, β-lactoglobulin, undergoes limited self-association, at milk pH values, to form a dimer with a geometry resembling two impinging spheres. The dimer dissociates in solution at 60°C, thus becoming susceptible to denaturation by unfolding of the tertiary structure. Alpha-lactalbumin has a homologous primary structure, resembling that of lysozyme, and has a highly compact, virtually spherical shape. The molecule is of greater heat stability than that of β-lactoglobulin.

Table 1.5 The whey proteins of milk

Fraction	Molecular weight[1]
Beta-lactoglobulin	18300
Alpha-lactalbumin	14000
Serum albumin	63000
Immunoglobulins	up to 1000000

[1]Molecular weight of monomer.

(c) Effect of heat on milk proteins

Casein micelles are remarkably stable at temperatures up to 140°C. In contrast, whey proteins are relatively heat-labile, extensive denaturation occurring at 80°C. Denaturation is accompanied by extensive breaking and randomization of the stabilizing disulphide bonds. Beta-lactoglobulin is more heat-labile than α-lactalbumin as a consequence of its one free sulphydryl group, which permits the initiation of autocatalytic disulphide exchange reactions.

Denaturation of β-lactoglobulin has further important consequences, since at 100°C and higher temperatures, interaction occurs between β-lactoglobulin and κ-casein. Interaction probably involves a thio-disulphide interchange, and while κ-casein remains on the surface, the micellar surface properties are altered. These changes affect the interaction of the micelles with calcium phosphate and, in turn, their stability. Preheating unconcentrated milk at 90°C, for example, reduces its stability during further processing at higher temperatures. In contrast, the stability of concentrated milk is increased by heating at 90°C. This probably results from the heat-induced reduction in calcium ion concentration, which is very marked in concentrated milk, outweighing the destabilizing effect of micellar reaction with β-lactoglobulin.

Continued heating leads to further changes including a general heat-induced proteolysis, formation of lysinoalanine and Maillard browning. Heating for periods in excess of 20 min at 140°C leads to destabilization of casein micelles and gel formation.

(d) Milk proteins and the behaviour of milk during processing

The properties of milk proteins relate to the behaviour of milk during processing both through the specific nature of the casein micelle and through the amino acid sequences which ultimately determine functional properties.

The most important reactions of the milk proteins are those which involve destabilization of the protein micelles. In some cases these are

---

* Alpha-lactalbumin serves an important function as a modifier during biosynthesis of lactose and the protein is present in all lactose-containing milks. Alpha-lactalbumin is bound to the enzyme galactosyl transferase and it has been suggested that its mode of action is to bring its monosaccharide binding site (specificity for glucose) into proximity to the monosaccharide binding site of the enzyme.

technologically desirable reactions, such as the formation of a gel either when the pH of milk is reduced (e.g. manufacture of fermented milks and acid-set cheese), or when κ-casein undergoes selective proteolysis (e.g. manufacture of Cheddar cheese). Under the correct conditions acidification may also be used to fractionate the milk proteins (e.g. manufacture of acid casein). In other cases, however, reactions involving destabilization of micelles are technologically undesirable. Examples include the various reactions involving aggregation of casein which occur during age thickening of concentrated milks.

The functional properties of both casein and whey reflect the sequence of amino acids and thus the conformational structure of the proteins. In the case of casein, the ampiphilic nature of the molecule, with the amino acids falling into hydrophilic and hydrophobic domains, imparts extremely good surface active properties and thus the functional properties of whipping/foaming and emulsification. These properties are, however, relatively dependent on pH value.

Whey proteins are not ampiphilic in nature and are generally of lower surface activity than casein. Foam stabilizing properties are, however, superior since a more rigid film is formed at the air/water interface.

(e) Nutritional properties of milk proteins

Milk protein is of very high nutritional value and both caseins and whey score very highly when compared to the United Nations Food and Agricultural Organization provisional scoring pattern for quality. The two types of milk protein are complementary with respect to content of essential amino acids and this obviates the slight deficiency in the sulphur amino acids, methionine and cysteine, in casein. Milk protein is also highly digestible, although digestibility can be reduced by processing. Processing can also lead to nutrient loss (see subsequent chapters).

1.4.2 Milk fat

Milk fat is generally regarded as being of complex composition. Triacylglycerols are dominant and constitute *ca.* 98% of milk fat, together with small amounts of di- and monoacylglycerols and free fatty acids. Measurable quantities of phospholipids, cholesterol and cholesterol esters and cerebrosides are also present (Table 1.6). Other ingredients are present only in very small quantities but may be of importance in determining the organoleptic character and nutritional status of milk. These are the fat soluble vitamins, mainly A, D and E, together with small

Table 1.6 Average lipid composition of milk

Lipid	% by weight
Triacylglycerols	97–98
Diacylglycerols	0.3–0.6
Monoacylglycerols	0.02–0.04
Free fatty acids	0.1–0.4
Free sterols	0.2–0.4
Sterol esters	trace only
Phospholipids	0.2–1.0
Hydrocarbons	trace only

quantities of vitamin K, flavour components, identified as aldehydes, ketones and lactones (see page 27) and carotenoid pigments.

(a) The fat globule

Lipid molecules in milk associate to form large spherical globules, which are surrounded by a phospholipid-rich layer, the milk fat globule membrane, derived during secretion through the apical plasma membrane (see page 8). The globules range in diameter from *ca.* 1 μm to *ca.* 12 μm, with a mean diameter of *ca.* 3 μm. The mean diameter is related to fat content of the milk and is greater in high-fat milk.

The milk fat globule membrane stabilizes the hydrophobic lipid in the aqueous environment of milk. Approximately 60% of milk phospholipid and 85% of milk cholesterol are located within the membrane, which also contains high concentrations of milk enzymes such as alkaline phosphatase and xanthine oxidase. The lipid composition of the milk fat globule membrane is similar to that of the plasma membrane, although composition changes during the ageing of milk. The membrane is often considered to be continuous over the entire globule surface, but there may be areas where the surface consists of components of cellular cytoplasm adsorbed to the fat before secretion.

The relationship between the fat globules and the aqueous phase of milk may change in a number of ways during storage, or processing, of milk (Figure 1.3). Creaming, for example, imparts the characteristic 'cream-line' appearance to non-homogenized bottled milk, but is undesirable in some products due to formation of a cream plug. Creaming is a consequence of the lower density of the fat phase and this density difference enables milk fat to be separated from milk by centrifugal centrifugation. Creaming is more rapid than would be expected from

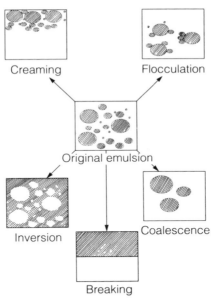

Figure 1.3 Possible changes to emulsions during processing and storage. Redrawn with permission from Fillery-Travis, A. *et al.* 1990. *Food Science and Technology Today*, 4, 89–93.

theoretical considerations, probably because of the clumping of globules under the influence of immunoglobulins. Homogenization (see Chapter 2, page 53) is used to reduce the average diameter of the fat globule to below 0.8 μm, so that the dispersive force of Brownian movement overcomes the tendency to float. The larger surface area of homogenized fat globules means that insufficient milk fat globule membrane is present for stability, but milk proteins, primarily the caseins, bind to the exposed fat and act as emulsifying agents to stabilize the globules. More drastic physical treatment, or thermal shock, can rupture the membrane, releasing free fat and 'breaking' the emulsion. This process is exploited in buttermaking (see Chapter 6, page 234).

(b) Fatty acid composition of milk fat triacylglycerols

The fatty acid composition of milk fat triacylglycerols is extremely complex and several hundred have been identified. It is, perhaps, fortunate for the understanding of milk fat that 15 fatty acids account for more than 95% of those present in triacylglycerols, the great majority being present only in very small quantities. All bovine milk fats comprise the same spectrum of 8 saturated fatty acids containing an even number of carbon atoms (4:0, 6:0, 8:0, 10:0, 12:0, 14:0, 16:0 and 18:0); two odd-numbered saturated acids (15:0 and 17:0); dienes and trienes (18:2 and 18:3) and three monounsaturated fatty acids (14:1, 16:1 and 18:1). The proportion of each fatty acid varies according to the stage of lactation and diet and it is possible to manipulate the relative proportion of the main fatty acids by alterations to the diet (see pages 18–19).

Under normal conditions saturated fatty acids account for *ca.* 70% of the total fatty acid content (by weight), monounsaturates account for *ca.* 27% and dienes and trienes for only *ca.* 3%.

(c) Distribution of triacylglycerols

Triacylglycerols containing an even number of carbon atoms are dominant in milk fat. This reflects the fatty acid composition, although the proportion of odd-numbered triacylglycerols is *ca.* 3 times that of the fatty acids.

The distribution of triacylglycerols is also a consequence of the fatty acid composition. Milk fat contains a range from a CN of less than 26 up to 54. Milk fat is not constructed on a random basis, fatty acids 4:0 and 6:0 being found on the glycerol moiety virtually exclusively at position $S_N 3$. For this reason, 40–50% (numerically) of the triacylglycerols are of CN < 42.

(d) Melting and crystallization of milk fat triacylglycerols

The melting and crystallization properties of milk fat are of major importance in determining the physical attributes of high-fat dairy products such as butter and ice cream. Melting behaviour is complex as a result of the complex nature of the fatty acid composition of the triacylglycerols. Melting commences at *ca.* −40°C and is complete at +40°C. At any intermediate temperature, both liquid and solid fat is present; the ratio of solid to liquid, at any given temperature, largely

---

* Medical concern over the adverse effect of consuming large quantities of dairy fat, usually stems from the high level of saturation. The monounsaturated fatty acid content, however, is dominated by 18:1 and it has been argued that the relatively high concentration of this fatty acid means that milk fat should more correctly be regarded as lacking in polyunsaturation (International Dairy Federation, 1991).

* The nomenclature of triacylglycerols is defined by reference to the number of carbon atoms present in the fatty acid moiety of the molecule. This is usually referred to as the carbon number (CN).

determining the rheological properties of the fat.

Crystallization commences as the temperature of liquid fat is reduced. The first stage, nucleation, requires a considerable degree of supercooling and, when fat is cooled in bulk, occurs at the surface of impurities (heterogeneous nucleation). The second stage, crystal growth, tends to be slow because of competitive inhibition, a process whereby a triacylglycerol not sufficiently similar to those already present is rejected at the surface of the growing crystal. Although significant in terms of slowing crystal growth, competitive inhibition is of less importance in milk fats than other crystal systems due to the general level of similarity between the triacylglycerol species. The formation of mixed crystals, containing more than one triacylglycerol species is common and the melting behaviour of milk fat has been explained in terms of the formation of a series of mixed crystals with extensively overlapping melting ranges.

Crystals of milk fat have a strong tendency to polymophism. Three, and possibly more, crystalline forms of fat exist and are of practical significance, α, β' and β. The α-form is least stable and tends to convert to the β' within a few minutes. In contrast, the β'-form is relatively stable in milk fat and further transformation to the β-form is a lengthy process which can take as long as several months depending on the nature of the triacylglycerols present.

The crystal structure of the fat describes the way in which a three-dimensional network develops from the individual crystals. To a great extent the solid and liquid fats may be considered to form a continuous phase and the crystal structure is of major importance in determining the plasticity of the fat.

It is possible to differentiate between a primary and secondary lipid structure. The primary structure is that which develops when the crystals form in the absence of agitation. Such a structure is of considerable strength but, if broken by application of mechanical force, does not readily re-form. In many instances the primary structure is replaced by a significantly weaker secondary structure, which readily re-forms after breakage. A large proportion of the structure is in

---

* In practice acylglycerols fall into three groups with respect to melting point.
1. High melting acylglycerols: melt at 20–40°C, comprise 10–15% total.
2. Middle (medium) melting acylglycerols: melt at 0–20°C, comprise 30–45% total.
3. Low melting acylglycerols: melt at below 0°C, comprise 30–55% total.

secondary form when crystallization occurs under conditions of vigorous agitation.

(e) Phospholipids

Phospholipids are primarily present in the milk fat globule membrane. The fatty acids of phospholipids differ in origin from those of other milk fats, a higher proportion tends to be unsaturated and chain lengths are longer.

(f) Effect of diet of cattle on composition of milk fat

The ability to modify the composition, and therefore the properties, of milk fat is of considerable potential importance. Modification of the physical properties permits the manufacture of products such as spreadable butter, while increasing the oleic acid content and reducing that of palmitic and myristic acids minimizes the increase in blood cholesterol level associated with consumption of milk fat.

The fatty acids of milk fat may be divided into three groups according to their origin.

1. Fatty acids arising only through *de novo* synthesis in the bovine mammary gland (4:0, 6:0, 8:0, 10:0, 12:0, 14:0, 14:1).
2. Fatty acids arising both through *de novo* synthesis in the mammary gland and from the blood supply (16:0, 16:1).
3. Fatty acids arising only from the blood supply (18:0, 18:1, 18:2, 18:3). Conversion of 18:0 to 18:1 occurs in the mammary gland through the activity of a specific desaturase.

Increasing the quantity of long-chain fatty acids reaching the mammary gland from feed, *via* the bloodstream, decreases the quantity derived by *de novo* synthesis and thus alters the composition of the milk fat.

Dietary modification of milk fat requires the feeding of concentrate in addition to grass and is most readily achieved during the indoor feeding system when grass intake is at a minimum. The addition of soya, or oats, to the diet, for example, results in milk fat containing up to 60% of C_{18} acids, with *cis* and *trans* 18:1 being dominant. Milk fat containing a high proportion of unsaturated C_{18} fatty acids produces a butter with similar spreading properties to margarine (see page 243).

An alternative means of dietary modification of milk fat is to produce a

The composition of milk

fat rich in 18:2 or 18:3 by protecting these fatty acids from hardening by rumen micro-organisms. This may be achieved by encapsulating unsaturated fat in sodium caseinate which is subsequently cross-linked by exposure to formaldehyde. The preparation passes through the rumen intact, but the cross-linkage is destroyed in the acid environment of the abomasum, enabling the digestion of the encapsulated fat. Under most conditions, 20–25% of the protected fat appears in milk and there is usually an increase in the total fat yield. Suitable protected feed has been available in Australia, but expense prevented wide-scale adoption and the product has been withdrawn from the market.

(g) Chemical changes during storage of milk fats

The importance of the fat phase in determining the flavour of many dairy products means that the rate and effect of chemical changes during storage are of major significance with respect to storage life. Two major modes of deteriorative change are involved, oxidation and lipolysis.

Oxidation is the major cause of chemical change and thus of major significance with respect to shelf life of many dairy products. Peroxides are formed by reaction between unsaturated fatty acid esters and oxygen. Peroxides have no influence on flavours, but readily decompose to yield carbonyl compounds which are the source of 'oxidized' flavours and detectable at very low levels, especially in bland products of high fat content such as some types of cream and butter.

Milk fat is relatively stable with respect to oxidation due to the high level of saturated fatty acids and the presence of natural antioxidants such as tocopherols. High-heat treatments involved in the manufacture of dairy products such as ultra-heat-treated (UHT) milk enhance the stability. The presence of pro-oxidants, especially copper and iron, however, accelerates oxidation which also occurs at a higher rate during storage at high temperature or at high concentrations of dissolved oxygen.

Lipolytic rancidity, the release of free fatty acids following hydrolysis of the triacylglycerol molecule is a lesser, but still a potentially significant, problem. Lipases are responsible and may be derived from milk or be produced by contaminating micro-organisms. Heat-stable enzymes of microbial origin are a major cause of lipolytic rancidity in processed milk products and improvements in raw milk hygienic quality have generally led to fewer problems. Despite this, the long storage lives now common with heat treated milk, especially UHT, does mean that problems may occur due to the heat-stable lipolytic enzymes derived from psychrotrophic bacteria (see page 84).

(h) Reduction of the cholesterol content of milk fat

In recent years much interest has been shown in the development of 'cholesterol-free' or 'low-cholesterol' dairy products. It should, however, be appreciated that in practice it is not possible to produce milk which is truly 'cholesterol-free' since skim milk contains residual small fat globules which are disproportionately rich in cholesterol, skim milk of 0.2% fat content containing 4.5 mg per serving. In the US, a level of below 2 mg per serving is proposed for milk designated 'cholesterol-free' and 2–20 mg per serving for milk designated 'low-cholesterol'.

At present two methods of reducing the cholesterol content of milk have been used, supercritical fluid extraction and steam distillation. In each case cholesterol is extracted from butter oil which is then blended with untreated skim milk (Figure 1.4). The use of bacterial enzymes to degrade cholesterol has also been examined but most cholesterol

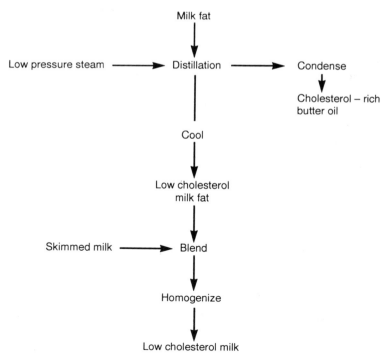

Figure 1.4 Manufacture of low-cholesterol milk by distillation.

dissimilating bacteria accumulate steroid intermediates. An exception is *Rhodococcus equi* and extracts of this organism have been successfully used to reduce the cholesterol content of milk on an experimental scale.

(i) Milk fat in the diet of man

The presence of fat is an important factor in determining the palatability of food. Dairy fats are particularly palatable due to their containing a large number of lipids of small molecular size, short chain fatty acids and their derivatives which contribute to flavour, aroma and, in the case of lipids, 'mouth feel'.

Triacylglycerols are an important energy source, yielding twice the energy per gram of carbohydrates. The energy value of short and medium chain fatty acids, which constitute *ca.* 12% of the total fatty acids in cow's milk is, however, relatively low.

Polyunsaturated fatty acids are metabolized in the rumen and milk is therefore a relatively poor source of the essential fatty acids linoleic and linolenic. Milk fats are, however, consumed in large quantities and contribute *ca.* 21 g per day to the average UK diet. This quantity contains 0.3 g linoleic acid out of a daily requirement of 5–10 g.

Milk fat contains substantial quantities of the vitamin A precursor β-carotene and, in the UK, provides *ca* 25% of the daily intake. Milk fat is also a minor, but significant, source of vitamin D, especially for pregnant and lactating women and children who have a high requirement.

Despite the important role of milk fats in human nutrition, high fat foods have come to be widely regarded in developed countries as being implicitly unhealthy. In many ways attitudes to dietary fats illustrate the very different problems faced by the developing and the developed nations. In the developing world the major problem is a shortage of dietary energy, which could be at least alleviated by increasing dietary fat intake. In contrast a major preoccupation in the developed world is the high intake of fats and the role, especially in the case of milk fat, in heart disease.

The unhealthy image associated with dairy products stems from three factors: the high fat content of whole milk and whole milk products, the cholesterol content and the saturated nature of the fatty acids.

A high dietary fat intake is associated with obesity and with hyperlipi-

Introduction

daemia (excessively high levels of blood lipoproteins). Although direct evidence is lacking, hyperlipidaemia is recognized as a risk factor predisposing to heart disease. For this reason preventive health strategies in a number of countries include as a major part reduction in fat consumption. In some cases the dairy industry has been able to respond to this requirement by the introduction of products such as skim milk and low-fat yoghurt, etc.

Both dietary cholesterol and saturated fatty acids have been associated with raised serum cholesterol levels, which have been, and still are, regarded as a risk factor for cardiovascular disease. There has, however, been criticism of some of the experimental work and doubt over what 'normal' and 'elevated' levels actually are. Further, the effect of milk fat on serum cholesterol levels has often been investigated in isolation from other dietary factors. Milk fat alone, for example, can cause an increase in serum cholesterol levels, whereas whole milk can result in a slight reduction.

The whole situation with respect to milk fat and disease is currently confused by conflicting claims and counter-claims with both the dairy industry and its competitors making use of highly selective scientific evidence. One of the major difficulties is that the 'debate' has tended to be centred on the properties of single dietary components and to ignore the importance of the overall diet and, indeed, lifestyle. There is no doubt that many individuals would benefit from a reduced level of fat intake, but other factors including greater exercise, reduction of smoking and alcohol consumption and removal of stress are likely to be of equal, or greater, importance.

1.4.3 Lactose

Lactose is the major solid constituent of milk. The concentration varies with yield between 4.2 and 5.0%, the lactose content usually being lowest in late lactation milk, or in milk from animals suffering from udder disease.

Lactose is a disaccharide and comprises α-D-glucose and β-D-galactose molecules. Three solid forms of lactose exist, α-lactose monohydride and anhydrous α- and β-lactose. The β-form is of markedly higher solubility but, through mutarotation, an equilibrium mixture of the two forms exists in solution. Lactose is one of the least soluble of the common sugars, having a solubility in water of only 17.8% at 25°C. This low solubility has consequences during the production of concentrated milk and frozen dairy products and it is sometimes necessary to induce crystallization to produce a large number of small crystals and thus avoid the defect 'sandiness'. The α-hydrate crystal form, which is most commonly formed, has a large number of shapes, of which the characteristic 'tomahawk' form is the main cause of 'sandiness'.

Lactose makes a major contribution to the colligative properties of milk; osmotic pressure, freezing point depression and boiling point elevation. Lactose, for example, accounts for *ca.* 50% of the osmotic pressure of milk. Changes in the lactose content of milk are associated with reciprocal changes in the content of other water-soluble constituents, especially sodium and chloride.

Lactose finds use as a food ingredient due to its protein stabilizing properties and low relative sweetness. Lactose may also be used as a partial replacer for sucrose in icings and toppings to improve mouth-feel without excess sweetness. It is also added to bakery products such as biscuits to impart a controlled degree of Maillard browning, a reaction considered undesirable in many food products.

Lactose is a significant source of dietary energy and may promote calcium absorption. Use of lactose as an energy source is limited, however, by the relatively high percentage of lactase-deficient persons. Lactase deficiency is most common amongst persons of African or Asian origin, but can affect members of any racial group. The degree of lactase deficiency varies and symptoms vary accordingly, from inability to digest even small quantities of dairy products to mild gastrointestinal upsets after consuming large quantities of lactose-containing food. Commercial processes for hydrolysis of lactose in milk and other dairy products have been developed in response to this problem (see Chapter 2, page 71).

1.4.4 Minerals

The minerals in milk consist principally of the bicarbonates, chlorides, citrates and bicarbonates of calcium, magnesium, potassium and sodium. All of the minerals are distributed between a soluble phase and a colloidal phase and while monovalent ions exist largely, or totally, in the soluble phase, as much as 66% of the calcium and 55% of the phosphorous may be in the colloidal phase (Table 1.7). The distribution of calcium, citrate, magnesium and phosphate between soluble and colloidal phases and their interactions with milk proteins have important consequences for the stability of milk and milk products.

Table 1.7 Distribution of milk minerals between the colloidal and aqueous phase

	Percentage	
	Soluble phase	Colloidal phase
Calcium total	33	67
Calcium (ionized)	100	0
Chloride	100	0
Citrate	94	6
Magnesium	67	33
Phosphorus (total)	45	55
Phosphorus (inorganic)	54	46
Potassium	93	7
Sodium	94	6

The role of the major milk salt, colloidal calcium phosphate in maintaining the integrity of casein micelles has been discussed earlier (see page 10). Calcium ions also bind to some caseins and complexes strongly with citrate. The soluble (non-micellar) calcium concentration in milk correlates closely with the soluble citrate concentration and the majority of the citrate exists as a complex with calcium ($CaCit^-$). Approximately half of the inorganic phosphate exists as colloidal calcium phosphate and only a small quantity (*ca.* 6.6%) of the calcium is present as soluble Ca^{2+} ion. Citrate also complexes much of the magnesium.

The citrate concentration in milk varies according to season and the diet of the cow, citrate concentration affecting, in turn, the soluble calcium content and milk stability. This has consequences for milk processing and may require the addition of anions to complex ionic calcium, to reduce that available for binding to casein and stabilize micelles against aggregation.

In addition to the importance of minerals in the stability of casein, the monovalent ions, together with lactose and other low molecular weight components, maintain the osmotic pressure at a value iso-osmotic with

---

* The total citrate content of milk is controlled by the effect of diet on *de novo* synthesis of fatty acids in the mammary gland. Citrate levels rise when *de novo* synthesis is inhibited, but are low during active *de novo* synthesis. Grass-fed cattle have low levels of citrate in milk since *de novo* synthesis is required to compensate for the low dietary intake. In contrast concentrated feed contains high levels of preformed fatty acids, obviating the need for *de novo* synthesis and citrate levels in milk are high. Seasonal variation in citrate reflects feeding practice, summer milk being produced by grass-fed cows, while winter milk is produced by cows fed on a high-concentrate diet.

that of blood. Changes in the concentration of milk salts or lactose are compensated for by reciprocal changes in the other component and the osmotic pressure of milk remains constant within narrow limits. The freezing point of milk also remains essentially constant (−0.53 to −0.57°C) and a freezing point depression of less than 0.525°C is indicative of adulteration of the milk with water.

Milk is an important source of dietary calcium and it has been suggested that the association with caseins may improve absorption in the gastrointestinal tract. Calcium is a key factor in determining healthy bone and tooth development in the young and an adequate intake is essential. Calcium status may also be a factor in the development of post-menopausal osteoporosis in women and the consumption of milk and dairy products (including calcium-enriched dairy products) has been recommended in this situation. It is doubted, however, that a high calcium intake in later life can reverse the deterioration in bones that is a predisposing factor to osteoporosis.

1.4.5 Minor components and micronutrients

Milk contains a large number of minor components, some of which are of relatively high biological or chemical activity and which may, therefore, affect the properties of milk.

(a) Urea

Urea is an important factor in determining stability of unconcentrated milk and seasonal variation in the concentration of urea in milk is reflected by variation in heat stability. The concentration of milk urea is controlled by blood urea, which in turn is largely determined by diet. Milk of high heat stability contains high concentrations of urea, although it is thought that isocyanate, formed from urea on heating, is responsible for stabilization through interactions with free sulphydryl groups of milk proteins.

(b) Enzymes

Milk contains a large number of enzymes (*ca.* 50). Although present in small quantities, some are of considerable importance in determining the stability of milk during storage. Of these proteases and lipases can affect the flavour and protein stability of milk, while oxidoreductases also affect flavour, especially in the lipid fraction.

Plasmin is the major proteinase in milk and is almost certainly identical with the blood protein of the same name. Plasmin concentration in milk varies according to the extent of leakage of blood components into milk, being high, for example, in very early, or very late lactation milk. High concentrations of plasmin in milk also arise from mastitis. In countries where standards of husbandry and udder health are high, such as Scotland, plasmin is only rarely present in significant quantities in bulk silo milk. Plasmin is a serine proteinase, with activity similar to that of trypsin. The enzyme is highly heat stable at the pH value of milk and retains substantial activity after pasteurization (see Chapter 2, page 84).

Lipoprotein lipase is the principal lipolytic enzyme in milk. The enzyme is present in large quantities in freshly drawn milk, but the protective effect of the milk fat globule membrane means that significant lipolysis due to lipoprotein lipase is rare. Spontaneous lipolysis, involving rapid production of free fatty acids and associated rancid and soapy flavours, does, however, occasionally occur. The causes are not well understood, but probably arise from a combination of factors, including diet and stage of lactation. Lipoprotein lipase is heat-labile and thus is of no significance in heat treated milk.

A number of oxidoreductase enzymes are present in milk, including catalase, lactoperoxidase and xanthine oxidase. Lactoperoxidase is present in high concentrations and is able to catalyse oxidation of unsaturated fatty acids with production of oxidized taste. Lactoperoxidase is also a component of the lactoperoxidase/thiocyanate/hydrogen peroxide system (LPS), which is a potentially important anti-microbial system.

The significance of other oxidoreductases in milk is not known, but is likely to be limited. Xanthine oxidase, for example, can cause off-flavour production through catalysis of non-specific oxidation, but this is not thought to be of importance in most circumstances.

Although alkaline phosphatase is of no importance with respect to milk stability, it is one of the most widely known enzymes in milk. This is due to its almost complete inactivation by pasteurization, but not by even marginally sub-pasteurization temperatures, and its widespread use as an index of adequate heat treatment.

(c) Vitamins

Milk is a source of the fat soluble vitamins, A (as the precursor, β-carotene), D and E and the water soluble vitamins C, B_1, B_2, B_6, B_{12}, pantothenic acid, niacin, biotin and folic acid. The dietary significance of milk as a source of fat soluble vitamins has been discussed earlier (page 21) and milk is also an important source of many water soluble vitamins. The vitamin content of milk is, however, strongly modified by losses during processing and storage.

1.5 THE FLAVOUR AND SENSORY PROPERTIES OF MILK

The bland mouth-feel of milk is a consequence of the emulsion structure, while the slightly sweet and salty taste results from the balance between lactose and milk minerals. The flavour and aroma are a consequence of a component balance involving a large number of compounds, many of which are present at sub-threshold levels. Many of the compounds identified as contributing to the flavour and aroma of milk are derived from the fat and the milk fat globule membrane. Compounds involved in determining flavour include carbonyls, alkanals, lactones, esters, sulphur-compounds, nitrogen-compounds and both aliphatic and aromatic hydrocarbons.

Certain feed components can enter the milk and cause taint. Tainting components usually partition into the fat phase and feed taint, therefore, is usually a greater problem with high-fat dairy products, such as cream and butter. It must be appreciated, however, that some feed-derived compounds contribute, at low levels, to the normal flavour of milk. Common causes of feed taint are lucerne (alfalfa), clover hay, silage, especially poor quality and brewers' grains. The problem can be controlled by withholding such feeds for 4–5 h before milking. Not all of the compounds responsible for feed taint have been identified, but silage is known to contain mixtures of methyl sulphide, aldehydes, ketones and esters. Fresh lucerne contains *trans*-2-hexanal, *trans*-3-hexanals and *trans*-3-hexenols, which impart a strong grassy flavour.

Taints in milk may also be caused by abrupt changes in feeding practices. The transition from a high concentrate diet during winter feeding to grass appears to be particularly important in this context.

Taints may also be present as a result of odourants entering the milk

---

* Unusual, or unacceptable, flavours in milk may result from an abnormal physiological condition in the producing cattle. Strong 'cowy' flavours, or an overtone of acetone, for example, are the result of ketosis or acetoneamia. The detection thresholds of the flavours are low and milk from a small number of ketotic cattle can adversely affect a relatively large volume of milk.

> **BOX 1.3 Buttercups and daisies**
>
> Opponents of modern farming methods sometimes claim that milk from cows grazing traditional meadows was of richer flavour due to a contribution from wild flowers and hedgerow plants. Such claims appear to be made on the basis of personal opinion and objective evidence is lacking. Weeds such as wild onion, land cress and buttercup, however, are well known as sources of taint and, possibly, of toxic substances (see page 33). A large number of buttercups amongst the grazing can also lead to an unnatural yellow coloration in the milk.

after drawing from the udder. Feed taints may be acquired in this way, especially where ventilation is inadequate and feed stored in close proximity to milk handling operations. Chemical flavours are usually the result of contamination with sanitizers derived from improperly rinsed equipment, chlorine- and iodine-based compounds usually being responsible.

Milk is only rarely consumed direct from the cow and consumer perception of desirable milk flavour is usually based on a product which has undergone heat processing and in which some deteriorative changes may have occurred. For this reason very limited changes resulting from both oxidation and thermal processing can be considered desirable. The oxidation product 4-*cis*-heptenol, for example, is associated with 'creaminess' at sub-threshold levels. At higher levels, however, an increasing number of persons consider changes undesirable. Flavour changes in milk during processing and storage, as well as those resulting from microbial spoilage are discussed in more detail in subsequent chapters.

1.6 POTENTIALLY HAZARDOUS SUBSTANCES IN MILK

Potentially hazardous substances may enter milk through transfer from the feed, or the environment and, in the case of some antibiotics and other anti-microbial drugs, through intra-mammary therapy. For obvious reasons attention has been focused on substances whose effects are well understood, but it has been suggested that there may be long-term hazards from very small quantities of environmental pollutants present on a continuing basis. Evidence for this contention is, however, currently lacking.

1.6.1 Antibiotics and other anti-microbial drugs

Antibiotics and other anti-microbial drugs are widely used in the treatment of mastitis. Intra-mammary administration is used, the drug entering the affected udder quarter *via* the teat canal. A wide range of drugs is used in treatment of mastitis including penicillin G, ampicillin, tetracyclines and sulpha drugs, such as sulphamethazine. Residues can persist in milk for as long as four days after administration in treatment of mastitis. Antibiotics are also widely used in prophylactic treatment of non-lactating cattle and, in this case, high levels of residues can be excreted, after calving, for longer periods. Anti-microbial drugs may also enter milk through medicated feed and through improper use of drugs for intramuscular administration, such as ceftiofur.

There is still some disagreement concerning the significance of drug residues in food for human consumption and, in particular, the long-term effects are not known. This is of particular significance with respect to children. Known short-term effects are the possibility of strong allergic reactions amongst sensitive persons, possible carcinogenicity if exposure is prolonged and the spread of antibiotic-resistance amongst micro-organisms. From a technological viewpoint the presence of antibiotic residues is undesirable because of interference with growth of starter micro-organisms during manufacture of cheese and fermented milks.

In many countries, creameries routinely test incoming milk for residues (see Chapter 2, pages 86–7). At the same time attempts are being made to educate farmers and, in some cases, veterinary surgeons, in responsible use of antibiotics and other anti-microbial drugs. It is, however, currently necessary to use financial penalties to discourage unacceptable practices. The nature of penalties and the means of imposition vary widely, but are illustrated by examples drawn from various parts of the UK (Table 1.8). In the US and some other countries, evidence of serious

---

* Intramuscular injection of suitable drugs, such as ceftiofur, is a popular means of mastitis treatment since, in many cases, it is not necessary to withdraw milk from sale after treatment. This is because residues in the milk are at a very low level (below 2 ppb) when the drug is correctly used. Detectable levels are usually the consequence of incorrect use, but it is possible that successive daily injections may permit residues to accumulate to detectable levels even when correct dosage is administered.

Table 1.8 Financial penalties in various parts of the UK resulting from antibiotic residues in milk

---

A. *England and Wales Milk Marketing Board*
  1 ppl

B. *Aberdeen and District Milk Marketing Board*
  5 ppl for 1st and 2nd failure within 12 months
  3 ppl for 3rd and 4th failure within 12 months
  1 ppl for 5th and subsequent failures
  Testing is repeated daily until a negative result is obtained

C. *North of Scotland Milk Marketing Board*
  2 ppl
  Testing is repeated daily until 2 consecutive negative results are obtained

D. *Scottish Milk Marketing Board*
  12 ppl for first failure
  3 ppl for any subsequent failures within 6 months

E. *Northern Ireland Milk Marketing Board*
  75% of month's price for 1st failure within 12 months
  60% of month's price for 2nd failure within 12 months
  10% of month's price for 4th failure within 12 months
  Contract suspended for 2 days for 5th failure
  Contract suspended for 4 days for 6th failure
  Contract suspended for 6 days for 7th or subsequent failures

---

misuse of antibiotics can lead to more draconian measures including prosecution. In some cases the producer may also be liable for any financial loss to the processor resulting from residues in milk.

1.6.2 Mycotoxins

Mycotoxins are toxic metabolites produced by a wide range of moulds. A wide variety of chemical compounds is involved, ranging from simple aromatics to complex polycyclics and toxic peptides. In many cases toxic effects are manifest as cancer or delayed organ damage due to long-term ingestion of subacute levels of mycotoxins.

Mycotoxins can enter milk indirectly through the consumption of mould-infected feed. Aflatoxins, produced by some strains of *Aspergillus flavus* or, less commonly, *A. parasiticus* are the most significant risk in milk. Cattle are more resistant to aflatoxins than most animals and may be fed on feed containing relatively high levels without development of symptoms of acute toxicity. Aflatoxin M_1 is excreted in the milk.

> BOX 1.4 **Physician heal thyself**
>
> Administration of drugs by farmers, without direct veterinary supervision, has contributed to the presence of residues in milk. In the US, for example, the widespread contamination of milk with sulpha drug residues, discovered during the mid-1980s, was at least partly due to the use of 'over the counter' drugs by producers. Despite educational programmes, it is probable that misuse through carelessness and the temptation of effecting a more rapid cure through overdosing can never be totally eliminated. There is no doubt, however, that gross abuse of antibiotics is now rare. Practices such as the addition of penicillin to milk in churns, to ensure adequate keeping quality in hot weather, are now, fortunately, part of agricultural folklore.

Aflatoxins, as a group, are toxic to the liver and also carcinogens of the liver in several animals, possibly including man. Analytical methods have been developed and may be used for determination of aflatoxins in milk. In the US a tolerance level of 0.05 ppb has been set for aflatoxin M_1 in milk.

1.6.3 Radioactive material

Following the 1952 nuclear accident at Windscale in the UK, it became apparent that radioactive material contaminating pasture was passed into milk. This led to a considerable amount of research into methods for the removal of radio-isotopes from milk. Over the years, however, interest tended to wane despite the growth of the nuclear power industry and an increasing number of nuclear power stations. The major nuclear accident at Chernobyl, in the Soviet Union, subsequently led to contamination of pasture land over wide areas of Europe and a consequent contamination of milk supply. The presence of radio-iodine amongst the isotopes present in milk placed children at particular risk of thyroid cancer.

In addition to large-scale radioactive contamination following a nuclear accident, concern is growing over the long-term effects of exposure to low-level radiation. All food must be considered a contributor to the total body burden, but it is not possible to accurately assess the specific role of milk.

> **BOX 1.5 A visionary brain**
>
> It is regrettable that quality assurance has, in many instances, become preoccupied with solving short-term problems, rather than with devising and implementing long-term strategies. A notable exception is the monitoring system for radioactivity in milk, developed at the Rowntree Company by the head of scientific affairs, the late John Colquhoun. The monitoring, which began after the Windscale accident, continued through years of negative results and finally proved its worth in the aftermath of Chernobyl.

1.6.4 Environmental chemicals

In recent years it has been recognized that many chemicals are degraded only slowly and that significant levels build-up in the environment. Inevitably these enter the food supply of man, including milk. Risk may be minimized by control of water supplies and protection of grazing land, although this can be difficult when the general environmental pollution load is high.

Particular concern has been expressed over two types of pollutant, agricultural chemicals and polychlorinated biphenyls. Usage of agricultural chemicals, such as herbicides and pesticides, should be strictly controlled to avoid contaminating the environment of dairy cattle. Careless, or irresponsible use (including disposal of waste chemicals into water courses) does, however, present a risk and provide a means by which milk is contaminated.

Polychlorinated biphenyls (PCBs) have been widely used in industry as insulators in electrical equipment and as fire retardants. Until the mid-1960s it was assumed that PCBs had remained confined within their intended usage, but subsequently it has been discovered that a high level of environmental pollution, involving both soil and water, has occurred. The compounds are highly resistant to degradation and lipophilic and have been frequently detected in milk fat. The widespread environmental contamination means that PCBs have also been detected in other foods, especially meat, fish and eggs and the relative importance of milk is not known. Levels detected are usually less than 2 mg/l and the toxicological implications are not clear. Polychlorinated biphenyls are known to cause illness when ingested at high levels, but the significance of long-term exposure to low levels is not known. There is, however, accumulation of PCBs in adipose tissue.

1.6.5 Poisonous plants

Toxic substances can enter milk as a result of cattle eating poisonous plants. In many cases, such as with yew, the cattle are themselves poisoned and ensuring pastures are free of harmful plants is considered to be a part of good husbandry. Modern pasture management has almost entirely eliminated the problem, although milk from traditional pastures containing buttercups is occasionally identified as the cause of stomach irritation. Many earlier reports, including allegations of milkborne poisoning through cattle consuming foxgloves and deadly nightshade, are apocryphal in nature.

1.7 THE MICROBIOLOGY OF MILK AT FARM LEVEL

1.7.1 Anti-microbial systems in raw milk

A number of anti-microbial systems are present in raw milk. These either form a part of the udder defence mechanism against infection, or confer disease resistance to suckling calves. Contrary to the assertions of the 'raw milk lobby', there is no evidence of any beneficial effect being conferred on human consumers of raw milk.

(a) Immunoglobulin

Milk frequently contains antibodies to potentially pathogenic microorganisms. These may enter milk from the bloodstream, or be produced within the udder. The primary function is protection of the calf in its early life, although immunoglobulins also contribute to udder defence mechanisms.

---

* Large-scale milkborne poisoning due to the compound polybrominated biphenyl (PBB), occurred in Michigan during 1973. The poisoning resulted, not from environmental contamination, but from the accidental substitution of a PBB-based flame retardant, Fire Master, for a magnesium oxide feed additive, Nutrimaster. Both products were manufactured in the same plant and, at the time of the substitution, similar packaging was used. More than 500 herds were affected and, while some cattle showed symptoms of illness, milk continued to enter the public supply. It has been estimated that every resident of Michigan ingested a significant quantity of PBB during this episode and, as Michigan is a net exporter of dairy products, it is likely that many people in other parts of the US were affected. The effects on the population, however, are still not clear.

(b) Phagocytosis

Phagocytosis, together with killing by polymorphonuclear leukocytes (PMNs), is the primary udder defence mechanism against micro-organisms causing mastitis. During infection the total cell count rises dramatically and affected quarters of the udder can excrete as many as 10^7 cell/ml, of which *ca.* 90% are PMNs. In many countries regular assessment of the cell count of bulk milk is used as a means of monitoring udder health within herds and, in the US, forms a part of milk quality payment schemes.

(c) Lactoperoxidase/thiocyanate/hydrogen peroxide system

The lactoperoxidase/thiocyanate/hydrogen peroxide system (LPS) is a powerful anti-microbial system against certain micro-organisms. In general the effect is greatest against Gram-negative organisms and LPS is bactericidal to *Escherichia coli* and *Salmonella*, as well as to the Gram-positive Group A streptococci. Most Gram-positive micro-organisms, including streptococci of other Groups and *Listeria monocytogenes*, are relatively resistant to LPS, growth being slowed but not prevented. The main *in vivo* role of LPS is thought to be protection of the calf against enteritis caused by organisms such as *E. coli*, but it may also form part of the udder defence mechanism.

Of the three components of LPS, only lactoperoxidase is a gene product of the mammary gland. Thiocyanate is derived from the diet, especially if clover or cruciferous crops are present, while hydrogen peroxide is derived from udder micro-organisms such as streptococci or from PMNs. Milk from healthy udders usually contains only very small amounts of hydrogen peroxide and this limits the activity of LPS. Proposals have been made to stimulate LPS as a means of providing a 'cold-sterilization' process either for bulk milk (see Chapter 2, page 47), or for specific situations, such as the preparation of infant formula in areas where safe water supplies are unavailable.

(d) Lactoferrin

Lactoferrin is an iron-binding protein similar to the transferrin present in serum. Lactoferrin is present in all milk, but the concentration is significantly elevated in unmilked animals, or animals with udder infection. Lactoferrin competes with bacteria for available iron and thus markedly reduces, or inhibits, multiplication. Anti-microbial activity in milk is considerably reduced by the high concentration of citrate and low concentration of bicarbonate present.

1.7.2 Micro-organisms in raw milk

(a) Micro-organisms derived from the udder

Milk drawn aseptically from the healthy udder is not sterile, but contains low numbers of micro-organisms, the so-called 'udder commensals'. These are predominantly micrococci and streptococci, although coryneform bacteria including *Corynebacterium bovis* are also fairly common. These bacteria are only rarely involved in mastitis and have no significant effect on milk yield, or quality.

The bacterial content of freshly drawn milk is significantly increased by mastitis. Causative organisms enter the udder through the duct at the teat tip. This is heavily keratinized and retains milk residues, although the keratin may possess anti-bacterial properties. The teat duct, especially the region adjacent to the orifice, can be colonized by organisms such as *Staphylococcus aureus*, which persist for many weeks, shedding into the outgoing milk, but not penetrating to the tip sinus. Machine milking has been implicated in propelling bacteria from the teat duct into the sinus and thus producing mastitis, but other mechanisms of entry exist, including growth through the teat canal.

A great many organisms can cause mastitis under specific circumstances but the most common, and the most significant cause of economic loss, are *Escherichia coli*, *Staph. aureus*, *Streptococcus agalactiae*, *Str. dysgalactiae* and *Str. uberis*. In addition *C. pyogenes* is responsible for 'August bag', a serious form of mastitis prevalent amongst unmilked cattle between July and September.

Staphylococcus aureus, *Str. agalactiae* and *E. coli* are all pathogenic to man, although the pathogenicity of *Str. agalactiae* and *E. coli* strains with a causal relationship to mastitis is uncertain. A proportion of strains of mastitic *Staph. aureus*, however, have been shown to produce enterotoxins. Other human pathogens are occasional causes of mastitis, these include *Str. pyogenes*, the causative organism of scarlet fever and pharyngitis, *Mycobacterium bovis*, or *M. tuberculosis*, *Nocardia* spp., *Actinomyces* spp. and *Cryptococcus neoformans* as well as the enteric pathogens, *Salmonella*, *Listeria monocytogenes*, *Bacillus cereus* and *Clostridium perfringens*. There is also some evidence to support the contention that *Campylobacter* is a cause of mastitis, but this is not generally accepted.

The mastitic udder appears to be of no importance with respect to food poisoning due to *B. cereus* or *Cl. perfringens* and its importance as the source of *Salmonella* and *L. monocytogenes* appears to be overemphasised. A feature of *L. monocytogenes* mastitis, however, is the lengthy period of carriage in the udder. Persistence for over a year, and thus into successive lactations, appears relatively common and persistence for as long as 3 years has been reported.

In addition to micro-organisms within the udder a contribution is made from the normal microflora of the teat exterior. The composition of this microflora is similar to that of the udder commensals, comprising micrococci, streptococci and a smaller number of coryneform bacteria, Gram-negative, rod-shaped bacteria and endospore-forming bacteria. Unless infected lesions are present, there are no public health implications.

(b) Organisms derived from the environment

The importance of the environment as a source of organisms in raw milk varies considerably. In the summer months, when most cattle graze open pastures, levels of contamination are relatively low, although micro-organisms may still enter milk from the exterior of improperly cleaned teats contaminated with soil, water or faeces. In the winter months, when cattle are at least partly fed and housed indoors, the incidence of environmental contamination is considerably high. Feed and bedding can be sources of micro-organisms and are of particular significance with respect to thermoduric spoilage organisms. The most important source of enteric pathogens, including *Salmonella* and *Campylobacter*, is faeces, which contaminate bedding and hence the teats and coat of the animal.

In earlier years when both hand and machine milking involved exposure of milk to the air within the milking shed, airborne contamination, especially from bedding and feed was a significant source of micro-organisms in milk. In modern systems the milk is fully enclosed and contamination arises from failure to adequately clean teats, or from accidents such as the dropping of a milking machine teat cluster. It will be appreciated that, in general, standards of housing have improved in

---

* Shedding of *Salmonella* in the milk has been reported in the absence of either clinical or sub-clinical udder infection by the organism. This condition is rare, but can occur in herds after outbreaks of acute salmonellosis. The number of salmonellas shed into the milk is variable and shedding can persist for some time after acute symptoms have disappeared. *Salmonella dublin* which, although pathogenic to man, is host-adapted to cattle is most commonly involved.

recent years and that the widespread use of separate milking parlours further reduces the risk of contamination. At the same time, financial pressures on farmers and the consequent need for increased productivity means that the milker may have insufficient time to adequately clean an unusually heavily soiled udder, or to sanitize a dropped teat cluster before replacing it on the cow.

(c) Milk handling equipment

Milk handling equipment is a notorious source of micro-organisms in milk, and is the major source of the Gram-negative, psychrotrophic spoilage bacteria. Contributory factors include poorly designed and constructed pipeline systems, etc., and inadequate cleaning and sanitization between milkings. Equipment needs to be very heavily contaminated to significantly increase the total number of micro-organisms in milk and the major problem lies with build-up at dead-ends, joints and fittings, such as tank outlet plugs, which are not amenable to cleaning-in-place.

In recent years the general standard of design and construction of milk handling equipment has improved. In many cases, however, installations are of increasing complexity. Many farms lack formal training schemes for workers even where large and sophisticated installations are operated and the need for increased productivity may again mean that procedures are skimped. As milk production becomes concentrated in units of increasing size, it is necessary to adopt policies to hygiene management which parallel those long adopted at processing plants.

(d) Personnel

Direct contamination of milk from the hands of personnel can occur during hand milking. It is also possible that milk may be contaminated through hands touching the milk contact surfaces of milking machines, etc. There would appear to be no practical possibility of introducing a

---

* Steam 'sterilization', involving the dismantling of milking machines and exposure to steam in a steam chest and the use of live steam to sanitize pipelines, is now obsolete in the UK due to the high cost and inconvenience. It was, however, a highly effective means of sanitization and the widespread introduction of detergent-sanitizers, led to a correspondingly widespread fall in standards. In the case of pipeline systems high standards can be obtained by circulating detergent-sanitizers at an initial temperature of *ca.* 80°C, or by using acidified boiling water. This requires, however, properly designed and maintained cleaning-in-place equipment, correct use of sanitizers and strict control of temperature.

significant number of spoilage micro-organisms by this route. The potential obviously exists, however, for the introduction of pathogens. These may be introduced either by a person suffering clinical symptoms of infection (or, less probably, in the carrier state), or by passive transfer from other sources of contamination, such as faeces. The risk is greatest from the enteric pathogens, *Salmonella* and *Campylobacter*, although this route may also be involved in transmission of the protozoan pathogen *Cryptosporidium*. It is difficult to assess the significance of the contamination of milk by personnel at farm level in the overall public health context. It is notable, however, that in Scotland, where the sale of unpasteurized milk is forbidden, a further reduction of morbidity due to *Salmonella* and *Campylobacter* resulted from introducing legislation to prevent farm workers being supplied with raw milk as part of their wages. This, and other evidence, suggests that where farm workers customarily drink raw milk cycles of infection and re-infection are established. It is, in any case desirable, that persons involved in handling milk at farm level should be made aware of their responsibilities as food handlers and to receive a full training with respect to personal hygiene, exclusion during sickness, etc.

1.7.3 Milk as a source of human pathogens

Raw milk is widely recognized as a source of human pathogens (see Chapter 2, page 44). In previous years tuberculosis and brucellosis were of major concern, but in many developed countries the causative organisms, *M. bovis* or *M. tuberculosis* and *Brucella abortus*, *Bruc. melitensis* or *Bruc. suis* have been eradicated from the national herd. Elsewhere, however, these organisms remain a serious problem, especially where milk is customarily consumed without heat treatment.

In developed countries, the major milkborne pathogens are those associated with enteric disease. As previously noted the major source is faeces and it appears impossible, under any conceivable commercial conditions, to entirely eliminate contamination from this source. At present there are no proposals to eliminate zoonotic pathogens from dairy cattle and it seems unlikely, at present, that any attempt would be

---

* Not all milkborne enteric pathogens have been identified. Brainerd diarrhoea is a chronic diarrhoeal syndrome first recognized following a point-source outbreak associated with raw milk. Sporadic outbreaks probably occurred on earlier occasions. Brainerd diarrhoea is a distinct clinical condition with an infectious aetiology, but the causative organism remains unknown (Archer, D.L. and Young, J.E. 1988. *Clinical Microbiology Reviews*, **1**, 377–98).

> **BOX 1.6 Triumph of the will**
>
> The eradication of *M. bovis*, *M. tuberculosis* and, at a later date, *Brucella* spp. from the UK national herd must be seen as a triumph for the agricultural and veterinary sciences, for bacteriologists and for the political will. In many ways the tuberculosis eradication scheme was a model of its kind involving, with the help of financial incentives, the establishment of 'tuberculin-tested' herds, regular testing of cattle and culling of those infected and improved herd management. This was supported by the general introduction of pasteurization. Although milk has been effectively eliminated as a vehicle of tuberculosis, the disease persists in developed countries, with the incidence increasing once more in cities such as New York. The disease is largely associated with immigrant communities living in very poor quality housing and it has been noted that the pioneers of the battle against tuberculosis would be shocked to find the disease still a significant problem today.

feasible. It is considered, however, to be a matter of good agricultural practice both to minimize the risk of initial infection and to take precautions against spread within a herd.

Cattle may be infected from a number of sources. Water has been identified as the main source of *Aeromonas* and *Campylobacter* and may also be involved in infection with *Salmonella*. Where possible water of potable quality should be supplied, although the animals themselves often show a distinct preference for natural, and potentially polluted, supplies.

Cattle feed has also been identified as a source of zoonotic pathogens and where animal protein is incorporated into compound feeds, the risk of recycling pathogens exists unless heat treatment is adequate. Infection with *L. monocytogenes* is often associated with feeding poorly made silage, especially of the 'big bale' type and there is also evidence of infection from naturally contaminated forage. Pastures may also be contaminated by fertilization with animal manure or sewage sludge, or by contaminated water through irrigation or flooding. The importance of this route of infection is, however, difficult to assess.

Spread of infection within a herd is usually by the faecal–oral route and can be rapid under intensive housing systems, or where hygiene is poor. Animals purchased from other herds present a risk, especially when bought at stock sales where the original vendor is unknown. Quarantine of incoming stock is considered essential.

Contamination of the fabric of buildings can serve as a long-term focus of infection and may require extremely rigorous measures to eliminate. Problems are generally less in buildings constructed to modern standards, but in one instance *Salm. typhimurium* persisted for many months in a calf-rearing unit of good construction, despite depopulation, cleaning and disinfection.

A number of other sources have been implicated as sources of zoonotic infection amongst dairy cattle. These include other farm and domestic animals, rodents, wild animals and birds and man. The relative importance of these sources is unknown, but some cases at least have been reported because of curiosity value.

Of the milkborne enteric pathogens, some strains of *E. coli*, including diarrhoeal strains, *Yersinia enterocolitica*, and *L. monocytogenes* are psychrotrophic and capable of growth at 4°C and possibly lower temperatures. *Salmonella* is not able to grow at such temperatures, but strains of some serovars are able to grow at less than 10°C, albeit slowly, and are thus capable of growth in many farm tanks, operating at 6–7°C. *Campylobacter* is not capable of growth in milk, but survival is prolonged at low temperatures and in the absence of significant lactic acid production by the spoilage microflora.

In the US, milkborne infections due to the rickettsial pathogen, *Coxiella burnetii* are common in some areas. In southern California, for example, the attack rate for consumers of pasteurized and unpasteurized milk, respectively, is 0.9 and 10.7%. Milkborne infection is also significant in some parts of Europe, but elsewhere, including the UK, the most common mode of transmission is by aerosol inhalation. Monitoring schemes to assess the incidence of *Cox. burnetii* in individual herds are established in some areas.

---

* Badgers are thought to be a natural reservoir of *M. bovis* and culling of badgers in the vicinity of dairy herds has been the cause of considerable controversy. There is also a well authenticated case of dairy cattle being infected with *Salm. agona*, through contamination of pasture in the vicinity of a badger set. The milk was bottled and sold without pasteurization, but the presence of *Salmonella* was detected by examination of milk collected as part of a routine sampling scheme and no cases of salmonellosis were reported.

1.7.4 Spoilage micro-organisms in milk at farm level

Micro-organisms can enter milk from a large number of sources (see pages 36–8). Despite the presence of anti-microbial systems, milk supports the growth of a wide range of micro-organisms and temperature is the major growth-limiting factor. In countries where milk is still stored on-farm and transported in churns, or other unrefrigerated containers the growth of mesophilic spoilage organisms can be rapid. Spoilage can occur, especially in hot weather, before the milk is delivered to the dairy, species of *Enterococcus*, *Streptococcus*, *Lactobacillus*, *Bacillus* and members of the *Enterobacteriaceae*, all being potentially involved. Milk is collected on a daily basis, usually after morning milking and processed as soon as possible after receipt at the dairy.

In developed countries, milk is stored on-farm in refrigerated bulk tanks, collected by refrigerated tanker and, if necessary, stored at the dairy in very large silos. Farm bulk tanks are designed for operation at temperatures below 7°C and while a relatively wide range of micro-organisms can grow between 5 and 7°C, the predominant spoilage organisms are Gram-negative psychrotrophic species, primarily *Pseudomonas* and *Alcaligenes*. Collection of bulk milk on alternate days is most common practice, although less frequent collection is sometimes employed. Some multiplication of psychrotrophs occurs under all conditions, although the increase in numbers is limited by dilution with fresh milk after each successive milking. At temperatures above 7°C, multiplication is rapid and, since a single bulk tank of high bacterial count can affect an entire tanker, tanker drivers are empowered to refuse milk stored at temperatures above 7°C, or which has an abnormal smell or appearance. Growth continues during storage in silos and the manufacturing properties of milk may be adversely affected by the proteolytic and lipolytic enzymes of psychrotrophs, even though no signs of spoilage exist. A more serious problem is the heat resistant nature of psychrotrophic enzymes, which cause deleterious changes in milk and milk products during storage after heat treatment (see Chapter 2, page 84).

---

* *Coxiella burnetii* is a member of the *Rickettsiae* and an obligate intracellular parasite. The organism forms endospores, which are resistant to dehydration and less rigorous pasteurization. The endospores are smaller than the parent cells and of reduced metabolic activity. Unlike the endospores of *Bacillus* and *Clostridium*, however, those of *C. burnetii* lack dipicolinic acid. *Coxiella burnetii* is the causative organism of Q-fever, symptoms of which are fever, severe headache and interstitial pneumonia. Epidemiologically, Q-fever may be regarded as a 'place' infection, causing outbreaks in localized areas which have undergone no change in exposure to the organism.

2

LIQUID MILK AND LIQUID MILK PRODUCTS

OBJECTIVES

After reading this chapter you should understand
- The difference between the various types of liquid milks
- The key roles of processing
- The basic technology of processing
- The major control points
- The nutritional effects of milk processing
- The nature of chemical changes associated with processing and subsequent storage
- Microbiological hazards and patterns of spoilage

2.1 INTRODUCTION

Liquid milk is the most important dairy food and supplies for this market generally receive precedence over those for processing operations such as milk powder production or cheese making. In the UK, for example, liquid milk sales account for approximately 50% of the total dairy market.

In recent years there has been considerable diversification of the types of milk available. Skimmed and semi-skimmed milk, which are legally defined by fat content (Table 2.1), may be considered as 'less unhealthy' milks while the development of products such as vitamin enriched milk is intended to promote a positive 'healthy' image. At the same time attempts are being made to increase milk sales by the development of milk-based beverages, which are essentially competitive with carbonated beverages such as colas, rather than with conventional milks. Such

Introduction

Table 2.1 Fat content of milk retailed in the UK

	Fat content (%)
Milk from Channel Islands, Jersey, Guernsey and South Devon breeds	minimum 4.0
Non-standardized whole milk from other breeds[1]	minimum 3.0
Standardized whole milk	minimum 3.83[2]
Semi-skimmed milk	1.5 to 1.8
Skimmed milk	maximum 0.3

[1] UK regulations apply only to cows' milk.
[2] Standardized whole milk is defined as milk imported into the UK from another EEC country which has a fat content not less than the guideline figure. This figure is fixed annually as the weighted average fat content of the whole milk produced and marketed in the UK during the previous milk year.

BOX 2.1 In sickness and in health

It is likely that the trend to 'healthy milks' will continue with the introduction of modified products such as low cholesterol milk and milk-based functional foods. To some extent, however, such products present a marketing dilemma since their development implies that unmodified milk is inherently 'unhealthy'.

---

products range from simple flavoured milk to sophisticated products formulated to resemble branded chocolate bars and to fully structured milk desserts. There is also increasing interest in milk analogues usually based on soya protein.

In industrialized nations most liquid milk is heat processed for safety

---

BOX 2.2 Plenty and want

In the affluent developed countries, substitute milk is seen as a means of avoiding animal protein for either dietary, or ethical reasons. In a wider context, however, the importance of milk substitutes should be seen as an efficient means of utilizing locally available raw materials as high quality protein in the non-industrialized nations of the third world.

but, where permitted, a small quantity is retailed raw. Consumption of raw milk carries a disproportionately high risk of infection by milkborne pathogens (Table 2.2). Of these, *Salmonella* and *Campylobacter* are of greatest importance, and it is agreed by *responsible* microbiologists that banning the sale of raw milk would result in an immediate and significant reduction in morbidity due to these and other pathogens. This may be demonstrated by reference to Scotland where a ban on raw milk sales resulted not only in a dramatic fall in the incidence of campylobacteriosis and salmonellosis but in a net economic saving after allowing for the cost of pasteurization equipment. Despite the overwhelming evidence of the hazards associated with raw milk consumption (those who require further proof need look no further than Dr James Chin's 1982 *Editorial* for the *Journal of Infectious Diseases*, **146**, 440–1), a vociferous minority continue to campaign for the continuing sale of this commodity. It is unfortunate that the proponents of raw milk often receive publicity disproportionate to the validity of their arguments and it is considered that, in many cases, the public is seriously misled as to its safety and alleged superiority.

Pasteurization is the most common means of heat-processing milk for safety and in most industrialized countries pasteurized milk is the largest selling liquid milk. The International Dairy Federation has defined pasteurization as "A process applied to a product with the object of minimizing possible health hazards arising from pathogenic microorganisms associated with milk, by heat treatment, which is consistent with minimal chemical, physical and organoleptic changes in the

Table 2.2 Infections[1] associated with raw milk consumption in industrialized nations

Micro-organism	Frequency
Salmonella	High
Campylobacter	High
Yersinia enterocolitica	Not known, possibly high in some areas
Escherichia coli 0157:H7	Not known, possibly high in some areas
Listeria monocytogenes	Probably low
Streptococcus (Group A and C)	Low[2]
Cryptosporidium parvum	High amongst occasional consumers
Poliovirus	Very low

[1] Occasional cases of illnesses such as tuberculosis, brucellosis and Q-fever, which were historically associated with raw milk, still occur.
[2] *Streptococcus zooepidemicus* (Group C) causes only mild, influenza-like symptoms in healthy adults and most cases are therefore unreported.

product". In practice, the process destroys vegetative microbial pathogens, but not endospores, and any beneficial effect on shelf life due to destruction of spoilage micro-organisms is incidental to the main concern of safety.

Minimum permitted treatments are defined by law and, in both the UK and the US are 62.8°C for 30 min (low temperature–long time: LTLT) or 71.7°C for 15 s (high temperature–short time: HTST). In recent years concern over the possible survival of *Listeria monocytogenes* (see pages 89–90) has led to some processors increasing the pasteurizing temperature to above the legal minimum.

Ultra-heat-treated (UHT) milk was developed to meet the demand for a milk which was stable for extended periods at room temperature and yet was free of the unpleasant taste associated with in-bottle sterilized

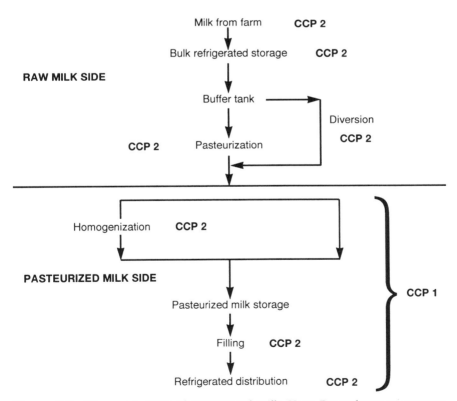

Figure 2.1 The production of pasteurized milk. *Note*: Procedures to prevent contamination of pasteurized milk with pathogenic micro-organisms collectively represent a CCP 1.

milk. UHT milk is particularly popular in continental Europe, and in countries such as France sales equal, or exceed, those of pasteurized milk. In the UK the continuing preference for creamline milk and the off-flavours associated with some UHT milk means that sales remain at a relatively low, but still significant, level.

The legal designation of UHT milk varies from country to country. In the UK, for example, it is defined as milk heated to no less than 135°C for at least 1 s and elsewhere processes vary from 130–150°C for 1–4 s. A more practical, and possibly more helpful, definition is that of Dr Harold Burton, a pioneer of UHT milk processing – 'a treatment in which product is heated to a temperature of 135 to 150°C in continuous flow in a heat exchanger for a sufficient length of time to achieve commercial sterility with an acceptable amount of change in the product'.

Although UHT milk is microbiologically stable at room temperatures, chemical reactions leading to deterioration are temperature dependent and ambient temperature storage is likely to reduce shelf life.

* From a technological viewpoint differences between UHT sterilized and in-bottle sterilized milk are best illustrated by plotting reaction kinetic data for inactivation of *Pseudomonas* proteases (a), destruction of thiamine (b) and inactivation of thermophilic (c) and mesophilic (d) endospores. The conditions for UHT and in-bottle sterilization are then superimposed on the resulting semi-logarithmic plot (Lewis, M.J. 1986. In *Modern Dairy Technology. vol.1. Advances in Milk Processing* (ed. Robinson, R.K.). Elsevier Applied Sciences, London, pp. 1–50).

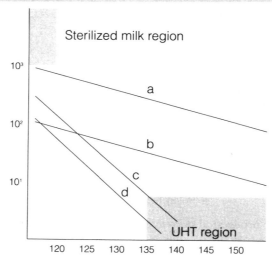

Technology

In addition to UHT sterilization, small quantities of milk are sterilized by an in-bottle process. In-bottle sterilized milk is legally defined in the UK as milk which has been maintained at a temperature above 100°C for a sufficient time to ensure that the 'turbidity' test, which detects undenatured whey proteins, is negative. In practice conventional in-bottle sterilized milk receives a heat treatment of 110–120°C for 20–40 min. Higher temperatures are used for production of sterilized milk by a combination of continuous and in-bottle heating (see page 69).

In the UK, consumption of sterilized milk is concentrated in its traditional markets, the industrial regions of the west Midlands and north west of England. Elsewhere the product is reviled and likely to be refused even as an accompaniment to tea or coffee.

2.2 TECHNOLOGY

In all cases heating is the key stage in liquid milk processing, but a number of other operations are involved. Each has a distinct technological purpose and each is described separately. It must be appreciated, however, that the plant should be considered as a whole with respect to correct and safe operation, particular care being necessary in the control of operations which involve a risk of recontaminating the heated product (see pages 51–54).

2.2.1 Incoming raw milk

Milk is commonly delivered to dairies in bulk and stored, before processing, in large refrigerated silos. According to local practice milk may be stored in the silos for up to 4 days and longer periods may be considered desirable to permit the most efficient utilization of plant. Growth of psychrotrophic micro-organisms and consequent production of heat-stable proteolytic and lipolytic enzymes occur and may cause

Table 2.3 Pre-treatment of raw milk

Thermization (*ca* 65° for *ca* 15 s)
Carbon dioxide addition
Deep cooling (*ca* 2°C)
Thiocyanate/lactoperoxidase/hydrogen peroxide system

Note: In addition to the above methods, direct addition of hydrogen peroxide has been employed in sub-Saharan Africa, but is not considered suitable for liquid milk elsewhere. Hydrogen peroxide is, however, permitted (at a concentration not exceeding 0.05%) as a sterilizing agent in cheesemilk in the United States.

spoilage of the processed milk. The problem is greatest with UHT processed milk but quality problems can also occur in pasteurized milk. A number of procedures have been described to permit extended storage of raw milk (Table 2.3) but of these only thermization is commercially employed on a regular basis.

2.2.2 Pasteurized milk

Stages in the production process of pasteurized milk and the critical control points for product safety and quality are illustrated in Figure 2.1.

CONTROL POINT: INCOMING RAW MILK CCP 2

Control

Initial quality determined at farm level (pages 35–8).

Formal scheme must be operated to ensure storage before processing is not excessive and that temperature is maintained at 2–4°C.

Monitoring

Chemical

Ensure compositional standards are met.

Ensure absence of antibiotic and other residues if required.

pH value is a useful predictor of storage instability of UHT milk.

Goat and sheep milk may be adulterated with cows' milk and species determination should be made if doubt exists.

Microbiological

Determine microbiological quality of incoming milk.

For UHT milk microbiological quality should also be determined after bulk storage, if prolonged. Direct assay of proteolytic enzymes may provide a more accurate assessment of suitability for processing than counts of psychrotrophic micro-organisms.

(a) Pasteurization

Although legislation permits two pasteurization processes, LTLT and HTST, the former is now rarely used in commercial practice and will not be discussed further.

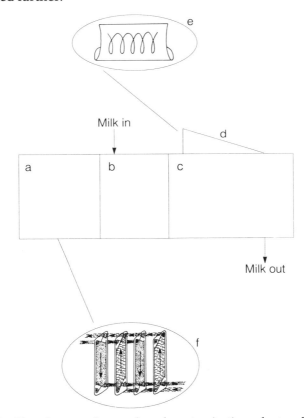

Figure 2.2 Plate heat exchanger-based pasteurization plant: schematic diagram. Raw milk enters the pasteurization plant and is preheated by outgoing milk in the regenerator section (b). The preheated milk then passes to the pasteurization section (c), where it is heated by hot water to the required pasteurization temperature. All heat exchange sections consist of a series of plates, milk flowing on one side and heating, or cooling medium on the other (f). The holding tube (d) ensures that the milk is held for the necessary period at pasteurization temperature. Holding period is determined by the length of the tube, which is fixed, and the flow rate, determined by a positive pressure pump. Plug flow represents an ideal situation in the holding tube, but is unattainable in practice. Turbulent flow (e) results in good mixing and a flat velocity profile. Pasteurizer design should take account of the flow characteristics in the holding tube and ensure that the minimum residence time is sufficient to ensure safety.

In theory many types of heat exchanger are suitable for application in the HTST process, but the plate heat exchanger (Figure 2.2) is now in virtually universal use. The detailed design of plant varies but the requirements for successful operation are common to all types (Table 2.4). Plate heat exchangers consist of a number of stainless steel plates clamped vertically into a rigid frame. Individual plates are separated by gaskets which form seals around the outer edges of the plates and round two of the ports so that the milk and the heating/cooling medium are distributed in thin films of large surface area on opposite sides of each plate. Plates are grouped into sections for heating, cooling and, in most cases, regeneration where heat is transferred from the hot, pasteurized milk to the cold raw milk. Regeneration is important with respect to economic operation of the plant and over 94% of the necessary heating/cooling can be achieved in this stage, permitting substantial savings in operating cost in relation to capital outlay.

An external holding tube is used to ensure the correct length of heat treatment. Holding time is a function of the length of the holding tube and the flow rate and it is necessary to strictly control the latter using a constant delivery pump or a flow control valve. The severity of the heat treatment, however, is dependent not only on the length of time in the holding tube but on the whole time-temperature characteristics of the process.

Pasteurization plant is fitted with a flow diversion device at the end of the holding tube which returns the heated milk to the raw milk side of the plant if the pasteurization temperature is not attained.

---

* Specified pasteurization processes were originally based on the heat treatment necessary for the destruction of *Mycobacterium tuberculosum*, then a common milk-borne pathogen. In the US a more rigorous treatment was subsequently specified to ensure destruction of the rickettsial pathogen *Coxiella burnettii*, which has a relatively high level of heat resistance attributed to the presence of endospore-like forms. As noted above, concern over the possible survival of *Listeria monocytogenes* has led to some processors voluntarily raising the pasteurization temperature.

* The positioning of the temperature sensor for the detector is a continuing cause of debate. If positioned at the beginning of the holding tube underheated milk leaving the heat exchange is detected and the flow diversion valve can operate in time to prevent any underheated milk entering the pasteurized stream. Such an arrangement, however, takes no account of any fall in temperature during transit through the holding tube. Conversely, positioning the sensor at the end of the tube means that temperature fall during holding is accounted for, but lag in operation of the diversion valve means that some underheated milk will pass into the pasteurized stream. Irregular flow and 'surging' of milk in the holding tube may also permit small quantities of underheated milk to pass undetected (Lewis, M.J. 1986. In *Modern Dairy Technology, vol. 1. Advances in Milk Processing* (ed. Robinson, R.K.). Elsevier Applied Sciences, London pp. 1–50).

Table 2.4 Requirements for successful operation of pasteurization plant

Requirement	Application of correct thermal process
Solution	(a) Use of thermostatic control to ensure heating medium at correct temperature
	(b) Use of positive control to ensure flow rate through holding tube correct
	(c) Use of long, thin holding tube to minimize short holding times due to turbulent flow
	(d) Fitting of automatic flow diversion device to return underheated milk to raw milk buffer tank
Requirement	Prevention of cross-contamination within pasteurizer
Solution	(a) Vent interspace between seals to atmosphere to provide an immediate visual indication of gasket failure
	(b) Maintain a positive pressure balance between pasteurized milk and raw milk in the regeneration section
	(c) Ensure correct positioning of flow diverter and associated pipework to avoid contamination of pasteurized milk when through-flow resumes after diversion
Requirement	Cleanability
Solution	(a) Fabricate milk contact surfaces from high grade stainless steel finished, preferably by electropolishing, to avoid crevices and consequent entrapment of soil
	(b) Welds, joins, etc., should be finished to the highest possible standard
	(c) All materials used in construction should withstand contact with cleaning fluids
Requirement	Limitation of heat damage
Solution	(a) Minimize temperature difference (1°C is desirable) between heating medium and milk
	(b) Minimize milk residence time in 'hot' section of pasteurizer
	(c) Ensure efficiency of cooling section
Requirement	Economic operation
Solution	(a) Ensure efficiency of regeneration section
	(b) Employ maximum possible ratio of heating surface to volume

Despite the high thermal efficiency of modern pasteurization plant, the increasing cost of energy has provided the impetus for research into alternative heating systems. The most promising of these is the use of microwave energy at 2450 MHz and plant has been developed for HTST pasteurization which has a comparable performance to conventional plant. Microwave heating, however, has not been adopted on a commercial scale.

CONTROL POINT: PASTEURIZATION CCP 1

Control

Thermal destruction of vegetative pathogens.

Monitoring

Ensure adequately trained operators available at *all* times.

Check condition of equipment (worn gaskets, etc.) on a regular and routine basis.

Monitor temperature of milk in holding tube and after cooling.

Verification

Periodic specialist examination and maintenance of plant by manufacturers. Ensure correct operation of flow diverter at start of each run.

Examination of thermograph records.

Phosphatase test.

Note: In most countries legislation specifies that a thermograph must be fitted and the record maintained for a given period

---

(b) Protection of pasteurized milk from recontamination

Prevention of recontamination is not a discrete process such as pasteurization or homogenization, but is of major importance in production of pasteurized milk that is both safe and of satisfactory storage life. Individual pieces of equipment such as plate heat exchangers are designed and constructed to minimize the possibility of the pasteurized

---

* Modern retailing practices require a long storage life for pasteurized milk. This requires a very low level of post-pasteurization contamination and a highly effective cold chain. Attempts to increase the storage life by raising the pasteurization temperature have not only been unsuccessful, but resulted in faster growth by any bacteria present. This 'pasteurization effect' is attributed to the inactivation of natural anti-microbial systems in milk at the higher pasteurization temperatures.

product being contaminated either with raw milk or from other sources, but it is necessary to view the potential for contamination from the viewpoint of the entire plant and its operations. Milk moves through the plant to the final container in closed pipes and tanks and 'hands-on' operations are, or should be, totally avoided. Despite this recontamination of pasteurized milk can occur with serious potential consequences both for public health and product spoilage. The possible routes of contamination are summarized in Figure 2.3 and means of control summarized in Table 2.5.

It is of particular importance to ensure that there are no cross-connections whatsoever between raw product equipment and piping and pasteurized product equipment and piping and that cleaning-in-place (CIP) systems are also entirely separate. The inherent safety of the plant should be reviewed after any change whatsoever in layout of equipment and alterations to pipe runs. Equally no change should be made, even on a temporary basis, that has not been authorized by technical management.

(c) Homogenization

Homogenization may be applied to milk to reduce the size of the fat globules and thus prevent, or greatly reduce 'creaming'. Piston driven valve homogenizers are used which subject the milk to high velocity turbulent flow fields (Figure 2.4). Homogenizers are sometimes incorporated in the pasteurizer and operated at, or near, pasteurization temperature. This has the advantage of permitting lower pressures to be used and reducing problems due to microbial contamination.

(d) Filling

Milk may be filled into a number of types of container including glass, or plastic, bottles and plastic-coated cardboard containers for retail sale, or plastic bag-in-box containers for bulk use. Re-usable glass bottles are

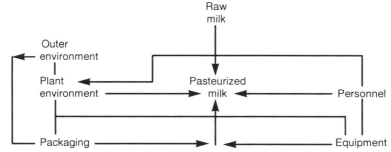

Figure 2.3 Possible routes of contamination of pasteurized milk.

Table 2.5 Precautions against re-contamination of pasteurized milk

Source	Raw milk
Pathway	(a) Direct
	(b) Indirect *via* contamination of plant environment, passive transfer on hands of personnel, etc.
Precautions	(a) Correct design (see text pages 49–51). Correct operation and maintenance of pasteurizer (see text pages 49–50).
	(b) Correct plant layout. Control of personnel movement and avoidance of 'hands-on' operations involving milk or milk-contact surface.
Source	Plant environment
Pathway	Usually indirect *via* contamination of equipment. Also possible *via* personnel and packaging
Precautions	Prevent contamination of plant environment from outer environment including animals and birds, soil and water. Eliminate contamination of pasteurized milk side of plant by drainage, etc. from raw milk side. Correct environmental sanitation.
Source	Personnel
Route	(a) Direct due to personnel suffering clinical illness or being convalescent or chronic carrier of pathogens.
	(b) Indirect due to introduction to plant of contamination from outside environment, etc. A particular hazard can exist in some rural areas where personnel are also smallholders.
Precautions	(a) Apply appropriate medical and exclusion policies.
	(b) Ensure good personal hygiene and correct use of protective clothing and footwear.
	(c) Prohibit raw produce such as eggs being brought into the plant by part-time farmers for sale to fellow workers.
Source	Equipment
Route	Direct following:
	(a) Contamination of equipment by raw milk, etc.
	(b) Development of biofilms and colonization of milk contact surfaces by micro-organisms.
Precautions	(a) Protect equipment (see above).
	(b) Restrict operating periods to not more than 8 h.
	(c) Utilize suitable cleaning and sanitization programmes.
Source	Packaging
Route	Direct following:
	(a) Failure to adequately clean and sanitize reusable bottles.
	(b) Contamination of packaging from plant environment, etc.
Precautions	(a) Institute correct cleaning, sanitization and inspection procedures.
	(b) Protect packaging from contamination.

Note: It is good practice to periodically review possibilities for contamination in each plant and to test precautions on a 'what if?' basis. During such exercises possibilities should not be ignored because they are considered 'unlikely', since it is 'unlikely' circumstances which lead to most catastrophes.

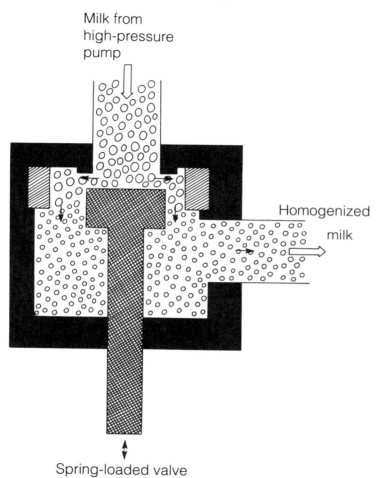

Figure 2.4 Principle of valve homogenizers.

now seen as an 'environmentally friendly' means of distributing milk at retail level. Such bottles do, however, cause considerable problems due to the need to ensure adequate cleaning and sanitization before re-use. The problem is exacerbated by the failure of consumers to rinse empty bottles or, even worse, to use the bottles before return as containers for paint and household chemicals. Modern bottle washers of the Hydro-type usually have five stages, involving prerinsing by both immersion and spray, before cleaning with a sodium hydroxide solution at *ca.* 62°C. The bottles are then rinsed with water at *ca.* 49°C, sanitized with a hypochlorite spray, before consecutive final rinses in warm (49 and 30°C) and cold water. Bottles leaving the washer should be inspected visually or by an automatic photoelectric cell device. Despite stringent

CONTROL POINT: HOMOGENIZATION CCP 2

Control

Reduce size of fat globules to eliminate creaming.

Sanitation to minimize microbial contamination.

Monitoring

Ensure correct pressure setting(s) and check periodically.

Ensure correct application of CIP or manual cleaning procedures.

Verification

Visual inspection.

Assessment of microbiological status.

---

precautions, improperly cleaned bottles do sometimes pass, undetected, to the consumer resulting in dissatisfaction and the possibility of legal action with attendant adverse publicity.

Single-use plastic bottles may be preformed or blow-moulded on site. Preformed bottles are usually made from polyethylene and most types can be used without prior rinsing. Blow-moulded bottles are made from high-density polyethylene or polypropylene. Moulding takes place at *ca.* 200°C. This sterilizes the bottles which may be used in programmes designed to minimize post-pasteurization contamination.

Polyethylene is most commonly used for coating cardboard, cartons being either prefabricated or, more commonly, fabricated from roll material directly before filling. Large capacity bags are usually preformed and protected by rigid cardboard outer packing. An outlet tap is incorporated in the bag.

---

* Migration of soluble compounds from the polyethylene lining of cardboard cartons can result in rapid tainting of the contents. The problem is greatest with small (0.5 l) cartons where taints may be detected after only 1 day of storage. The nature of the compounds is not known.

Technology

Large capacity filling machines are mechanically complex, but essentially employ pistons to deliver a predetermined quantity of milk into each container. Bottles are sealed with an aluminium foil cap, while other containers are heat-sealed. A recent innovation is a retail milk carton which can be re-sealed after initial use.

CONTROL POINT: FILLING CCP 2

Control

Cleaning and sanitizing glass bottles.

Monitoring

Ensure correct operation of bottle washer.

Visual/automated inspection of *all* bottles before filling.

Verification

Check operation of automated inspection equipment.

Microbiological analysis of bottle rinses.

Control

Filling of predetermined volume.

Correct sealing to prevent leakage and contamination.

Monitoring

Operator to ensure correct functioning.

Verification

Measurement of volume filled.

Examination of seals including tear-down tests (heat-sealed cartons).

(e) Separation and 'sterilization' processes

Separation and 'sterilization' processes are designed to produce pasteurized milk of extremely high microbiological quality and thus of extended shelf life. The milk is also highly resistant to fat oxidation while lacking the cooked flavours associated with high temperature treatments. More than one approach is possible, but the underlying principles involve the concentration of most of the bacteria into a fraction of small volume. This bacteria-rich fraction is then 'sterilized' at a temperature sufficient to kill endospore-forming bacteria such as *Bacillus cereus* and then recombined with the milk before conventional pasteurization. Pasteurization is required to inactivate any bacteria which were not partitioned into the bacteria rich fraction, or which were derived from the plant.

A number of approaches have been made to developing successful processes including the use of gravity, or centrifugal separation, ultrafiltration and microfiltration. The technology of a commercial-scale plant, the Alfa-Laval Bactocatch™, which combines centrifugation with microfiltration is illustrated in Figure 2.5.

(f) End-product testing

End-product testing is required for all types of pasteurized milk. It must be appreciated, however, that end-product testing is essentially a means of verifying that correct processing procedures have been observed and not a means of imposing control.

2.2.3 Ultra-heat-treatment of milk

Stages in the production of UHT milk are summarized and the critical control points for safety and stability indicated in Figure 2.6.

Key stages for product safety and stability are sterilization and aseptic packaging, although homogenization is also essential to avoid fat separation and hardening during storage. It must also be appreciated that all plant downstream of the sterilization stage must be capable both of

---

* It is usually inappropriate to take action on the basis of a single set of results of end-product tests and judgement is based on performance over a period of time. In the US, further sampling is undertaken if the standard plate count exceeds the standard. Under the 'three out of five' system of compliance, no less than three of the next five samples examined must have standard plate counts below the standard (*Grade A Pasteurized Milk Ordinance*. USPHS, FDA, 1978).

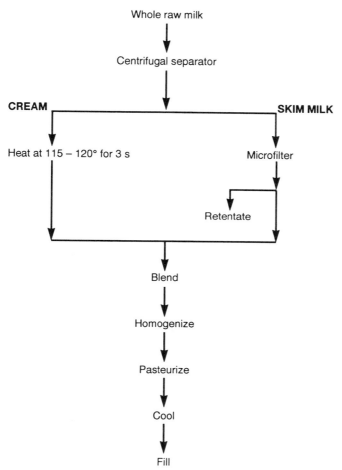

Figure 2.5 The Bactocatch separation and 'sterilization' process.

being sterilized and maintained in a sterile condition for the duration of the production run.

(a) Sterilization

Plant for continuous UHT sterilization of milk may be placed in two categories: indirect heating and direct heating. In common with pasteurization plant, each type must be fitted with flow diversion devices to prevent underheated milk entering the UHT milk stream and recording thermographs.

Indirect heating is carried out using tubular, or plate, heat exchangers

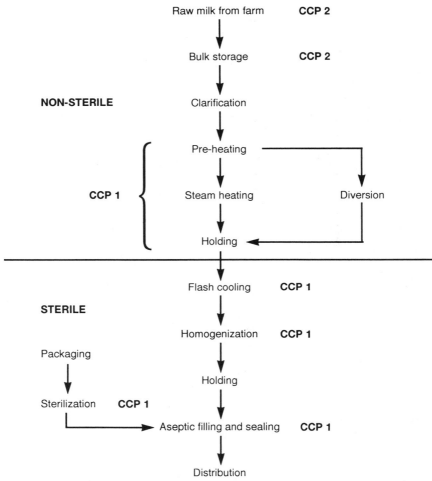

Figure 2.6 UHT sterilization of milk by direct heating.

similar to those used for pasteurization. It is necessary, however, to employ higher operating pressures to prevent the milk boiling at the higher temperatures. Pressurized hot water, or steam is used as heating

---

* A novel indirect method, the Alfa-Laval 'Achilles and the tortoise' system dispenses with a conventional heat exchanger and holding tube and is further unusual in that both milk and packaging pass through the same sterilization zone. Milk is fed into a plastic film formed continuously into a tube, the tube then passing through a bath of heating liquid and acting as a moving heat exchanger through which the milk passes at 10 times the speed of the tube. After heating and cooling, the tube/milk is separated into individual packages. The system avoids problems of fouling, avoids the use of sterilizing chemicals or steam and permits a lower nominal heat process.

medium and there is a trend to high regeneration efficiencies of 85–90%. Plant of high regeneration efficiency usually has large heat exchange systems and thus longer total transit times. Improvements in operating efficiency are therefore accompanied by increases in the amount of chemical damage.

The production of UHT milk by direct heating systems involves the mixing of superheated steam with milk, which is consequently heated almost instantaneously to the required sterilizing temperature.

Incoming raw milk is heated by a regenerative heat exchanger to 70–80°C before passing to the mixer. Much effort has been expended in the design of systems for mixing steam and milk, it being desirable from the viewpoint of both product quality and operating efficiency that the temperature rise should not only be very rapid, but that all of the milk should be heated at the same rate. Two basic systems are used – injection (steam into milk) and infusion (milk into steam). The APV Uperiser™ plant is a widely used example of an injection system, a specially designed injector being used to promote turbulence in the milk stream. Infusion systems include the Pasilac™, where droplets of milk are mixed with steam, and the DaSi™, where thin sheets of milk fall into an atmosphere of steam.

Heated milk passes to a holding tube and then to a vacuum vessel. At this stage the temperature of the milk falls rapidly, the resulting loss of energy causing some of the water and other volatiles to vaporize. This process, 'flash cooling', has three technological objectives:

1. Very rapid cooling which reduces the extent of thermal damage.
2. Removal of water to restore the milk to its original composition.
3. The removal of low molecular weight volatile compounds which have a deleterious effect on product quality.

The degree of cooling and the quantity of water removed is determined by the level of vacuum. The quantity of water removed must equal that added during steam injection and, as the thermal characteristics vary from plant to plant, individual calibration is necessary.

When installing UHT plant the choice between direct and indirect heating is influenced by both economic and quality considerations (Table 2.6).

Fouling, a major problem in indirect plants, is caused by deposits on the

Table 2.6 Comparison of indirect and direct heating for UHT processing

A. Economic aspects

Capital expenditure
Indirect plant less complex and less expensive than direct.

Operating efficiency
Operating efficiency and costs less in indirect plant due to difficulty of recovering energy lost during flash cooling in direct plant.

Run time
Reduced fouling in direct plant means run times of up to 20 h in direct heated plant compared to 6–10 h in low regeneration efficiency indirect plant and 14–16 h in high regeneration efficiency plant. Longer run times at least partly offset the lower capital and running costs of indirect plant.

B. Quality aspects

Organoleptic quality
Direct plant produces milk of superior flavour and odour. Flakes of deposited material are absent.

---

milk side of heat exchange surfaces which increasingly impair plant efficiency to a point at which operation is not viable. The deposits are difficult to remove and product quality may be affected by small flakes entering the milk. Although deposits may form either in the regenerator section of the plant or during the heating of the milk to sterilization temperatures the length of production runs is limited by fouling of the high temperature section. These may be minimized by reducing the temperature differential between the heating medium and the milk and by ensuring that only good quality milk is processed. pH value is of particular significance in this context and build-up of deposits is particularly rapid when processing milk of a pH value below 6.6. However, natural variation in milk composition can lead to twofold variation in the efficient run time of plant.

The higher quality of direct heated UHT milk is attributable to the shorter period of time it is held at the higher end of the temperature range. Comparisons are complicated, however, by the fact that different plants have different time–temperature profiles (Figure 2.7) and that chemical damage, in some plant, results primarily from excessive heating and cooling periods. The most satisfactory method of comparing plant was devised by the German food engineer, Dr H. Kessler, who introduced a microbiological parameter, B^* and a chemical parameter C^* (see Figure 2.8 for definition and derivation of these parameters). It is highly desirable that the value of B^* should exceed 1, while the value of

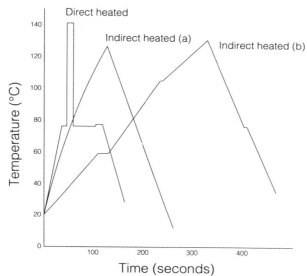

Figure 2.7 Kinetic reaction data for different sterilization procedures. Based on Lewis, M.J. 1986. In *Modern Dairy Technology, vol. 1, Advances in Food Processing* (ed. Robinson, R.K.). Elsevier Applied Science, London.

C^* should be as low as possible. Typical values for indirect plants are $B^* = 1.25$; $C^* = 0.49$ and for direct plants $B^* = 2.18$; $C^* = 0.30$.

(b) Homogenization

As noted above, homogenization is essential for the stability of UHT milk. However homogenization conditions, and consequently emulsion properties, have wider implications for product quality. Factors to be considered include homogenization pressures, single or double stage operation and positioning upstream or downstream of the sterilization process. Information on the best conditions is not readily available, but it is generally recognized that, from the viewpoint of plant design, the homogenizer should be upstream of sterilization to avoid the need for sterile operation. However subsequent heating, particularly by direct processes, may destabilize the emulsion and it is often necessary to place the homogenizer downstream.

(c) Maintenance of sterile conditions downstream of sterilization

Sterility is achieved by circulating hot water through the plant and ensuring that all parts in contact with the milk attain a temperature of 130°C. Valves, seals, etc., must be capable of withstanding this tempera-

CONTROL POINT: UHT STERILIZATION CCP 1

Control

Inactivation of vegetative micro-organisms and endospores by heating for approved temperature–time combinations.

Monitoring

Measurement of temperature, pressure and flow rates.

Ensuring correct operation of control and safety devices.

Inspection of the physical condition of the plant.

Verification

Sterility testing.

Examination of plant records.

CONTROL POINT: HOMOGENIZATION CCP 1 OR 2

Control and monitoring of the homogenization process itself is the same as that during pasteurized production. Where homogenizer is downstream of the sterilizer it is necessary to use equipment designed and constructed for aseptic operation. The cleanliness of the homogenizer should be checked before assembly and operation monitored throughout use to ensure aseptic operation.

---

ture and special consideration must be given to the design of ancillary equipment such as pumps. Downstream homogenizers present a special problem and in operation sterility is maintained by a sterile block, a steam chest through which the piston rods pass.

(d) Aseptic filling

Aseptic filling is an integral part of the production of UHT milk and all plant and associated equipment must meet basic conditions:

Microbiological parameter

A B^* value of 1.0 represents a process of 135°C[1] for 10.1 s (Z value = 10.5°C).
B^* may then be evaluated for any plant for which the temperature-time profile is known according to the formula

$$B^* = \int \frac{10^{(135-\theta)/10.5}}{10.1} dt$$

where θ is the temperature and
t is the time in seconds

Chemical parameter

The C^* value is based on the conditions required for a 3% reduction of thiamine, which are 135°C for 30.5 s (Z value = 31.4°C).

$$C^* = \int \frac{10^{(135-\theta)/31.4}}{30.5} dt$$

[1] 135°C is used as reference temperature instead of 121°C to reduce the extent of extrapolation required for UHT processing

Figure 2.8 Definition and derivation of microbiological parameter (B^*) and chemical parameter (C^*) for assessment of UHT plants. [1]135°C is used as reference temperature instead of 121°C to reduce extent of extrapolation required for UHT processing. Based on Kessler, H.E. 1981. *Food Engineering and Dairy Technology*. Verlag A. Kessler, Friesing.

CONTROL POINT: MAINTENANCE OF STERILITY CCP 1

Control

Prevention of recontamination of sterilized milk.

Monitoring

Assessment of efficiency of cleaning.

Measurement and recording temperature of circulating hot water.

Verification

Sterility testing.

1. Transfer of milk from the sterilizing plant to the point of filling must be made under aseptic conditions.
2. Plant must be capable of being sterilized before use.
3. A means must exist of sterilizing packaging.
4. Filling, sealing and critical transfer operations must be carried out in a 'sterile' environment.

A wide variety of aseptic packaging systems is available (Table 2.7), but the vast majority of UHT milk is packaged in cartons using either the Tetra Brik™ or Combibloc™ system. Each of these employs cartons of plastic:paper:aluminium foil laminates but they differ in that cartons for Tetra Brik™ are formed direct from the reel and sealed below the product surface, while Combibloc™ cartons are performed and sealed above the product surface. Milk packed in Combibloc-type packaging is more susceptible to oxidative deterioration, although in many plants this problem is minimized by flushing the atmosphere above the milk with an inert gas.

Plastic bottles and pouches have been introduced more recently for packing UHT milk and, like bottles for pasteurized milk, may be either blow-moulded on site or preformed. Bottles, however, must be manufactured to more rigorous specifications to meet the requirements of the long shelf life of UHT milk. Triple-laminated high density polyethylene is widely used for bottle manufacture, while pouches are made from co-extruded polyethylene and polyvinylidene chloride, or ethyl vinyl acetate.

Table 2.7 Examples of systems for aseptic packaging of UHT milk

System	Pack type and material	Sterilization	Filling environment
Tetra Brik	Web-fed paper/plastic/Al foil laminate carton	H_2O_2 and heat	H_2O_2 and steam
Combibloc	Pre-formed paper/plastic/Al foil laminate carton	H_2O_2 and hot air	Sterile air
Liquipack	Pre-formed paper/plastic/Al foil laminate carton	Synergistic H_2O_2 and ultraviolet light	Sterile air
Freshfill	Pre-formed thermoplastic cup	H_2O_2 and hot air	Sterile air

Adapted from Hersom, A. C. 1985. *Food Reviews International*, 1, 215–70.

(e) Verification of plant performance

Verification that the plant is capable of consistently producing a sterile end-product should be made at commissioning, or after major maintenance work. Challenge tests are used which verify the performance of the entire plant as well as that of critical stages such as heat treatment of the milk and sterilization of the packaging material. *Bacillus stearother-*

CONTROL POINT: ASEPTIC FILLING CCP 1

Control

Presterilization of plant.

Installation of filler in 'clean' area with control of air flow and pressure.

Sterilization of packaging material and of air or gases used for flushing pack.

Correct formation of pack and making of seals.

Monitoring

Ensure physical condition of plant acceptable.

Ensure correct air flow and pressure in 'clean' area.

Ensure correct operation of systems for sterilization of packaging, etc.

Ensure correct formation of pack and integrity of seals.

Verification

Tear-down testing of seals.

Sterility testing.

Note: Quality control of packaging material is primarily the responsibility of the manufacturer, but material should be inspected prior to use to ensure freedom from dust and other contaminants, creasing and other visual defects.

---

mophilus is widely used as challenge organism for milk sterilization, while the choice of challenge organism for packaging sterilization varies according to the method of sterilization used. *B. subtilis*, for example, is the common choice where H_2O_2 is used as sterilizing agent. Sealing equipment should also be subject to rigorous checks at this stage and either dye or biotest methods may be used.

(f) End-product testing

No index test exists for the UHT sterilization process and end-product testing for sterility is used to verify the performance of the whole system. The numbers of micro-organisms present in non-sterile packs are very low unless gross underprocessing has occurred and incubation is required before microbiological testing. Methods used are discussed in more detail in pages 98–99.

To be effective any end-product testing must be based on a properly devised and implemented statistical sampling scheme. During commissioning it is usual to sample 100% of production, but during subsequent operations the number of packs is gradually reduced to 0.1% and finally as low as 0.01% of production, or a fixed number of packs per batch.

2.2.4 In-bottle sterilized milk

Milk for sterilization is filtered or clarified and homogenized before heat treatment. In conventional production the milk is filled into thick-walled, narrow-necked bottles which are closed by machine-crimped Crown™ caps.

Batch retorts were originally used for sterilization but these have now been largely replaced for large-scale processing by continuous retorts of the hydrostatic and rotary types. These retorts are designed to agitate the milk during processing and the resulting improvement in heat transfer permits a reduction in the overall thermal treatment. From the viewpoint of flavour changes and nutritional losses, this leads to an improved quality, although normal criteria for judging milk quality do not apply in the traditional sterilized market and severity of processing may be determined by the amount of caramelization required.

More recent processes for in-bottle sterilized milk have involved processing in a continuous-flow heat exchanger. Milk sterilized in this way may be aseptically filled into sterile bottles, but it is more common to combine continuous heat treatment with an additional retorting period after filling. The severity of this heat treatment is considerably less than that used for conventional in-bottle sterilization, but is sufficient to complete the denaturation of the whey proteins and thus satisfy the requirements of the turbidity test. A recently developed two-stage process (Stork International) involves continuous UHT processing in a tubular heat exchanger at 138–140°C for 2 s followed by filling, under clean conditions, into bottles. The second stage heating takes place in a continuous hydrostatic retort, the sterilization process being 10–12 min at 117–123°C. The process may be used with plastic bottles, but in this case compressed air is used to maintain the total pressure in the sterilizing chamber at 0.3–0.5 bar higher than the pressure of saturated steam alone to prevent the bottles, weakened at high temperature, bursting.

Sterilized milk should be cooled as quickly as possible after processing to avoid further thermal damage and, in extreme cases, outgrowth of surviving endospores of thermophilic bacteria. The potential for contamination at this stage is less than with a double seam can, but cooling water must be of potable quality.

2.2.5 Modified milk, added-value milk and milk analogues

(a) 'Less harmful' and 'more healthful' milks

The simplest form of 'less harmful' milk is skimmed or semi-skimmed milk in which the fat content is reduced by centrifugal separation in accordance with legal criteria (see Table 2.1; page 43).

Fat soluble vitamins are removed from milk with the fat and this leads to a dietary significant loss of vitamin A. Pasteurized skim milk contains only a trace of the vitamin, while in pasteurized semi-skimmed milk the level is 25 µg/100 ml (retinol equivalent) compared with 55 µg/100 ml in pasteurized full cream milk. Addition of vitamin A to restore the content to that of whole milk is mandatory in the US. Vitamin A is added as retinyl palmitate in concentrated butter oil and in the presence of an emulsifier. Photo-oxidative losses are greater for added Vitamin A than that naturally occurring and result in an oily, 'hay-like' flavour, although some protection against this is obtained by reducing the concentration in the butter oil.

Skimmed milk lacks the smooth body and texture of whole milk and several products are available in which skim milk powder is added to produce a low-fat milk with similar organoleptic properties to whole milk. Such products also contain added vitamin A at levels equal to, or above, those in whole milk.

CONTROL POINT: STERILIZATION CCP 1

Control

Inactivation of vegetative and endospore-forming bacteria.

Only correctly calibrated retorts should be used and correct operation, appropriate to the type of retort in use, is essential.

Management systems must ensure that all bottles receive heat treatment.

Retort operators must be fully trained and be aware of the safety implications of improper thermal processing.

Monitoring

The initial product temperature should be checked and any delay between filling and thermal processing recorded. The correct functioning of the retort, its controls and recording devices should be checked visually or, in the case of computerized instrumentation, by test routines. Monitoring requirements vary according to retort design. However temperatures and pressures should be monitored in all cases. A calibrated and certified mercury in glass or platinum resistance thermometer should be used, other types are not suitable. Temperature–time recorders should be checked against the certified thermometer. These instruments are not sufficiently accurate for process control, but provide a record of temperature profiles. The operator should also maintain manual records of all aspects of retort operation.

Verification

Sterility testing.

Examination of process records.

Note: Other control procedures will be necessary where all, or part, of the processing involves a continuous flow heat exchanger.

A number of 'more healthful' milk based products have been proposed, although not all patents have become commercial reality. These include various vitamin-enriched formulations, high calcium and low-sodium milks and a semi-skimmed milk containing 11% soluble dietary fibre.

In recent years much interest has been shown in the development of 'cholesterol-free' or 'low-cholesterol' milks. The methods of reducing the cholesterol content are the same as those applied to other dairy products and are discussed in greater detail in Chapter 1, page 20

The necessary technology for the production of low-lactose milk has been available for some time, enzymatic hydrolysis using microbial β-galactosidase now being the preferred method. Lactose may either be hydrolysed before heat treatment using an immobilized β-galactosidase, or by the post-heating addition of enzyme before packaging. The small size of the market means that only UHT sterilized lactose-hydrolysed milk is currently commercially viable due to its long storage life.

Low-lactose milk may offer only a partial solution to the problems faced by lactase-deficient individuals since oligosaccharides are formed if less than 80% of lactose is hydrolysed. Oligosaccharides are poorly adsorbed in the small intestine and are fermented in the colon with possible resulting digestive problems similar to those of lactose intolerance.

Milk is one of a number of non-fermented dairy foods which is used as a vehicle for *Bifidobacterium*. Refrigeration is required to prevent growth of the organism and product life is restricted by the requirement that at consumption 10^6 viable cells/ml must be present (see pages 349–51 for a discussion of probiotic and therapeutic properties associated with *Bifidobacterium* and its use in fermented milks).

BOX 2.3 **Too much of a good thing**

Over-fortification of milk with vitamin D at a small Massachusetts dairy resulted in at least seven cases of vitamin D toxicity and one death over a 4 year period. In some cases milk contained 600 times the stated quantity of vitamin D. Over-fortification resulted from a combination of unsatisfactory procedures at the dairy and incorrect labelling of the concentrate, which contained 3.6 times the stated quantity of the vitamin (*Food Chemical News*, July 8, 1991).

> **BOX 2.4 So deeply sweet**
>
> Hydrolysis of lactose yields glucose and galactose and the resultant milk has a sweet taste which is advantageous in production of sweetened flavoured milk. Rather surprisingly, there is evidence that many teenagers and young adults preferred the sweet taste of lactose-hydrolysed milk. A market for 'sweet' milk may therefore exist amongst persons for whom dietary considerations are unimportant. The market for lactose-hydrolysed milk has also been extended to cats, many of whom are lactase deficient.

(b) Added-value milks

The simplest added-value milks are those intended for immediate consumption which are prepared by adding an appropriately flavoured and coloured syrup to a fresh milk base and 'shaking' to mix the contents and provide a degree of aeration. Flavoured milks are also produced for retail distribution and usually contain a stabilizer such as carrageenan to prevent separation of the ingredients during storage and to improve the product texture. A thickening agent may also be present and products may be weakly gelified. In such cases the milk may be structured to permit formation of a stable foam when the refrigerated product is shaken in its carton and this principle is employed in a brand extension exercise to produce a drink related to aerated chocolate bars. Flavoured milks are usually sweetened by sucrose, but increasing use is being made of intense sweeteners such as aspartame™

The increasing market for added-value milks has led to many approaches to product development. A process has been described, for example, for carbonation of UHT sterilized milks. Carbonation not only produces an effervescent drink, likely to be popular with children and adolescents, but results in improved perception of flavours.

---

* Instability of chocolate milk is a particular problem to the industry and is of three types, sedimentation of cocoa particles, formation of large flocs and segregation into light and dark layers. Sedimentation and flocculation are controlled by stabilizers which initiate formation of a weak network of protein and protein-covered cocoa particles. Segregation probably results from uniaxial compression of the total network due to gravitational force and ceases when gravitational force is counterbalanced by the elastic modulus multiplied by the deformation gradient of the network (van den Boomgaard, T. *et al.* 1987. *International Journal of Food Science and Technology,* 22, 279–91).

Added-value milks may be based on whole, skimmed or semi-skimmed milk, a minimum milk content of 85% being required in the UK if the product is to be described as 'milk-based'. Additional milk solids are often provided by dried skim milk powder or concentrated skim milk but do not legally contribute to milk content.

Added-value milks are heat processed for safety and stability. Pasteurization and UHT sterilization are most commonly used although some flavoured milks are processed by in-bottle sterilization. Sugar and some other ingredients have a protective effect on micro-organisms and a more rigorous heat treatment is required. Allowance must also be made for the higher viscosity and consequent slower heating of some products. Minimum heat treatments in the UK are 72°C for 15 s for pasteurization, 140°C for 2 s for UHT treatment and 108°C for 45 min for in-bottle sterilization. Thermoduric organisms have, however, been reported to survive a UHT process of 140°C for 4 s and considerably more rigorous heat treatment than the stipulated minimum may be required, especially where the product is to be marketed in hot countries. Addition of all ingredients before the final heat processing is desirable, but some are heat-labile and for optimal organoleptic quality are added after heat treatment. Aseptic addition using the Tetradosing^{TM} system is required for UHT products and a high standard of practice is necessary when making post-process additions to pasteurized products. In the case of sterilized products it is important that the flavouring should either mask, or complement, any cooked flavours, and that both flavouring and colouring should remain stable over the product storage period. Additional buffering capacity may be required where natural flavouring derived from acidic fruit is used to prevent coagulation during UHT processing.

It is also necessary to be aware that additives may themselves be the source of micro-organisms, including pathogens, and are a critical control point. Particular care is required when chocolate, or cocoa, is used as flavouring, or gelatin as stabilizer since these commodities have been identified as sources of *Salmonella*.

---

* The reduced availability of calcium due to binding by some stabilizing gums and, in chocolate milk, by cocoa is a matter for dietary concern in the case of high-risk persons who consume added-value milk drinks in preference to milk itself.

(c) Gelified (structured) milks

Gelified milks are related both to weakly gelified milk drinks and to highly structured, aerated products such as mousse (see page 389). In each case stabilizers and thickeners play a major role in determining product characteristics. Gelified milks are primarily prepared as ready-to-eat dairy desserts, but despite their structure are a special type of liquid milk product.

Gelified milks are a relatively recent development and vary in nature according to the properties of the gel. The product is made from pasteurized or sterilized milk, sucrose, or other sweetener, flavouring and colouring, stabilizers, thickeners and gelifying agents. Manufacture involves the addition of ingredients to cooled, homogenized milk, and cooking at temperatures between 60 and 90°C. Gelified milks may either be pasteurized for sale from refrigerated display cabinets, or UHT sterilized for ambient temperature storage. In the latter case aroma is usually added as a final step process before aseptic packaging.

The rheological properties of the finished product are largely dependent on the nature of the stabilizing, thickening and gelifying agents and the cooking these receive during manufacture. Starch, carrageenan, agar-agar, alginates, pectins and, less commonly, gelatin are used depending on the required properties of the product. Starch is probably the most important of these ingredients and structurally significant changes to starch during cooking are summarized in Table 2.8.

The ability to define the properties of the finished product by choice of ingredients and manufacturing conditions means that gelification can be used to manufacture a wide range of products. These range from pour-over sauces and custard of relatively low viscosity to flans with a tender gel structure, which can be pushed out of a cup (demouldable) to heavier puddings and multi-layer desserts.

---

* Starch is considered particularly useful in formulating gelified milks since native starch is capable of modification to provide specific properties. Waxy maize starch, for example, may be stabilized by the introduction of ester cross-linking groups between chains. A high level of cross-linking results in a starch which remains of low-viscosity and easy to handle until UHT processed and which has a short and creamy texture. Where storage stability of the end-product is of particular importance, cross-linking may be combined with esterification and etherification.

Table 2.8 Structurally significant changes to starch during the cooking process in gelified milk production

1. Swells under combined effect of water and heat
2. Depolymerization of amylose and amylopectin
3. Thermal energy permits passage of water through molecular network
4. Starch granules swell with a corresponding decrease in density
5. Density ultimately decreases sufficiently to allow granules to remain in suspension thus increasing viscosity of the solution

(d) Milk analogues

Milk analogues have been manufactured from soya beans for many years using a relatively simple extraction technique involving soaking the beans, grinding in water, filtering to remove sediment and heating. This process has a high protein extraction efficiency but the resulting soya milk has a pronounced and unacceptable flavour due to the lipoxygenase-mediated oxidation of linoleic and linolenic acids, esters and triacylglycerols and formation of aldehydes, ketones and alcohols which impart the characteristic 'beany' flavour.

Various attempts have been made to overcome the acceptability problems caused by poor flavour. These include alkali soaking, acid grinding and hot water extraction, but while these improve flavour there is an adverse effect on the efficiency of protein extraction. A process, the Illinois process, has however been developed which produces a smooth, bland flavoured milk with good colloidal stability and a high efficiency of protein extraction (Figure 2.9). Residual lipoxygenase activity may be significantly reduced by anti-oxidants of which propyl gallate, especially in the presence of citric and ascorbic acid, is most effective.

Other legumes such as cowpea (*Vigna unguiculata*) have been used for preparation of milk analogues in countries such as Nigeria where soya

---

* Although legume-based milk analogues represent an important source of protein in developing regions, the biological value of the protein is considerably reduced by the presence of proteinase inhibitors. Eight proteinase inhibitors have been identified in soya beans, of which the Kunitz and Bowman-Birk inhibitors are of greatest importance. Proteinaceous inhibitors do not act by directly affecting intestinal proteolysis but, possibly, by stimulating pancreatic secretions leading to excessive endogenous losses of essential amino acids. Soya bean trypsin inhibitors may also be directly involved in the metabolism of methionine and may thus be responsible for apparent methionine deficiency. Although proteinase inhibitors may be inactivated by heat, it may not be possible to achieve total inactivation without a detrimental effect on protein nutritional quality and processes must be carefully designed and controlled to strike a balance between these two factors.

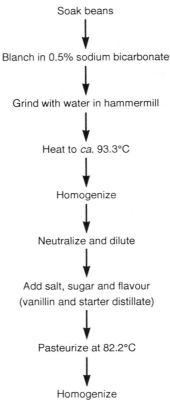

Figure 2.9 Illinois process for manufacture of soya milk analogue. Data from Nelson, A.W. *et al.* 1979. *Journal of Food Science*, **41**, 57–62.

beans are not readily available. Problems, however, remain with digestibility and overall acceptability.

2.3 CHEMISTRY

2.3.1 Nutritional changes during processing and storage

Nutritional changes during HTST pasteurization and UHT sterilization are, in most cases, small and of limited significance, although the effects of indirect UHT processing are greater than those of direct UHT processing. Post-process nutrient losses, however, can mean that the nutritional status of pasteurized milk and, under some conditions, UHT sterilized milk at the point of consumption is much reduced. Major changes occur in the nutritional status of milk during the in-bottle sterilization process, although the extent varies considerably with the severity of the process.

(a) Vitamins

Pasteurized and UHT sterilized milk have a similar spectrum of vitamin loss, although the extent is likely to be greater in UHT milk. The fat-soluble vitamins A, D and E and the water-soluble vitamins biotin, nicotinic acid, pantothenic acid and riboflavin are relatively heat-stable and there is no detectable loss during pasteurization or during most UHT processes. Losses of less than 10% of folic acid, thiamine, vitamin B_6 and vitamin B_{12} occur during pasteurization, losses of vitamin B_6 being slightly higher during UHT processing. The most significant loss is of vitamin C, total vitamin C content (ascorbic acid plus dehydroascorbic acid) being reduced by 10–25% during pasteurization and by 25%, or more, during UHT processing. Losses in total vitamin C content are almost entirely due to the heat instability of the oxidized form, dehydroascorbic acid, and are thus minimized by handling procedures which limit the dissolved oxygen content of the milk. Loss of vitamin C continues during storage and while photodegradation (see below) is involved, this may again be minimized by use of processing which excludes oxygen. UHT milk produced by direct heating, or by indirect heating followed by deaeration, has a very low oxygen content and storage losses of vitamins are insignificant. In the case of conventional indirect heated UHT milk, however, significant storage losses occur, not only of vitamin C, but also of folic acid, vitamin B_6 and, to a lesser extent, vitamin B_{12}. In the case of folic acid, 100% of activity can be lost within 14 days, while 50% of vitamin B_6 activity can be lost within 3 months. Folic acid can be protected by addition of vitamin C to the milk before processing. This practice leads, however, to markedly increased losses of vitamin B_{12}.

Light-mediated losses of vitamins are of importance in pasteurized milk, but in the case of UHT milk the use of aluminium foil-lined cartons provides total protection. As much as 75% of the riboflavin content can be lost during exposure to direct sunlight for 1 h and significant losses also result from the use of high intensity lighting in retail display cabinets. There is also a significant, but lesser loss of vitamin A.

The extent of vitamin loss during in-bottle sterilization varies according to the process, but in all cases major loss occurs. The vitamin C content is reduced by 30–100%, thiamine by 20–50%, vitamin B_6 by 15–50%, vitamin B_{12} by 20–100% and folic acid by 30–50%. Losses are signifi-

> **BOX 2.5 Through a glass darkly**
>
> In the UK, where delivery of milk to the home is common, bottles may remain on the doorstep from delivery in the morning until the householder returns in the evening. Loss of light-sensitive vitamins under these conditions is rapid and some years ago nutritionists, concerned about the consequences especially for children, suggested that milk should be filled into brown bottles. During trials, however, it was found that while brown bottles were effective in reducing light-mediated vitamin loss, the contents warmed very rapidly on sunny days leading to equally rapid spoilage.

cantly lower when continuous flow heating is used as part of the process than when the entire heat-treatment is in-bottle.

(b) Proteins and amino acids

The effect of both pasteurization and UHT sterilization is limited. The Maillard reaction is initiated, but loss of available lysine is only 1–2% in pasteurized milk, *ca.* 4% in direct heated UHT milk and *ca.* 5.5% in indirect heated. Such losses are not biologically significant. In pasteurized milk significant light-mediated losses can occur during storage. Losses are most significant with respect to methionine, tryptophan and tyrosine. Significant denaturation of whey proteins occurs during UHT processing but this has no effect on the biological value.

Available lysine losses of up to 13% can occur during processing of in-bottle sterilized milk. In contrast to pasteurization, or UHT sterilization, there is also significant formation of lysinoalanine which is present at levels of 170–570 mg/l after conventional processing. Lysinoalanine is formed by the reaction of free ε amino groups of lysyl residues with dehydroalanyl residues derived from the decomposition of cystine–cysteine and serine by β-elimination. Lysinoalanine is toxic in the diet of rodents but not monkeys or, by extrapolation, humans. Its formation is,

---

* Homogenization has no direct effect on the nutritional status of milk fat, but it has been claimed that the smaller size of the fat globules increases the digestibility of fat by the healthy adult. This in turn leads to xanthine oxidase, absorbed to the smaller fat globules, being carried to the artery walls and heart muscles where histochemical changes and, ultimately, disease states result from attacks on plasmalogens. Many nutritionists, however, treat these claims with considerable scepticism.

however, accompanied by a reduction in the Net Protein Utilization.

(c) Lactose

Lactulose formation increases with the temperature of heat treatment, typical values being *ca.* 50 mg/l in pasteurized milk, 100–500 mg/l in UHT sterilized milk and 900–1380 mg/l in conventional in-bottle sterilized milk. Differences in lactulose content may be used to distinguish in-bottle sterilized from UHT and pasteurized milk.

Variable quantities of galactose are formed in UHT and in-bottle sterilized milk as a consequence of thermal degradation. In very severe in-bottle treatment small quantities of epilactose are formed by epimerization and there is some degradation of galactose to tagatose.

(d) Minerals

Milk salts are of two types with respect to heat treatment; those unaffected such as sodium, potassium, chloride and sulphate, and those affected including calcium, magnesium, citrate and phosphate. Heating affects the equilibrium distribution of calcium salts which results in a decrease in soluble calcium and precipitation of solid calcium phosphate. Losses during pasteurization are significant only in exceptional circumstances despite the role of milk as a major dietary source of calcium. In the case of UHT milk, however, as much as 40–50% of soluble calcium appears in the colloidal phase and work based on rat feeding experiments suggests that due to a decrease in adsorption of calcium, UHT milk should not be depended on as a source of calcium for children and pre-menopausal women.

The formation of colloidal calcium during UHT processing is not fully understood. It is thought, however, that the newly formed calcium phosphate either builds up on that already present in casein micelles or

---

* Two pathways have been suggested for lactulose formation, alkaline epimerization of lactose, catalysed by the amino groups of casein or by the Lobry de Bruin–Alberda van Ekenstein transformation catalysed by phosphate and citrate. Lactulose is not hydrolysed by mammalian enzymes, but may be fermented in the large intestine with consequent flatulence. This is unlikely to be a significant problem with the quantities present in pasteurized, or UHT sterilized, milk but flatulence is a possibility if significant quantities of in-bottle sterilized milk are consumed. Lactulose formation can continue during storage of UHT sterilized milk, but only at temperatures in excess of 25°C, which effectively act as a secondary heat treatment (Jiminez-Perez, S. *et al.* 1992. *Journal of Food Protection*, 55, 304–6).

forms a new colloidal phase in association with denatured whey proteins.

2.3.2 Changes affecting structure and quality

(a) Proteins and amino acids

Heat treatment of milk results in denaturation of the whey proteins. The extent varies according to the severity of heat treatment from partial during pasteurization to total during in-bottle sterilization. Immunoglobulins are the most labile, serum albumin, β-lactoglobulin and α-lactalbumin having greater stability. However, denatured β-lactoglobulin interacts with κ-casein in a reaction involving a thiol–disulphide interchange, which alters the micellar surface structure and stability. The size distribution is also affected, the relative number of both large and small casein micelles increasing. The increase in the number of small casein micelles increases the reflectance of milk leading to a whiter appearance. This is enhanced by changes resulting from homogenization, but non-enzymatic browning reactions have an opposite effect, lowering reflectance and increasing the green and yellow components of the colour system.

The denaturation of whey proteins plays an important role in the development of 'cooked milk' flavour. This is insignificant in HTST pasteurized milk, but forms part of the characteristic taste of sterilized milk, albeit of secondary importance to the bitter and acrid flavours resulting from the Maillard reaction and from changes to lipids.

In the case of UHT sterilized milk 'cooked milk' flavour is a serious quality defect. It is readily detectable both in the milk itself and in other foods and drinks containing milk as an ingredient and is probably largely responsible for the relative unpopularity of UHT milk. The defect may be described by other terms including 'cabbage', 'sulphur' and 'caramel' flavours. Sulphydryl groups, particularly free or active sulphydryls, are generally considered to be responsible for 'cooked milk' flavour. These are derived from whey proteins, the sulphydryl groups of which are exposed during thermal denaturation. Particular attention has been paid to β-lactoglobulin, each 36 000 Dalton dimer of which contains two sulphydryl (–SH) groups and four disulphide (–S–S) groups and which accounts for over 50% of the whey proteins. Sulphydryl and disulphide groups are also present in blood serum albumin and disulphide groups in α-lactalbumin but casein contains only very few sulphydryl groups and makes little contribution to development of 'cooked milk' flavour.

During UHT processing the number of active sulphydryl groups increases, probably at the expense of disulphide groups with production following first order reaction kinetics. At the same time the total amount of sulphydryl decreases as a consequence of volatilization and at higher temperatures this also limits the total amount of active sulphydryl present.

Detailed studies have been made of the low molecular weight volatile sulphur compounds present in UHT milk. 'Cabbage' flavours showed a close correlation with total volatile sulphur and it is possible that volatile sulphur components rather than active sulphydryl groups are responsible for the very strong 'cooked' and 'cabbage' flavour of some types of freshly processed UHT milk. The major components have been identified by gas chromatography and mass spectrometry as hydrogen sulphide, carbonyl sulphide (COS), methanethiol (CH_3SH), carbon disulphide (CS_2) and dimethyl sulphide (($CH_3)_2S$).

'Cooked milk' flavours are most apparent immediately after processing but diminish with time to a normal acceptable flavour which in turn is supplanted by unacceptable oxidized or rancid flavours. Changes in flavour during storage have been shown to occur in two phases, each of which has a number of distinct stages (Figure 2.10). Reduction in 'cooked milk' or 'cabbage odours' occurs most rapidly in milks of high oxygen content and these are preferred for up to 13 days after manufacture. A high oxygen content, however, also leads to greater vitamin loss and a more rapid onset of fat oxidation and consequent oxidized or rancid flavours resulting from production of methyl ketones and aldehydes. Changes occur irrespective of the temperature of storage but are more rapid at higher temperatures.

Development of 'cooked milk' flavour is much greater in indirect heated milk than in direct heated due to the removal of volatiles during flash cooling. Direct heated milk may be considered to be at the end of the primary phase immediately after manufacture and often has organoleptic properties very similar to those of pasteurized milk.

The major whey proteins are also involved in light-mediated reactions during storage of milk. Riboflavin catalyses a number of reactions which result in the formation of new low molecular weight fractions and an increase in amino groups. Whey proteins can also serve as oxidizable substrates for the photogeneration of superoxide anion, riboflavin again serving as catalyst.

In addition to nutritional consequences, light-mediated degradation of

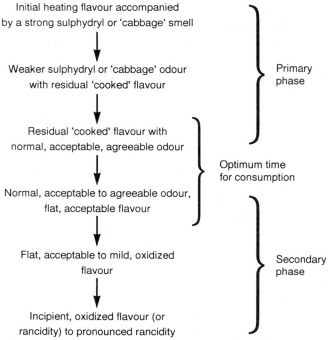

Figure 2.10 Changes in 'cooked milk' flavours during storage of UHT milk.

methionine to methional is of importance in the development of off-flavours and curtailment of storage life. Methional has been shown to be responsible for the development of off-flavours variously described as 'activated', 'oxidized' and 'sunlight'. Light-mediated reactions may also lead to discoloration of milk by degradation products of tryptophan and tyrosine.

(b) Lipids

Homogenization has a number of important consequences for the chemistry of milk fat. The major technological objective of homogenization, prevention of gravity separation, is achieved not only by reducing the size of the fat globule from 3 to 10 μm to less than 2 μm, but also by stabilization against cluster formation due to denaturation of immunoglobulins and interaction of the globule surface with casein. The fat surface is increased four to sixfold and the globule stabilized by adsorption of milk proteins including casein micelles and subunits. Casein-associated lipases are inactivated by heating, but

> **BOX 2.6 No noxious thing**
>
> In recent years there has been considerable concern over the mutagenic potential of compounds produced in heated foods. In-bottle sterilized milk has been identified as a possible high risk food with respect to mutagen production, but investigations suggest that fears are unfounded under commercially used processing conditions. Potential mutagens may, however, be formed in milk during domestic cooking (Berg, H.E. *et al.* 1990. *Journal of Food Science*, **55**, 1000–3).

the fat remains susceptible to microbial lipases and to light-mediated lipolysis.

During heating lactones and methyl ketones are formed from fat and have a deleterious effect on flavour. Quantities present in pasteurized milk are small in comparison with those in more rigorously heated milk and in the case of methyl ketone, quantities present are little higher than in unheated milk (12 nmol/g of fat versus 10 nmol/g of fat). Higher levels are present in UHT milk (*ca.* 21 nmol/g), but formation of lactones and methyl ketones is of major significance only in in-bottle sterilized milk, where levels typically exceed 100 nmol/g of fat.

Deteriorative changes in fats during storage result primarily from light-mediated lipid oxidation or heat-resistant microbial lipases. Copper, or iron, induced oxidation of unsaturated fatty acids is of less importance where modern stainless steel equipment is used but contributes to overall off-flavour development.

(c) Enzymes

Pasteurization inactivates a number of milk-derived enzymes and enzyme inactivation tests such as the phosphatase test have been used for many years as an index of adequate pasteurization (see page 88).

---

* Homogenization reduces the concentration of the fat globule membrane in relation to the surface area of the fat globule and this may have important consequences with respect to fat autoxidation. The situation is not straightforward and it is probable that both anti-oxidant and pro-oxidant factors coexist in the fat globule membrane. Further some components may be able to act as either anti- or pro-oxidants depending on environmental conditions (Chen, Z.Y. and Nawar, W.W. 1991. *Journal of Food Science*, **56**, 398–41; Berg, H.E. *et al.* 1990. *Journal of Food Science*, **55**, 1000–3).

As noted above, milk-derived lipases are also inactivated although if the fat globule membrane has been badly damaged there may be some residual activity. In contrast, the major milk proteinase, plasmin, is largely resistant to pasteurization retaining 70–80% of activity after HTST treatment. Plasminogens appear to be less affected than active plasmin and the rate of activation of plasminogens may increase after pasteurization due to inactivation of compounds which usually act as inhibitors. Pasteurization therefore has no apparent effect on the overall level of proteolytic activity, even if some inactivation of plasmin occurs. Some 30–40% of plasmin activity remains after UHT sterilization but, according to the heat treatment applied, the enzyme may be totally inactivated by in-bottle sterilization.

Proteases and lipases produced by psychrotrophic bacteria (see pages 92–5) are markedly heat resistant and readily withstand pasteurization and UHT sterilization. The relatively short refrigerated storage lives mean that heat-stable enzymes are of limited importance in the spoilage of pasteurized milk, but are associated with age thickening, or gelation, which is a serious problem with UHT milk. The cause is limited hydrolysis of casein by proteolytic enzymes, followed by physical aggregation of the modified casein micelles to form a gel structure which entraps whey proteins and fat globules. There is also evidence for the involvement of plasmin. The UHT process itself plays a role in predisposing milk to gelation by producing a high proportion of small-sized micelles.

Thickening, or incipient thickening, may be accompanied by the development of off-flavours, the 'astringent' defect being attributed to γ-casein like polypeptides resulting from breakdown of casein. Bitter favours also develop as a result of the activity of heat-stable microbial proteases. Heat-stable lipases are not involved in structural changes but activity leads to an increase in fatty acid content of the milk and acidic off-flavours.

(d) Non-enzymatic browning

Non-enzymatic browning *via* the Maillard reaction is initiated during both pasteurization and UHT sterilization. The use of low temperature storage for relatively short periods means that the reaction remains incipient in pasteurized milk. In UHT milk the reaction continues during storage, although there is no significant increase in the end-product 5-hydroxymethyl-2-furaldehyde (hydroxymethylfurfural: HMF) at moderate ambient temperatures. This has been attributed to loss of HMF through oxidation, or other transformations. At temperatures of *ca.* 35°C, however, UHT milk takes on the characteristics of in-bottle sterilized milk. In general browning reactions are more significant in indirect heated milk.

Non-enzymatic browning occurs to a considerable extent during in-bottle sterilization and yields, among other end-products melanoidins, which are responsible for the brown coloration and HMF, which is responsible for acrid flavours. Partial Maillard degradation of lactose also leads to formation of organic acids, principally formic acid, which lowers the pH of the milk and can cause protein destabilization.

(e) Sedimentation

Sedimentation is a relatively common fault which shortens the life of UHT sterilized milk through deposition of a layer of proteinaceous material. The problem involves the transport of intact micelles from the bulk of the solution to the bottom of the container. The phenomenon has been poorly understood and various theories developed, many of which concern aggregation of casein micelles as a result of heat-induced changes. More recently sedimentation has been shown to be a physical phenomenon, unavoidable when the milk is left undisturbed. Native micelles sediment to some extent, but the sedimentation rate is faster after UHT processing due to the higher weight of heated micelles. Some aggregation may occur, but is not necessary for sedimentation.

2.3.3 Chemical analysis of milk

Chemical analysis of milk is required to ensure that minimum compositional standards are met and that the milk is free from adulterants and other contaminants. Analysis is usually carried out on milk both before and after processing. Commonly applied tests, and their function, are listed in Table 2.9.

Table 2.9 Function of chemical analyses applied to milk

Protein	Compositional quality: some payment schemes
Fat	Compositional quality: some payment schemes
Lactose	Compositional quality: some payment schemes
Cell count	Udder health: payment schemes in United States
Sediment	Production hygiene
Anti-microbials	Good veterinary practice: suitability for sale/processing

Chemical criteria are often used as part of bonus payment schemes to encourage good standards of milk production (See Chapter 1, page 5). Historically fat content has been the main criterion, but protein and, in some countries, lactose content are now increasingly considered to be of importance and, in addition to fat content, form the basis of bonus payment schemes. Somatic cell counts, previously used to monitor udder health and the suitability of milk for processing, now also form the basis for bonus payment schemes in the US, while determination of sediment is of lessened importance following widespread improvements in farm hygiene.

Analytical methods for the commonly used tests are very well established. In recent years there has been a growing trend towards instrumental methods (Table 2.10), although in some cases the traditional method remains definitive. Infrared measurements, for example, are acceptable as a screening method for excess water in milk, but legal action for adulteration must be based on traditional methods of measuring freezing point depression.

In recent years increasingly sophisticated analytical techniques have been required to enable testing for antibiotic and drug residues and, in the case of milk from animals such as goat, for detecting adulteration with milk from other species such as cow.

Beta-lactam antibiotics have previously been of greatest concern in milk and a sensitive disc assay using *Bacillus stearothermophilus* was

Table 2.10 Traditional and instrumental analysis of milk

Parameter	Traditional methodology	Instrumental methodology
Protein	Kjeldahl	near-infrared spectroscopy
Fat	Gerber/Rose–Gottlieb	Milko-tester
Lactose	Polarimeter	near-infrared spectroscopy
Adulterants		
water	Freezing point depression using cryoscopy	infra-red analysis[1]
other species	None available	Enzyme-linked immunoassay
Somatic cell count	Microscopy	Coulter counter™ Fossmatic™[2]

[1] Infrared analysis is suitable only as a screen and freezing point depression must be used for definitive analysis.
[2] Both Coulter counter™ and Fossmatic™ can show unacceptable variation when used in bonus payment schemes and greater standardization is required.

developed for the detection of residues. This test, however, is relatively insensitive to other antibiotics and the 3 h test time required means that individual tanker testing, the most effective means of control, is not possible. In response to this requirement a number of rapid tests providing results in *ca.* 10 min have been developed. Among the best known are the Charm II™ test, the Cite™ test and the Penzyme™ test. The Charm II™ test is a competitive radioimmunoassay using binding sites on microbial cells or, in the case of tetracyclines and chloramphenicol, antibodies as binding sites. The test detects a wide range of antibiotics as well as sulpha drugs and shows good correlation with standard disc assays. It does, however, require the use of radiolabelled material.

Screening tests cannot identify individual residues and high performance liquid chromatography has been used for confirmatory purposes. The Charm II™ test system may be used as detector.

The increasing popularity of milk of other species, especially goat, which is sold at a premium price, has led to some instances of adulteration with cheaper cows' milk. This problem is generally greater with cheese, but has also occurred with milk. Serological assays have been developed, usually based on antisera to whey proteins, which permit detection of adulterating milk at a low level. The enzyme-linked immunosorbent assay (ELISA) has been widely used.

The quality of the raw milk is generally considered to be more critical for UHT processing and pH value is an effective predictor of instability in the end-product. Determination of pH value is most usefully made on the milk after bulk storage. It is also common practice to determine the numbers of psychrotrophic bacteria at this stage, but viable counts may not reflect levels of heat-stable enzyme activity. Direct assays for proteolytic enzymes are more valuable analyses and a number of methods, usually based on ELISA, have been described.

Chemical assays also play an important role in verifying the efficiency of pasteurization and in-bottle sterilization. Laboratory testing for verifica-

---

* The use of ELISA can cause difficulties in smaller quality control laboratories which lack experience of serological techniques. In this situation an alternative approach to assay of proteolytic enzymes may, when fully developed, be more appropriate. This assay is based on the degradation of luciferase by the proteolytic enzymes, residual luciferase then being determined by the ATP-driven bioluminescent luciferin–luciferase reaction (Rowe, M. *et al.* 1991. *Dairy Industries International*, **56**(12), 35–7).

tion of correct pasteurization is classically based on inactivation of the enzyme alkaline phosphatase. Use of this test is a statutory requirement in many countries and standard methods, usually based on colourimetric determination of residual phosphatase, are well established. Problems of false-positive reactions occur due to reactivation of phosphatase after heating and the presence of heat-resistant phosphatases produced by psychrotrophic bacteria during growth in raw milk. False-negative reactions, of major significance with respect to safety, have been ascribed to operator error, but it must be appreciated that colourimetric assay methods are relatively insensitive and will not detect the small quantities of raw milk which may be present as a result of 'surging', or delayed response of the flow diversion device. More sensitive fluorimetric assays such as the Fluorophos ™ are now available and offer a greater degree of protection. The turbidity test is used as verification that in-bottle sterilized milk has received the legal minimum heat processing.

2.4 MICROBIOLOGY

2.4.1 Liquid milk products and foodborne disease

(a) Pasteurized milk

Pasteurized milk has been responsible for a number of outbreaks of foodborne disease in recent years including an outbreak of salmonellosis in Chicago, USA which involved 16 284 known cases during 1985.

Campylobacter has also been implicated as the cause of foodborne disease associated with pasteurized milk. Underprocessing appears to

---

* Alkaline phosphatase is not suitable as an index of heat treatment at the higher pasteurization temperatures now increasingly used and it is necessary to consider other enzymes. Lactoperoxidase has been considered as an index for milk heat treated at 78°C for 15 s for international commerce while γ-glutamyl transpeptidase has been proposed for in-plant use as an index of pasteurization at the same temperature (Patel, S.S. and Wilbey, R.A. 1989. *Journal of the Society of Dairy Technology*, **42**, 79–80).

* This is the largest outbreak of salmonellosis ever recorded and the actual number of cases may have been as high as 250 000. Both this and a smaller outbreak in Cambridge, UK resulted from contamination of pasteurized milk by raw milk. The underlying causes were very similar, involving a raw/pasteurized interface consisting of a loop pipe 'safeguarded' by valves. This inherently unsafe design, which in the Cambridge plant also involved a common cleaning-in-place circuit, was compounded by poor plant maintenance, poor operator training, incorrect plant operation and poor supervision. An alarming aspect of the report of the Outbreak Control Team at Cambridge was that the situation which had led to contamination was likely to be similar in plants throughout the country (Romney, T. 1988. *Food Science and Technology Today*, **2**, 268–71).

have been involved rather than post-process contamination, although the latter route is a serious potential hazard. Further, many underlying factors were similar to those of the *Salmonella* outbreaks; unsatisfactory plant design, inadequate operator training and poor supervision.

The first of two published reports of *Campylobacter* infection associated with pasteurized milk involved more than 2500 children attending school in two English towns. It is probable that raw milk by-passed the pasteurization plant, although evidence was lacking. In a smaller outbreak involving a boarding school in the US, *C. jejuni* survived batch pasteurization. The pasteurization plant was privately operated by the school and the severity of the process had been arbitrarily lessened to reduce 'burnt flavours' in the milk.

Pasteurized milk was the vehicle in an outbreak of *Listeria monocytogenes* infection which affected 49 people in the US. This outbreak was of particular significance in that epidemiological evidence failed to fully resolve the cause of the outbreak and the suggestion was made that *L. monocytogenes* had survived correctly applied HTST pasteurization. This has led to an extensive investigation of the heat resistance of the organism in the context of commercial HTST pasteurization.

Listeria monocytogenes is present within phagocytes in milk drawn from infected udders and it has been postulated that the apparently high degree of heat resistance is due to protection by the phagocytes. The concept that phagocytes provided a thermal shield was dismissed on the basis of the thermodynamics of pasteurizer operation and attention became focused on the possibility of *L. monocytogenes* acquiring enhanced thermal tolerance through the mediation of induced stress proteins. A possible role for heat-stable phagocytic superoxide dismutase has also been discussed.

It has been thought that *L. monocytogenes* may enter a heat resistant state as a result of exposure to the rise in body temperature (up to 42.8°C) of cows in response to infection. Heat induced thermotolerance

---

* Birds (jackdaws and magpies) have been implicated as vectors of *Campylobacter* in an outbreak which affected up to 59 people in Gateshead, UK, during 1990. The birds contaminated bottled pasteurized milk when pecking through the foil caps, *Campylobacter* being isolated both from beaks and milk in pecked bottles. Persons at risk lived in new housing close to open country with large bird flocks, but residents were sceptical of risk and responded poorly to a health education scheme. Earlier outbreaks of campylobacteriosis in which birds were implicated as vehicles of infection had been reported in South Wales (Hudson, S.J. *et al.* 1991. *Epidemiology and Infection*, **107**, 363–72).

has been demonstrated but the significance is disputed. Workers who used strictly anaerobic (Hungate technique) incubation to enhance recovery of the damaged cells, consider that survival of *L. monocytogenes* during HTST pasteurization presents a finite risk, but other experts have stated categorically that no risk of survival exists. A significant development was the application of risk analysis to survival of *Listeria* during pasteurization. All relevant factors were included in the analysis and it was concluded that there is no justification for raising the pasteurization temperature, or increasing the holding period. The current situation appears to be that the majority of informed opinion considers that standard HTST pasteurization is adequate to inactivate *L. monocytogenes* at levels conceivably present in milk. The safety margin may, however, be lower than during LTLT pasteurization. Doubts do, however, persist amongst professional microbiologists, especially in view of the possible inadequacy of recovery methods for sub-lethally stressed cells.

Yersinia enterocolitica has been implicated in three large outbreaks of illness associated with pasteurized milk in the US including the first reported outbreak of foodborne yersiniosis in Oneida County, New York. Chocolate-flavoured milk was involved, the organism being introduced with chocolate syrup which was added after pasteurization.

The largest known outbreak of yersiniosis resulted from consumption of pasteurized milk and involved several thousand people in Memphis and other areas of the US. The cause was unusual in that the organism entered the milk at point of consumption due to contamination on the outer surface of cartons. The contamination was derived from improperly cleaned crates which had been used to deliver waste milk to a pig farm where there had been direct contact with pig manure. This outbreak illustrates the need for extreme caution where milk processing operations have close contacts with farms.

Two milkborne outbreaks of yersiniosis occurred among hospitalized children in the UK. Pasteurized milk was involved in each outbreak, one

---

* Diagnosis of yersiniosis can be complicated by the variability of symptoms. In some cases, especially those involving teenaged children, the symptoms may be confused with those of acute appendicitis and during the Oneida County outbreak unnecessary appendectomies were carried out on no less than 18 of 38 known victims. Diagnosis was also complicated in the Memphis outbreak by the extraintestinal nature of symptoms amongst adults, the predominant symptom being pharyngitis with positive throat cultures.

of which continued over a period of several months. In the latter case *Y. frederiksenii* was isolated from some patients as well as *Y. enterocolitica*.

Staphylococcus aureus has only rarely been involved in food poisoning associated with consumption of pasteurized milk, although enterotoxigenic strains may readily be isolated from raw milk. An outbreak involving chocolate-flavoured milk in California, however, involved at least 500 school children and was attributed to growth of the organism in raw milk and survival of the enterotoxin (A) during pasteurization.

Although *Bacillus cereus* and other potentially pathogenic species may be readily isolated from pasteurized milk, and can be involved in spoilage (see below) there have been no fully authenticated reports of food poisoning.

(b) UHT and in-bottle sterilized milk

The safety record of these milks is extremely good and there have been no published reports of foodborne disease associated with their consumption.

2.4.2 Spoilage of liquid milk products

(a) Pasteurized milk

The spoilage microflora of pasteurized milk is of two types; post-process contaminants which have entered the milk after heating and heat resistant bacteria which have survived heating. Under current circumstances post-process contaminants are of greatest spoilage significance. A number of surveys in the UK, the US and elsewhere have shown that the level of such contamination is often unacceptably high.

Post-process contamination is dominated by Gram-negative rod-shaped bacteria. These often have a degree of resistance to commonly used sanitizers and are able to colonize milk contact surfaces including stainless steel and Buna™ rubber gaskets. Examinations made directly

---

* Some strains of *Citrobacter*, *Enterobacter* and *Serratia* have been implicated as causes of diarrhoeal disease, although in no case has there been an association with pasteurized milk. Environmental strains may also be distinct from clinical isolates. In this connection it should be appreciated that environmental species of *Yersinia*, some of which were previously identified with *Y. enterocolitica*, but which are generally considered non-pathogenic are also fairly common in pasteurized milk.

after processing commonly indicate that members of the *Enterobacteriaceae* including *Serratia*, *Enterobacter*, *Citrobacter* and *Hafnia* are numerically dominant. Such bacteria are of environmental origin, probably derived originally from water and, unlike *E. coli*, have no role as an index organism. Despite the numerical dominance of the *Enterobacteriaceae* in newly pasteurized milk, the ultimate spoilage microflora usually consists of psychrotrophic Gram-negative rods, primarily *Pseudomonas*, *Alcaligenes* and, to a lesser extent, *Flavobacterium*. Species of *Acinetobacter* and *Psychrobacter* ('*Moraxella*-like bacteria') are occasionally present in large numbers and bacteria such as '*Alteromonas putrefaciens*' ('*Pseudomonas putrefaciens*'), which cannot readily be assigned to existing genera, may also be isolated. During storage at temperatures below *ca.* 8°C, the competitive advantage of psychrotrophic genera over the *Enterobacteriaceae* permits relatively rapid development from an initially small sub-population. The *Enterobacteriaceae*, however, may remain dominant at storage temperatures above 8°C and, occasionally, in other circumstances.

Spoilage by psychrotrophic Gram-negative rod-shaped bacteria is usually proteolytic and lipolytic in nature and involves various off-tastes, clot formation and, in some cases, virtually complete digestion of the protein. Phospholipases and glycosidases may also be involved in spoilage and where fluorescent pseudomonads are present in large numbers, visible quantities of pigment may be present when spoilage is advanced. *Acinetobacter* and *Psychrobacter* are of limited spoilage potential, although there may be some lipolysis. Occasional strains also produce a capsular polysaccharide and growth results in the classic 'ropy' milk. Growth of *Acinetobacter* and *Psychrobacter* at low temperatures tends to be slower than that of other psychrotrophs and these genera are often overgrown by *Pseudomonas*. *Pseudomonas fragi* has been associated with a specific defect, 'fruity-flavour' due to production of ethyl butyrate and ethyl hexanoate from butyric and caproic acid. The defect is of particular prevalence in milk from cows fed large quantities of silage.

Spoilage due to members of the *Enterobacteriaceae* varies according to the biochemical properties of the various genera. Lactose-fermenting genera, for example, produce acid clotting, possibly with gas formation, while spoilage by other genera primarily involves proteolysis. Off-taints are produced which may be 'faecal' in nature.

Other micro-organisms are occasionally present as post-process contaminants. Acidification due to species of *Lactobacillus* and *Lactococcus* is rare, but can occur if milk is held at ambient temperatures. Lactic acid formation is accompanied by a sour odour due to trace amounts of acetic acid, propionic acid and other volatiles. In some cases the sour aroma is detected before acidification. Some strains of *L. lactis* sub-sp. *lactis* are capable of producing 3-methylbutanal from leucine and imparting a malty flavour to milk. In the US, at least, 3-methylbutanal-producing strains show a geographical distribution, being of significantly greater prevalence in north-eastern states than in Pacific coastal states. A strain of *Lactobacillus* has also been implicated in production of malty flavours.

Heat resistant bacteria present in pasteurized milk are of two types, endospore-forming genera and genera of high vegetative heat resistance. Endospore-forming genera are of greatest importance and endospores are readily isolated in small numbers from pasteurized milk, the numbers and types reflecting those in the raw milk. Species of *Bacillus* are of greatest importance, although clostridial endospores are also commonly present.

Although *Bacillus* species are the most important members of the heat resistant microflora of pasteurized milk, their role in spoilage is limited and even though psychrotrophic species are able to grow at temperatures below 5°C, they are almost invariably overgrown by Gram-negative, post-process contaminants. There are, however, exceptions and temperature abuse can permit rapid development of species such as *B. cereus*, which can grow rapidly at temperatures above *ca.* 8°C. *Bacillus cereus* is classically associated with 'bitty' spoilage of the cream layer due to production of extracellular phospholipase (lecithinase), although in practice soft clotting may be a more common form of spoilage. Milk churns (cans) were a notorious source of *B. cereus* and the incidence of spoilage due to the organism is usually lower where bulk collection is used. Bulk tanks may themselves, however, become a source of psychrotrophic *Bacillus* species. In common with other *Bacillus* species, the incidence of *B. cereus* shows seasonal variation and spoilage is more common in the summer months.

Psychrotrophic species of *Bacillus* become the dominant spoilage organisms at storage temperatures below 5°C when competitive Gram-negative microflora are present only in low numbers, possibly as a result of the imposition of severe precautions against post-process contamination in attempts to extend storage life. *Bacillus circulans* is able to grow at 2°C and *B. cereus* and similar species at 4–5°C and there is also some evidence that mesophilic species can adapt to psychrotrophy. Spoilage patterns vary according to the species present, but the problem is such that only limited life extension is possible. The heat resistance of endospores of psychrotrophic species is relatively low and attempts have been made to extend the HTST process to inactivation of these endospores either by increasing the severity of the process, or by introducing a double heat treatment (tyndallization). These attempts have met only limited success and indeed may enhance the subsequent growth rate of spoilage micro-organisms due to inactivation of naturally occurring inhibitors (see pages 33–4) and activation of endospores. Other methods of reducing the number of endospores including removal by centrifugation (bactofugation) and separation–sterilization treatment may be more succesful but involve reduced throughput and greater capital expense.

Although endospores of *Clostridium* species are present in pasteurized milk, the redox potential is almost invariably too high to permit germination and growth.

A number of vegetative bacteria are able to survive HTST pasteurization. Most of the genera involved are Gram-positive and have only a minimal role in spoilage, especially during refrigerated storage. Common isolates include species of *Microbacterium, Micrococcus, Enterococcus, Streptococcus, Lactobacillus* and 'coryneform' bacteria. Survival rates vary from 100% in the case of *Microbacterium lacticum* to less than 1% in the case of some strains of *Enterococcus, Streptococcus, Lactobacillus* and 'coryneform' bacteria. A single Gram-negative bacterium, '*Alcaligenes tolerans*', is able to survive, although at a level of only 1 to 10% and the spoilage significance is not known.

(b) UHT sterilized milk

A target level of 0.02% failure (leading to spoilage) is generally accepted by UHT processors, investigations and remedial action being initiated in response to a trend towards 0.05–0.07% failure, or single instances of failure in the order of 0.1%.

The majority of post-process contamination is related to packaging problems. These usually involve the aseptic filling process, or faulty seams or pin-holes in the packaging itself. The nature of contamination is random and many types of micro-organism may be involved. Spoilage is often rapid and may occur before the pack is acquired by the consumer.

Spoilage of UHT milk is usually the result of post-process contamination.

Instances of contamination due to failure of equipment sterilization downstream of thermal processing have occurred, but such instances are now increasingly rare. Contamination of equipment leads to spoilage of an entire batch, or batches, and indicates major shortcomings in operation and control.

Spoilage due to the survival of heat-resistant endospores is less common. The thermophilic *B. stearothermophilus* has the greatest survival potential, but is unable to grow below *ca*. 30°C and is only a major problem in hot climates. In temperate climates *B. coagulans*, *B. subtilis* and *B. licheniformis* are the most important spoilage species, although more heat resistant strains of *B. cereus* have also been implicated.

Heat-resistant enzymes produced by psychrotrophic bacteria growing in the milk before heat treatment are a further important cause of both proteolytic and lipolytic spoilage. Both proteolytic and lipolytic enzymes are produced, spoilage by proteolytic enzymes involves gelation and bitter flavours, while lipolytic enzyme spoilage results in rancid flavours (see page 84). Heat-stable enzymes are produced by a wide range of psychrotrophic bacteria including *Pseudomonas*, *Alcaligenes* and *Flavobacterium*. Although growth temperature is important in determining the quantity of enzyme produced, the excreted enzyme has a similar heat stability irrespective of growth temperature. There has been disagreement concerning the kinetics of inactivation, but it is generally agreed that residual activity after UHT sterilization can be as high as 40% of that in the raw milk. In some instances it is not possible to directly relate numbers of psychrotrophs in the raw milk to thermostable enzyme activity in the end-product, but it appears that UHT milk prepared from raw milk containing more than 5×10^6 cfu/ml psychrotrophs is subject to rapid spoilage.

(c) In-bottle sterilized milk

Leakage through caps and consequent post-process contamination can lead to spoilage of in-bottle sterilized milk, but this is rare and most problems result from the survival of heat-resistant endospores. Endospores of *B. stearothermophilus* are present in varying numbers, but germination and outgrowth do not occur under usual conditions of storage. Some older processes were, however, marginal with respect to commercial sterility and while modern processes have a higher safety margin, spoilage does occasionally occur as the result of the survival of endospores of mesophilic species of *Bacillus*. The pattern of spoilage is determined by the properties of the causative organism and typically involves gas production and various types of curdling.

2.4.3 Microbiological analysis

(a) Incoming raw milk

It is common practice to monitor the microbiological quality of incoming raw milk and in some countries, including the UK, quality bonus payments are partly based on this criterion (see Chapter 1, page 6).

For many years the quality of incoming raw milk was assessed on the basis of a mesophilic plate count (standard plate count) together with, in some countries, a dye reduction test. The standard plate count is still widely used for legislative purposes but it is now recognized that the widespread use of on-farm refrigeration has significantly reduced the value of these earlier tests and that the psychrotrophic plate count is the most reliable indicator of conditions on the producing farm. Like all viable counts, however, the psychrotrophic plate count is imprecise, labour intensive and requires 10 days' incubation at 7°C to obtain a result. The direct psychrotrophic count has, therefore, largely been replaced by counts involving preincubation at *ca.* 6°C for 5 days before a colony count incubated at 21–25°C. Although an improvement, such techniques remain unsatisfactory and there has consequently been considerable interest in developing alternative, more rapid, methods. These are usually based on preincubation at an elevated temperature, in the presence of an inhibitor of the Gram-positive (non-psychrotrophic) microflora, followed by enumeration of the dominant, Gram-negative (psychrotrophic) microflora. Enumeration may be made by conventional colony count techniques, or by rapid or automated methods. The direct epifluorescent technique (DEFT) and measurement of microbial adenosine triphosphate are probably the most widely used rapid methods, although the *Limulus* lysate test is also used. Electrical measurement based on impediometry/conductimetry is most advanced with respect to automation and data handling, but cannot be considered a truly rapid technique. In many countries milk testing is on a centralized basis and the very large scale of operations has led to a

---

* The ideal test for microbiological quality has been defined as being rapid and economical and reflecting the total number of organisms in the milk sample, the number of psychrotrophic organisms, conditions of production on the farm and the time and temperature of storage. Meeting these objectives has been a preoccupation of dairy microbiologists since the earliest days of the science (Bigalke, D. 1984. *Dairy and Food Sanitation*, 4, 189–90).

demand for automation even where cultural techniques are used. A wide range of approaches to automation have been adopted including the use of laboratory robots, rapid plating devices such as the Spiral plate maker™ and Autoloop™ and laser colony counters.

A comparison of the standard European Economic Community test (preincubation for 5 days at 6°C, enumeration at 21°C for 25 h) with an accelerated test (preincubation for 24 h at 21°C in the presence of sodium deoxycholate, enumeration at 30°C for 24 h) showed both methods to correctly classify the majority of samples. Despite this, it was considered that the need for a simple, truly rapid means of determining psychrotrophic bacteria still exists.

It has been argued that the number of heat-resistant endospores should be determined in raw milk destined for sterilization. Numbers of endospores are conventionally determined after heating at 80°C for 10 min, but a more severe treatment of 100°C for 10 min is considered to be more relevant to milk for sterilization. An upper limit of 10 spores/ml has been suggested, but a more stringent standard may be required where the milk is to be packed into a large volume container.

(b) Pasteurized milk

Microbiological tests applied to pasteurized milk may be placed in three broad and overlapping categories.

1. Tests on freshly pasteurized milk to ensure that general milk quality and post-process hygiene are satisfactory.
2. Tests to ensure that keeping quality is likely to be satisfactory.
3. Tests to ensure that handling during retail distribution is satisfactory.

Various approaches may be taken and two, the legislative requirements in England and Wales (EEC regulations) and the voluntary milk quality scheme introduced in Scotland are compared in Table 2.11. The Scottish scheme is more extensive with respect to freshly pasteurized milk and also takes a different approach to keeping quality and handling during retail distribution.

In England and Wales determination of keeping quality is a retrospective test based on the number of psychrotrophic bacteria present after low temperature storage. In contrast the Scottish scheme is predictive and based on enumeration of the Gram-negative bacteria present in milk

Table 2.11 Examples of criteria for defining acceptable quality of pasteurized milk

England and Wales	
Freshly pasteurized	
total plate count (30°C for 72 h)	<3 × 10^4 cfu/ml
coliforms	<1/ml
After storage for 25 h at 21°C	
total plate count (21°C for 25 h)	<1 × 10^5 cfu/ml
During retail distribution	
methylene blue test	decolorization >30 min
Scotland (Scottish pasteurized milk quality testing scheme)	
Freshly pasteurized milk	
freezing point depression	>0.530°C
total plate count (30°C for 72 h)	<2 × 10^4 cfu/ml
coliforms	<1 cfu/ml
thermoduric count[1]	<1 × 10^4 cfu/ml
Keeping quality	
Griffiths' technique	>4 days
During retail distribution	
coliforms	<100 cfu/ml

[1] After heating milk at 63°C for 30 min.

after preincubation at 21°C for 25 h. Enumeration is on a selective medium containing crystal violet, penicillin and nisin to prevent growth of Gram-positive bacteria. Alternatively the inhibitors may be added to the milk itself prior to preincubation followed by enumeration on non-selective medium, or by a rapid method such as determination of microbial ATP. Preincubation at 15°C for 25 h has been proposed as a means of improving the predictive value of the test. Preincubation techniques have been used in other countries including the US where impedance detection proved most effective.

Testing to ensure satisfactory handling during retail distribution has been introduced relatively recently and reflects the increasing level of sales through supermarkets. The use, in England and Wales, of the methylene blue dye reduction test is of dubious value and the coliform test used in Scotland is considered more appropriate.

(c) UHT sterilized milk

In contrast to both pasteurized and in-bottle sterilized milk there is no index test for UHT treatment. End-product testing is therefore required to verify correct heat treatment as well as ensuring the absence of post-process contamination.

Table 2.12 Destructive methods of determining sterility of UHT milk

Visual-organoleptic evaluation
Microbiological methods
    Conventional colony counts
    Automated colony counts
    Direct epifluorescent technique
    ATP measurement
    Impediometry
    Turbidometry
Gas chromatographic determination of microbial metabolites
pH value
Titratable acidity
Stability to ethanol
Resazurin reduction
Nitrate reduction

Unless circumstances are truly catastrophic the number of micro-organisms present in non-sterile packs of UHT milk is very small. End-product testing involves incubation of some, or all, of the output at a temperature permitting rapid growth of any micro-organisms present. All of the output is tested on initiation of production at a new plant, but if operation proceeds without significant problems the sampling rate is progressively reduced to 0.01%. A temperature of 30°C is most commonly used for a minimum of 5 days, although incubation at 25 or 37°C has also been used and packs for export to tropical countries are incubated at 55°C to detect thermophilic endospore-forming bacteria. Visual examination for swelling due to gas production is the simplest means of detecting non-sterile packs but is unreliable since microbial growth frequently takes place without gas production. Until recently destructive testing has been the only means of assessing growth. Various methods have been employed (Table 2.12) but in a comparative study a resazurin dye reduction test, carried out after 3 days' incubation at 30°C, was found to be most reliable.

Considerable efforts have been made towards developing instrumental methods for non-destructive testing. The first instrument developed, the Electester™, detects changes in viscoelastic properties resulting from microbial growth. Its application is limited to cardboard carton packed milks. More recently the use of ultrasound has been proposed and methodology developed which permits a high level of testing and which is suitable for use with plastic and foil laminate packs as well as cardboard.

(d) In-bottle sterilized milk

Incubation testing is used to supplement the turbidity test. Detection of contamination usually involves visual examination and viable counts.

Exercises

EXERCISE 2.1.

Consider the post-process contamination of pasteurized milk. What are the main types of micro-organism likely to be introduced from each source of contamination and what are the consequences for product safety and shelf stability?

Are regular medical examinations, including stool testing, effective in reducing the risk of milkborne disease?

Useful discussion on the role of the carrier in foodborne disease may be found in *Epidemiology and Infection* (1987), **98**, 223–32; *The Lancet* (1987), **II**, 865 and the *British Medical Journal* (1990), **300**, 207–8.

EXERCISE 2.2.

Design in outline schemes to investigate and correct the following situations in a plant producing UHT milk.

1. Sporadic problems of spoilage due to endospore forming, Gram-positive, rod-shaped bacteria. Spoilage is reported only during summer months.
2. Continuing problems of spoilage due to non-endospore-forming, Gram-negative, rod-shaped bacteria. Affected packs are those produced on a Monday after plant shutdown on Sunday.

EXERCISE 2.3.

Your company produces UHT processed, flavoured milk drinks containing semi-skimmed milk, sucrose, flavouring and colour. It is planned to expand this range into the following markets:

1. A lactose-reduced dairy drink.
2. A low-fat and low-calorie sweetened drink.
3. A 'luxury' product of increased viscosity and 'creamy' texture, but of low fat content.

Develop, qualitatively, recipes for each of the products specified. Are any modifications likely to be necessary to the UHT process?

Note: The lactose-reduced dairy drink may be based on functional ingredients other than milk providing these are of dairy origin. Additional information is available elsewhere in this book, especially in Chapter 4 (Milk proteins) and Chapter 9 (Ice cream).

EXERCISE 2.4.

Recent research using a new gas chromatographic method has shown that 3-deoxypentulose, formed as an intermediate in lactose degradation, is present in in-bottle sterilized milk and may also be formed in UHT sterilized milk during storage at high (above 30°C) temperatures.

1. Consider the likely pathway of 3-deoxypentulose formation and relate this to the formation of tagatose, galactose and other minor sugars.
2. What are the nutritional and technological consequences of lactose degradation during the heating of milk?

Further information may be obtained from Troyano, E. *et al.* 1992. *Journal of Dairy Research*, **59**, 507–15; Andrews, G.R. 1986. *Journal of Dairy Research*, **53**, 665–80 and Olono, A. *et al.* 1989. *Food Chemistry*, **31**, 259–65.

EXERCISE 2.5.

Staphylococcus aureus may readily be isolated from raw milk but, while there have been other reports of *Staph. aureus* growing and elaborating toxin in raw milk, the risk to consumers of pasteurized milk appears low. Predict the likely behaviour of *Staph. aureus* (with respect to both growth and enterotoxin) in raw milk stored under the following conditions:

1. In churns at *ca.* 15°C for a total of 15 h before processing.
2. In a bulk tank at *ca.* 8°C for a total of 48 h before transfer to creamery silos at 5°C for a further 16 h.

It has been stated (in relation to foods as a whole) that the presence of large numbers of *Staph. aureus* is of no consequence providing that enterotoxin is not also present. Discuss this statement with respect to food safety, the epidemiology of *Staph. aureus* and the adequacy of current methods for detection of enterotoxins.

Further information may be obtained from Varnam, A.H. and Evans, M.G. 1991. *Foodborne Pathogens: An Illustrated Text*. Wolfe Publishing Ltd, London and Doyle, M.P. (ed.) 1989. *Foodborne Bacterial Pathogens*. Marcel Dekker, New York.

3
CONCENTRATED AND DRIED MILK PRODUCTS

OBJECTIVES

After reading this chapter you should understand
- The differences between the various types of concentrated milk products
- The various means by which milk may be concentrated
- The design of commonly used equipment
- The drying of milk
- The design of spray dryers
- The relationship between the processing of dried milk and its end-uses
- The major control points
- The nature of chemical changes associated with concentration, drying and subsequent storage
- Microbiological hazards and patterns of spoilage

3.1 INTRODUCTION

Very large quantities of concentrated and dried milk products are manufactured. Although consumer products of this type, such as canned evaporated and sweetened condensed milk and instantized spray dried milk, are well established, much of the production is utilized in ingredient form. Major advantages lie in a considerable reduction in transport and storage costs as a result of the reduction in bulk, and convenience of use during formulation. In the case of sweetened condensed milk, there is also a significant extension of the ambient temperature storage life, while functional properties, especially of dried milk, may be 'tailored' to specific end-use.

> **BOX 3.1 Wigan Pier revisited**
>
> The main concentrated milk consumer products, canned sweetened condensed milk and canned evaporated milk were initially used as a shelf-stable and, in the case of skim milk products, cheap substitutes for fresh milk in tea making, etc. Subsequently these products became used as a pour-over accompaniment for canned fruit and other desserts. The situation with respect to consumer acceptability is similar to that of in-bottle sterilized milk, in that the strong and distinctive flavour and colour of each product is reviled by many younger people, but considered desirable by older, long-term consumers.

3.2 TECHNOLOGY

3.2.1 Basic technology of concentration

The manufacture of concentrated milk products has an obvious prerequisite, the removal of water. Current technology permits three approaches:

1. Evaporation: removal of H_2O as a gas (vapour).
2. Reverse osmosis/ultrafiltration: removal of H_2O as a liquid (water).
3. Freeze concentration: removal of H_2O as a solid (ice).

(a) Evaporation

Evaporation is both the oldest established method and the most widely used. The process is, however, relatively expensive in terms of energy and inevitably involves a level of heat damage to the product. Evaporation is a simple heat transfer process which, in less developed countries still involves boiling in open pans. Elsewhere thermal damage is minimized by boiling at temperatures of *ca.* 40 to *ca.* 70°C under reduced pressure. Under such conditions the temperature of the boiling milk is determined by pressure (always a vacuum) and, to a much lesser extent, the concentration of the milk (boiling point elevation) and the hydrostatic pressure of a column of liquid. The milk temperature during evaporation is thus effectively a pressure controlled value, the vapour and liquid are in equilibrium and the temperature at any time is a function of the saturated vapour pressure.

A modern evaporator installation can appear of fearsome complexity but the basic requirements are straightforward and the same for each plant (Figure 3.1).

1. A vacuum evaporator acting as a heat exchanger.
2. A separator for the separation of vapour and concentrate.
3. A condenser for the vapour.
4. Equipment for the production of a vacuum and the removal of the condensate.

The evaporator is the most important part of the system with respect both to the quality of the concentrate and the operating efficiency. Over the years many designs have been used including the simple batch-pan, various types of circulation evaporators and the more recent high efficiency type such as the plate, rising and falling film designs. Special designs such as scraped surface heat exchangers have also been proposed for evaporating milk to very high solids contents. In recent years the falling film evaporator has achieved predominance in the dairy industry and while other types remain in use, their application is largely limited to small scale, specialist situations. The advantages of the falling film design lie in its simple construction and relatively low capital cost, low level of thermal damage and high operating efficiency. The very small liquid content of falling film evaporators also means that the plant is highly responsive to changes in operating parameters such as energy input, vacuum and feed rate. This means that the plant is easily controlled and permits fully automatic operation using appropriate feed-back devices and microprocessors.

The falling film evaporator consists of a bundle of tubes down which the milk flows (Figure 3.2). The tubes are surrounded by a steam heating jacket and maintained under vacuum. The tubes usually range from 4 to 10 m in length, although 15 m is not uncommon, with a diameter of 25–80 mm. Milk is evenly distributed over the heating

---

* Heat flow may be defined (ignoring changes in boiling point elevation during the operating cycle of a batch evaporator) as:

$$q = UA \, \Delta T = UA \, (t_e - t_s)$$

where q is the heat flow (joules/s), A is the heat transfer area (m^2), U is the overall heat transfer coefficient (watts/m^2/°K), Δ is the temperature difference (°K), t_e is the temperature in the evaporator (°K) and t_s is the temperature of steam (°K).

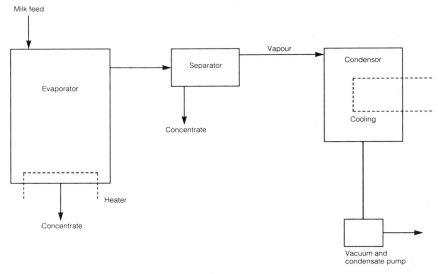

Figure 3.1 The basic components of an evaporator.

surfaces using a perforated plate, or a double cone valve distributor. During operation milk flows, by gravity, down the tubes. A high vapour velocity is an important design feature and is influenced by the length and diameter of the tubes, temperature difference and the level of vacuum. High velocities limit the thermal damage to milk by ensuring a short residence time and also increase the heat transfer coefficient and consequently the thermal efficiency of the evaporator.

In falling film evaporators, the necessary degree of concentration is usually achieved in a single pass through the evaporator. However, conventional single unit evaporators have an unacceptably high consumption of steam and cooling water and it has been common practice for several years to build multiple effect evaporators in which the vapour from one evaporator unit (effect) is used to heat a

---

* The vapour formed influences flow by increasing turbulence, decreasing the thickness of the laminar layer close to the tube wall and increasing the velocity of the film by the 'wind over water' effect. A maximum vapour velocity of *ca.* 60 m/s is obtained in the lower part of the tubes close to the discharge, but in practice the maximum velocity is limited by the risk of the vapour tearing the milk film away from the tube walls. Increasing product viscosity, which reduces the heat transfer coefficient, acts to counter the high vapour velocity during the later stages of evaporation. Viscosity can be reduced by use of high preheat temperatures, but the extent to which this is possible is often limited by the need to avoid thermal damage to the product.

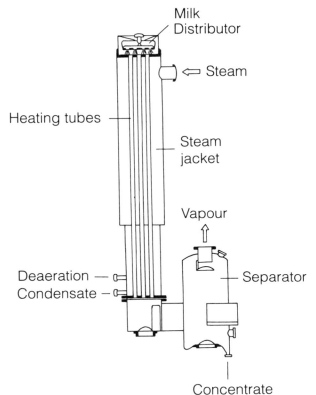

Figure 3.2 Falling film evaporator. Redrawn by permission of G.E.A. Wiegand GmBH, Ettlingen, Germany.

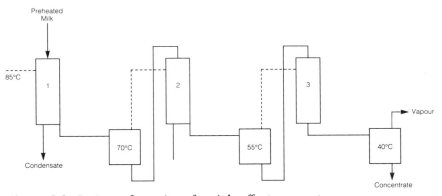

Figure 3.3 Basic configuration of a triple effect evaporator.

second effect. The effects are linked in series with a common condenser and vacuum source, the vacuum being highest in the lowest temperature effect and milk flow is usually from high to low temperature (Figure 3.3). In operation the temperature difference (usually *ca.* 15°C) between effects is the same and the amount of water removed in each effect is approximately equal. Energy savings, however, are offset by higher capital cost. The total temperature difference in the plant is that between the maximum heating temperature in the first effect and the lowest product boiling temperature in the last effect. The temperature difference is distributed evenly between the individual effects and therefore decreases with increasing number of effects. Larger heating surfaces become required to achieve a given evaporation rate and energy savings decrease markedly when more than four effects are installed. Most plants use five or six effects and it is rare that more than seven are installed. The heat transfer coefficient is lower in later effects due to the increase in viscosity and this further increases the heat exchange surface required. This, in turn, further increases the capital cost and also leads to operational problems due to a higher level of burning-on and deposit formation. These difficulties may be overcome by countercurrent flow of milk, so that the most viscous product is evaporated in the highest temperature effect, or by recycling part of the product in the high viscosity part of the plant to ensure that the heating surfaces are covered. The latter system, however, results in uneven retention times and the possibility of lower quality end-product.

Vapour recompression is used to obtain improvements in thermal efficiency beyond those possible with conventional multi-effect plants. Vapour recompression involves compressing part, or all, of the vapour from an effect to raise its temperature, and thus recover the condensation heat. The compressed vapour is then used to heat the same effect. When applied to the first effect, the evaporative capacity of that effect is increased considerably. In a double effect plant, for example, *ca.* 66% of the evaporation occurs in the first effect,

---

* In a single effect plant the enthalpy (heat content) of the evaporated vapour is approximately equal to the heat input. Reduction in steam consumption arises from the use of the enthalpy (condensation heat) of one effect to heat a second effect. Savings in heat consumption are proportional to the number of effects and may be illustrated by comparing the theoretical specific heat consumption (SHC):

$$SHC = \frac{\text{Amount of steam used for heating}}{\text{Amount of water evaporated}}$$

Theoretical values are; single effect, 1.0; double effect, 0.5; triple effect, 0.33 and quadruple effect 0.25.

obviating the necessity for large heating surfaces in the second effect handling viscous product.

Recompression equipment is of two types, thermal or mechanical. Thermal systems are the cheapest to install and of relatively simple construction, compression being effected by a steam-jet. The steam-jet recompressor acts as a heat pump which utilizes the pressure energy of the driving steam. Thermal vapour recompression is most efficient when temperature differences are small (*ca.* 6°C) and this is reflected in low steam and cooling water costs and in reduced scaling of the heating surfaces. Modern thermal recompression evaporators typically have five or six effects, a theoretical specific heat consumption of *ca.* 0.12% and a total energy requirement of 360 kJ/kg water evaporated. It is only possible, however, to compress part of the vapour and the enthalpy of the motive steam is discharged as residual heat into the cooling water.

Mechanical vapour recompression (MVR) offers even greater thermal efficiency since all of the vapours are recompressed, although the capital cost is significantly higher. Simplicity and ease of operation are important and single-stage centrifugal recompressors are widely employed. Various designs are used including high-pressure fans and turbo-compressors. Single-stage recompressors typically provide compression ratios of 1:1.2 to 1:2, but multi-stage compressors can be used where higher compression ratios are required. Use may also be made of positive displacement recompressors, such as the piston or sliding vane rotary types. The vapour is superheated, with poor heat transfer properties due to its dryness, and is cooled to its saturation point by injection of condensate. Mechanical recompressors are often driven by electric motors, but very high efficiency can be obtained by using internal combustion engines or steam turbines and recovering waste heat. Incoming milk must be heated to the boiling temperature of the first effect, but this may be achieved with very little (*ca.* 10%) external heat input providing that full use is made of waste heat recovery.

Mechanical vapour recompression evaporators only rarely have more than three effects, a three effect plant having a total energy requirement of 125 kJ/kg compared with 750 kJ/kg for a conventional three effect plant and 360 kJ/kg for a five to six effect plant with thermal recompression. Mechanical vapour recompression offers further advantages in that the maximum evaporation temperature may be as low as 69°C, while a minimum evaporating temperature of 63°C may be used, which prevents growth of most strains of the thermophilic *Bacillus stearothermophilus* and minimizes heat transfer problems due to high product viscosity. Only small quantities of steam and cooling water are required to stabilize operation and to maintain the correct heat balance.

Separators are necessary for the separation of entrained milk from the vapour. Various types are available, but in most instances, the simple gravity type is preferred.

Vapour condensation and production of a vacuum is an important aspect of evaporating plant, production of a vacuum requiring the removal of:

1. Steam, or vapour, produced during evaporation.
2. Gases dissolved, or dispersed, in the incoming milk.
3. Air leaked into the plant.

Vapour constitutes by far the greatest volume and in conventional, or thermal vapour compression plants, is condensed either by direct contact with water or by contact with a cooled surface. The condenser thus also acts as the main source of vacuum. Additional vacuum pumps are required to remove non-condensable gases such as leaked air. Water and condensed vapour must be removed by suction using either pumps or a barometric tube. In many modern plants the condensate is of sufficient purity for use as boiler feed or cleaning-in-place.

Evaporators and ancillary equipment are of high capital cost and must be utilized as fully as possible. Long production runs are necessary and in some cases the evaporator may be operated continuously for 20–22 h. The length of runs is limited by fouling and the need to avoid growth of thermophilic bacteria. Fouling is an inevitable process, the rate and nature of which depends on the nature of the product being concentrated, heat flux and operating conditions. Cleaning and sanitization are achieved by in-place circulation of appropriate materials including sodium hydroxide and nitric acid.

(b) Reverse osmosis and ultrafiltration

Reverse osmosis (RO) and ultrafiltration (UF) are both pressure-activated membrane separation techniques in which solutes of different molecular weight are separated from solution. There is considerable overlap between the two techniques and distinctions are somewhat

CONTROL POINT: EVAPORATOR OPERATION CCP 2

Control

Operation of evaporator under optimal conditions to assure product quality and process economy.

Prevent build-up of thermophilic micro-organisms.

Monitoring

Evaporator must be equipped with appropriate instrumentation to monitor operating parameters such as temperature and vacuum.

Monitor degree of concentration of product leaving evaporator.

Where long production runs necessary, operate evaporator at temperatures above maximum growth temperature of thermophilic micro-organisms. Cleaning routines to be correctly implemented.

Employment of experienced staff is necessary at *all* times.

Verification

Examination of plant records.

Inspection of plant to ensure adequate cleaning.

Periodic specialist examination and maintenance of plant by manufacturers.

---

arbitrary. There are, however, differences in performance in that RO requires an operating pressure of 100–2000 lb/in^2 and retains solutes with molecular weights generally less than 500, while UF operates at lower pressures of 10–100 lb/in^2 and retains solutes with molecular weight generally over 1000. The separation mechanism also differs and while UF can largely, but not totally, be explained as a relatively simple screen filtration process, RO involves diffusive transport and molecular screening.

In both UF and RO the properties of the membranes are of major importance. General properties are common to all membrane processes:

1. Capable of giving the required degree of separation (molecular weight cut-off) at high permeate flux rates and over extended time periods.
2. Must be cleanable and withstand treatments used in cleaning and sanitizing.
3. Must have sufficient strength to give a long *in situ* life.

Membranes were originally constructed of cellulose acetate, but this material has been largely superseded by non-cellulosics, especially polysulphone and, more recently, ceramic materials such as zirconium dioxide and aluminium oxides. Two types of membrane structure exist, 'asymmetrical' and 'composite'. Asymmetric membranes are cast as a single material of total thickness 1–2 mm. This comprises an ultra-thin (0.1–1.5 μm) layer of dense polymer supported by a relatively thick substructure of porous polymer. Composite membranes consist of a very thin film of active polymer on top of a highly porous sublayer made from a different polymer. In either case a backing layer may be present to increase mechanical strength.

In current usage, membranes may be cast in one of four configurations; tubular, flat sheet, spiral wound and hollow fibre, the latter being primarily used for water treatment. Each configuration has advantages and disadvantages and no single combination of membrane and configuration is suitable for all applications.

One of the most important features determining the efficiency of membrane systems is the rate of permeate passage, the flux. Flux is a function of membrane properties, operating parameters and the physico-chemical properties of the solution being concentrated. Membrane fouling, which results in an irreversible decline in flux, can be a major problem and may be caused by growth of microorganisms, blocking of membrane pores by particles, or physicochemical reactions such as a gel formation. Cleaning and sanitizing programmes are thus an essential part of the operation of membrane processes.

Membrane separation processes take place without phase change and, in theory, energy requirements are much reduced. The degree of concentration possible is, however, relatively low and membrane separation systems are usually operated in conjunction with an evaporator. Usually *ca.* 60% of water is removed at the membrane stage and 40% by the evaporator and under these conditions the saving in energy is insufficient to offset the capital cost of the membrane equipment and its operating cost. The inclusion of an evaporator also limits other theoretical advantages including reduction of thermal damage and avoidance of flavour stripping. Use of reverse osmosis and ultra filtration is therefore limited to specific situations such as the preparation of starting material for manufacture of fermented milk and cheese.

(c) Freeze concentration

Freeze concentration involves the cooling of milk to below the freezing point and the subsequent removal of the ice crystals formed. At present freeze concentration plants have not been developed beyond the pilot scale. Work suggests that a concentration of 36–38% is possible for both whole and skim milk in a single-stage plant. The concentrate is of good organoleptic quality and undergoes no significant undesirable changes. The loss of milk components of *ca.* 100 mg/l is also very low. Freeze concentration is, however, an expensive process, current costs being three to four times greater than thermal evaporation or reverse osmosis. The potential for microbial contamination is also considerable and very careful cleaning is required.

3.2.2 Manufacture of concentrated milk products

A number of concentrated milk products are made for both the industrial and consumer markets (Figure 3.4). The key processing stage in all cases is the removal of water, usually by evaporation, although ancillary processes, such as the thermal processing (retorting) of canned evaporated milk, are themselves of considerable technological importance.

(a) Concentrated milk

Concentrated milk was devised for direct consumption, after appropriate dilution, as a fluid milk. Advantages stemming from reduced bulk have been small and concentrated milk has not become an important dairy product.

Concentrated milk is prepared from high quality (Grade A) milk which is pre-heated at temperatures approximating to pasteurization. The milk is concentrated, usually by a ratio of 3:1, standardized, homogenized and

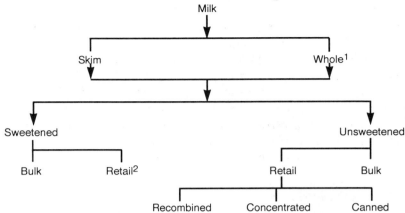

Figure 3.4 Production of various concentrated milk products. [1]Small quantities are semi-skimmed or contain vegetable fat as a substitute for milk fat. [2] Sold in cans, tubes or cartons.

pasteurized before packaging. A higher temperature of pasteurization is required to offset the protective effect of the higher solids content, a common process being 79.4°C for 25 s.

Although the water activity level of concentrated milk is lowered by the higher total solids content, the effect on microbial growth is minimal and refrigeration is required during storage and distribution.

(b) Bulk condensed milk

Large quantities of bulk condensed milk are made by evaporation and are used primarily as a source of milk solids in confectionery, bakery and other manufactured foods. Considerable interest has been shown in the use of milk concentrated by ultrafiltration or, to a lesser extent, reverse osmosis for manufacture of cheese and fermented milks, while retentates have also been used in the manufacture of ice cream.

Bulk condensed milk is usually made from manufacturing-grade milk, although Grade A milk is used where the condensed product is to be used to standardize liquid milk, or in preparation of milk drinks. Either skim, fat-reduced, or whole milk may be used according to the end-use and for particular uses non-milk fat may be substituted. Skim milk may be blended with whey (5:1) before concentration for use in ice cream manufacture.

Where appropriate, the milk is separated before concentration, but

CONTROL POINT: MILK AND MILK PRETREATMENT CCP 1

Control

Incoming milk to be of suitable quality.

Heat processing to be adequate for destruction of vegetative pathogens.

Other pretreatments to be correctly applied.

Monitoring

Incoming milk: in general, as for liquid milk (Chapter 2, page 48), even where manufacturing grade milk is used.

Heat treatment: equipment should be fitted with a thermograph and temperature monitored by operatives.

Other pretreatments: control parameters should be monitored by operatives and recorded either manually or automatically as appropriate.

Verification

Examination of plant records.

Note: Heating to ensure safety is a CCP 1, all other processes are CCP 2s. Control measures described here are generalized and applicable to all types of concentrated milks.

---

standardization for fat content may be delayed until final standardization for total solids content after concentration. Homogenization of the milk (other than skim milk) is normal at this stage. In all cases the milk is preheated, either in a continuous heater, or in a 'hot well'. The degree of heating varies but is commonly in the range 65.5–76.7°C. More severe treatments of up to 82–93.5°C for up to 15 min may be used for special applications to increase product viscosity, etc. Vegetative pathogens are destroyed at the upper end of normal preheat treatments, but may survive at the lower end. This is clearly undesirable and a supplementary heat treatment equivalent to HTST pasteurization should be introduced.

The degree of concentration depends on the product end use, but is normally in the range 2.5:1 to 4:1. The product leaving the evaporator is not sterile and further opportunities for contamination occur during post-evaporation handling operations such as standardization and packaging. A high standard of hygiene together with rapid and efficient cooling is seen as an integral part of processing and is necessary for an adequate storage life.

(c) Canned evaporated milk

Canned evaporated milk may be seen as the consumer equivalent of bulk condensed milk but, while the end products are similar, there are differences in pre-concentration treatment resulting from the need to stabilize the milk against instability (coagulation) induced by retorting and thickening associated with long-term storage (see pages 141–145).

Canned evaporated milk is made from high quality milk. Whole milk is most common, but smaller quantities of skim milk and milk filled with non-dairy fats are also used. The milk is separated, where applicable, and standardized, precautions are taken at this stage against instability, although strictly these are concerned with producing optimum viscosity rather than optimum heat stability. Stabilizing salts are added, those permitted in the UK including NaH_2PO_4, Na_2HPO_4 and Na_3PO_4, $NaHCO_3$ and $KHCO_3$, Na_3citrate and H_3citrate and $CaCl_2$. Both phosphate and citrate ions have a specific stabilizing effect, but pH value is also of major importance. It has been suggested that choice of stabilizer should, therefore, depend on the milk pH, Na_2HPO_4, Na_3PO_4, Na_3citrate or $NaHCO_3$ being added to milk of acid pH and NaH_2PO_4 or $CaCl_2$ to milk with a pH value approaching neutrality. The amount of stabilizing salt added depends on the amount needed to bring the pH value to that of optimum stability, 6.6–6.7. Summer milk is most prone to instability but seasonal variations are much less than with unconcentrated milk. The effectivity of stabilizing salts, however, varies markedly with season, phosphates being most effective during June to August and least effective in the winter months.

Preheating is also of importance in determining stability and must be of sufficient severity to denature and aggregate whey proteins, most modern plants employing temperatures of 120–122°C for a few minutes. Time and temperature of preheating have a major effect on stability, especially at intensities above that required to denature 90% of the β-lactoglobulin. Preheat parameters must, therefore, be carefully determined and rigorously controlled.

CONTROL POINT: STABILIZATION CCP 2

Control

Ensure stability of final consumer product by application of correct preheating process and addition of stabilizing salts.

Monitoring

Calculation of quantity of salts required and supervision of addition by trained and experienced personnel.

Preheat equipment to be fitted with thermograph.

Verification

Redetermine stability before canning and retorting.

Examination of plant records.

Following preheating the milk passses to the evaporator where the water content is reduced to give a total solids content higher than that desired in the end-product. The concentrate is homogenized and cooled. Two stage homogenization is usual, a typical process involving pressures of 17.5 and 3.5 MPa on the first and second stages, respectively. Standardization is also possible at this stage, but preheated skim milk concentrate must be used. Samples are tested for stability and further stabilizing salts may be added, providing that the legal limit is not exceeded.

In conventional practice, filled and sealed cans are retorted in either a batch or continuous retort, but in recent years there has been a tendency to combine ultra-high temperature processing of the milk with aseptic filling of presterilized cans, bottles or cartons. Thermal processes must be sufficiently severe to ensure microbiological safety and stability and higher retort temperatures are also desirable to minimize age-

---

* In small-scale production the concentrated milk may be stored at 4°C before the canning process, but this practice promotes subsequent age-thickening and should be avoided (de Koning, P.J. *et al.* 1992. *Netherlands Milk and Dairy Journal*, 46, 3–18).

CONTROL POINT: STERILIZATION CCP 1

Control, monitoring and verification procedures are basically the same as those applied during production of in-bottle or UHT sterilized milk (Chapter 1, pages 64 and 70) depending on the type of process applied. Where double seam cans are used, a procedure must be applied to check, on a regular basis, that seam parameters are correct. This requires trained and experienced personnel.

---

thickening. Excessively severe processing, however results in an undesirable level of Maillard browning and other deleterious reactions. Typical processes for conventional retorting involve heating at 120°C for 10 min, or 115°C for 15–20 min. More severe processes may be necessary where the product is to be stored in a hot climate and in some countries addition of nisin is permitted to control thermophilic spoilage. Batch retorts are usually of the rotary type to ensure uniformity of heating and cooling and continuous retorts impart a rolling motion to cans for the same purpose.

Ultra-high temperature processing may vary from 136°C for 30 s to 150°C for less than 1 s. Ultra-high temperature processed evaporated milk undergoes significantly less heat-induced change, but viscosity is lower and may permit fat separation. There may also be a greater tendency to age-thickening due to the lower total heat load. If necessary, however, these problems may be overcome by holding at 100–115°C. Sedimentation, however, remains a problem especially with skim or low-fat milk.

(d) Sweetened condensed milk

Sweetened condensed milk may be made from either whole or skim milk and may either be produced for bulk industrial use or packaged into containers (cans or tubes) for retail sale. The extended shelf life

---

* Proposals have been made to exploit ultrafiltration to produce a new generation of sweetened, in-can sterilized concentrated products. Such products are stabilized by removal of salts during a combined ultrafiltration and diafiltration process. Stable skim milk products were successfully produced but had a 'chalky' mouth-feel. Organoleptic problems were subsequently overcome by re-addition of cream, although the homogenization required itself led to instability (Muir, D.D. *et al.* 1984. *Journal of Food Technology*, **19**, 369–76).

is derived not from thermal processing, but the reduced water activity (a_w) level resulting from the high solute concentration. Refrigeration is, however, required for many types of bulk sweetened condensed milk.

The milk used is commonly of 'manufacturing grade', although the relatively lax standards previously permitted in some countries have now largely been replaced by more rigorous criteria. The milk is preheated and, in some cases, superheated, before passing to the evaporator. Sugar may be dissolved in the preheated milk before evaporation and this is general practice for bulk sweetened condensed milk, which is usually of relatively short shelf life. The much longer shelf life necessary for retail packs means that age-thickening may be a significant problem (see page 144) which can be exacerbated by addition of sugar before evaporation. In this case normal practice is to introduce the sugar as a *ca.* 65% solution in the later stages of evaporation. Where sugar is added at this stage, it is necessary to pass the concentrate through a 'finisher', a small separate evaporator specially designed for the removal of relatively small volumes of water to a high total solids content and for handling viscous material.

The sugar usually added is sucrose, but for industrial use other sugars may wholly, or partly, replace sucrose. The concentration of sugar in water is expressed as the sugar ratio (sucrose in aqueous phase) and may be calculated as:

$$\text{Sugar ratio} = \frac{\% \text{ sugar in condensed milk}}{100 - \text{total solids in condensed milk}} \times 100$$

The sugar ratio is usually 63.5–64.5 for the retail product, but can be as low as 42 for bulk, whole milk product to be stored under refrigeration.

Milk leaving the evaporator is partially cooled to *ca.* 30°C and seeded with very fine lactose crystals. These form nuclei as the milk lactose crystallizes out during further cooling and induce the formation of a large number of small, undetectable lactose crystals. Inadequate seeding leads to the formation of a small number of large crystals which impart a sandy texture to the product. The high viscosity of sweetened condensed milk can cause problems where conventional coolers are used and flash cooling in a vacuum-chamber may be preferred.

Sweetened condensed milk is not sterile and while the a_w level is

CONTROL POINT: SUGAR ADDITION CCP 2

Control

Quantity of sugar added must be correct with respect to sugar ratio required.

Sugar to be of acceptable quality and not to introduce micro-organisms or other contaminants.

Monitoring

Calculation of sugar required by trained and experienced person.

Addition of sugar supervised by experienced person.

Sugar to be obtained from reputable supplier and to meet pre-agreed standards.

Sugar to be stored under correct conditions.

Verification

Scales, flow meters, etc., used for measuring sugar to be checked on a regular basis.

Examination of plant records.

Inspection of storage facilities on a regular basis

---

sufficiently low to inhibit growth of many bacteria, other micro-organisms, especially yeasts and moulds are able to develop. It is necessary to apply stringent control both of general hygiene and of specific operations and detailed measures should be incorporated into master manufacturing schedules. Special account should be taken of the viscous nature of sweetened condensed milk and cleaning schedules adjusted accordingly.

Filling of sweetened condensed milk into retail containers is a relatively complex process due to the product viscosity and positive-action plunger-type fillers are used. Containers should be presterilized by gas flame or steam and the airspace should be minimal to prevent mould growth. Filling equipment is sometimes protected by directed air flow and ultraviolet light. Tanks for bulk storage of sweetened condensed milk should also be fully filled and may be equipped with ultraviolet lights to prevent surface mould growth. Refrigeration should be applied where necessary.

A special type of sweetened condensed milk, block milk, is used to a limited extent in Europe and the US in the manufacture of confectionery. Block milk is of higher solids content than conventional skim milk and has a paste-like consistency. The product is made in batch cookers, in which milk is heated to *ca.* 60°C before addition of a sugar syrup. Water evaporates under vacuum, a stirrer being used to work the paste. The composition of block milk varies according to end-use, most types having a total solids content of 85–90%, fat of 9–20%, sugar of 20–50% and a moisture content of 10–15%. Shelf life is short due to the rapid onset of age thickening and sugar crystallization.

(e) Recombined concentrated milk

Recombined concentrated milk, prepared from milk powder and anhydrous milk fat, is used as a substitute for whole milk in areas where fresh milk is scarce or highly seasonal in availability. Manufacture involves preparation of a reconstituted milk concentrate of solids-non-fat content *ca.* 18% and fat content *ca.* 8%. The viscosity of the concentrate should be in the range 13–18 centipoise. The concentrate is heated to 25°C and blended with molten milk fat at a temperature of 40–45°C. The mixture is then homogenized in two stages at pressures of 16 MPa in the first stage and 3.5 MPa in the second stage. The recombined concentrated milk is then UHT processed and aseptically packaged.

BOX 3.2 **A free man on Sundays**

Before the development of expensive energy replacement foods, sweetened condensed milk, packed in tubes, formed an important part of the diet of many hill walkers and climbers. The manufacturers were able to exploit this niche market by sales through Youth Hostels. Indeed sweetened condensed milk was sometimes a staple of Youth Hostel diet being used as a spread in sandwiches in addition to its more conventional culinary roles.

Recombined concentrated milk is subject to instability, the extent of the problem depending on the preheat treatment applied before the evaporation stage in manufacture of the milk powder. A special, high-heat, heat-stable skim milk powder is now available.

(f) End-product testing

End-product testing is required for all types of concentrated milk. Actual tests applied vary (see page 148). However the role of chemical analysis is to ensure compliance with compositional standards and product quality. Microbiological analysis ensures that general standards of manufacturing hygiene have been satisfactory. In the case of in-can, or UHT sterilized evaporated milk, sterility testing is required.

3.2.3 Basic technology of drying

The drying process may be seen as a continuation of concentration, with the intention of producing a stable, low moisture content product with minimum organoleptic change and functional attributes appropriate to the end-use. The concentration stage and its associated processes such as pre-heating are an integral part of the production of dried milks and play a major role in determining the properties and functional attributes.

Industrial-scale drying of milk involves the removal of water through application of heat at temperatures above ambient. Water in foodstuffs is held in varying degrees of bonding and the concept of a clear demarcation between free and bound water is a potentially misleading oversimplification. A substantial part of the water in concentrated milk is, however, only loosely bound and for drying purposes may be considered as 'free'. There are two possible means of removing water in a 'free' state:

1. Evaporation, the means in roller drying, where the saturated vapour pressure of the surface giving up moisture equals atmospheric pressure ($P''_{vo} = P$).
2. Vaporization, the means in spray drying, where the saturated

---

* The addition of buttermilk to the unconcentrated milk has been suggested as a means of controlling instability. Stabilization is possible during production of the recombined concentrate by control of homogenization and by addition of phosphates or citrates as stabilizers. Stabilizers are only effective, however, when used in conjunction with control of pH value, maximum stability being obtained at values of *ca.* 6.7 (Augustin, M.A. and Clarke, P.T. 1990. *Journal of Dairy Research*, **57**, 213–26).

vapour pressure of the surface giving up moisture is less than atmospheric pressure ($P''_{vo} < P$).

It is thus necessary to supply the latent heat of vaporization (or evaporation) and this involves two major process-controlled factors:

1. The transfer of heat to provide the necessary latent heat of vaporization.
2. The movement of water, or water vapour, through the food material, and then away from it, to effect the separation of water from the foodstuff.

Drying by vaporization may be considered to occur in three stages. In the first stage the surface of the product being dried contains 'free' water and vaporization proceeds from this surface. This stage is referred to as constant rate drying during which

$$\frac{d_w}{d_o} = \text{constant}$$

where w is the mass of material being dried.

Shrinkage of the material occurs as a consequence of the drying of the 'free' liquid surface (initial moisture content X_{oo}) and continues until a solid structure is formed (X_o). No further shrinkage occurs after this point, but moisture is transported from the interior to the surface by capillary forces. The drying continues at virtually constant rate until the maximum hygroscopic water content ($X_{hyg.max}$) is reached. At this point the end of the first stage is indicated by a sharp break in the drying stage line (Figure 3.5).

The second stage of drying is characterized by a decreasing drying rate. 'Free' moisture recedes into the product interior and the surface moisture content decreases from the maximum hygroscopic moisture content to the equilibrium moisture content ($X_{equ.}$). The average depth of the moisture level progressively increases and heat transmission now involves heat transfer to the product surface. Conductivity of the dry surface layers is very low and conduction within the product becomes of increasing importance in determining drying rate. An exception occurs with dried products of relatively high bulk density which have a small cavity volume and very small pores, where drying rate is determined more by resistance to moisture diffusion than by heat conduction. As the second stage proceeds, removal of water more tightly bound by sorption commences.

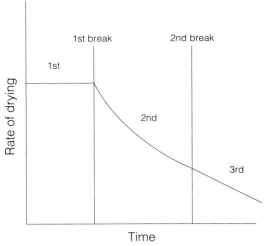

Figure 3.5 The three stages of drying.

During the third drying stage the maximum hygroscopic moisture content is present in the deepest part of the product interior. The difference between atmospheric pressure and the saturated vapour pressure of the surface giving up water progressively falls and this is reflected in a decreasing drying rate. As the vapour pressure of the residual moisture in the centre of the product enters equilibrium with the atmospheric moisture the drying rate approaches zero and the process is effectively complete.

3.2.4 Spray drying

Spray drying is now of overwhelming importance in the manufacture of dried milks (and other dried dairy products) and, except for special purposes, has largely supplanted the earlier process of roller drying. Roller drying is not, therefore, discussed further. At its simplest, spray

---

* Heat transfer during the first stage of spray drying may be expressed as:

$$q = h_m A(t_a - t_m)$$

where q is the heat transfer rate (kJ/kg/s), h_m is the surface heat transfer coefficient (J/m²/s/°C), A is the area through which heat flow occurs (m²), t_a is the air temperature (°C) and t_s is the drying surface temperature (°C). Dryer efficiency may be defined as the ratio of the heat theoretically required to supply the heat of vaporization of the water removed to the actual heat used by the dryer. It is necessary to take account of losses of heat before the supply into the dryer and calculations of efficiency must be based on the total heat available from the fuel burned to supply dryer heat (Earle, R.G. 1983. *Unit Operations in Food Processing*, 2nd edn. Pergamon Press, Oxford; Evans, A.A. 1989. *Australian Journal of Dairy Technology*, 44, 97–100).

Technology

drying involves mixing in a drying chamber atomized milk, of droplet diameter less than 300 μm (preferably 10–100 μm), with hot air at an inlet temperature of 150° to 220°C. The air gives up heat and moisture is correspondingly removed from the droplets. The vaporization temperature at commonly used inlet temperatures is between 40 and 50°C, the temperature of the dry particles never exceeding that of the cooled outlet air. Dry particles are separated from the air, either before or after removal from the drying chamber, and cooled before packaging.

In many ways the apparent simplicity of spray drying is deceptive and it is therefore possible to seriously underestimate the complexities involved in design, construction and operation of even the most unsophisticated dryers. Further the dryer must be considered in a broader context, not only of accompanying plant such as evaporators, but also of the design and construction of the factory building and of all operational procedures.

The design of a basic two-stage spray dryer is illustrated diagrammatically in Figure 3.6. Concentrated milk enters the top of the dryer *via* the atomizer. The viscosity of the feed is a major factor affecting the properties of the dried powder and may be controlled, within limits

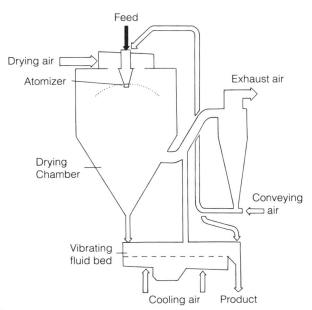

Figure 3.6 Basic two-stage spray dryer. Redrawn by permission of Niro Atomizer A/S, Copenhagen, Denmark.

which vary according to product type, by altering the degree of concentration by the evaporator.

Atomizers are of three types, rotary (disc), pressure nozzle and two-fluid nozzle, the rotary or the pressure nozzle type being most common. Rotary atomizers vary in design, but consist essentially of a rotating disc driven by a high speed electric motor. Feed enters the centre of the disc and is directed to peripheral exit points, consisting of slots or holes, by curved vanes incorporated into the body of the disc. The atomizer is fed by a simple pump and capable of dealing with feed containing solid particles. Disc atomizers are capable of handling high viscosity feed and have a very high throughput, but the 'umbrella'-shaped spray means that a large drying chamber is required to prevent excessive burning-on of product to the chamber walls. The rotary atomizer produces a range of droplet sizes according to the design of the atomizer and the speed of rotation, the behaviour of the largest droplets having the controlling influence. The extent of air inclusion within the droplet is relatively high and the powders of medium particle density. The extent of air inclusion can be reduced by flushing the atomizing disc with steam but this is an expensive procedure which may also cause heat damage to the product. Alternatively a sealed disc has been developed which minimizes air entrapment before atomization proper.

The pressure nozzle atomizer is fed by a high-pressure, positive displacement pump, usually of the piston-type. Operating pressures of 2–6 MPa are required for low viscosity liquids and *ca.* 20 MPa for high viscosity. Liquid feed enters the nozzle tangentially, the resulting spin atomizing the liquid as it leaves the nozzle. The throughput of a single pressure nozzle is low in comparison with a rotary atomizer and large dryers require multiple nozzles fed *via* a ring manifold. The installation cost for a large dryer is therefore higher, although the downward flow of product into a narrow cone means that a large diameter drying chamber is unnecessary. The diameter of the droplets falls within a narrow range depending on the pressure applied, nozzle design and capacity and the viscosity and surface tension of the feed. Nozzle atomizers produce a powder of high bulk density which, in general has better reconstitution properties, longer shelf life where fat-containing and reduced stack losses.

---

* In the past, choice of atomizer has tended to vary according to country of installation, the nozzle type being more common in the USA and the rotary type in Europe. The properties of the powders produced do, however, differ and choice of atomizer type is today likely to be based on this and other technological considerations such as the properties of the product to be dried.

The addition of dry ingredients at the time of atomization has advantages in the manufacture of products such as infant formula which are based on spray dried milk supplemented with crystalline sugars, dried starch flakes, etc. Dry ingredients are introduced *via* the fines return system and up to 70% can be incorporated.

Agglomeration is the major step in the manufacture of instant products and involves the formation of a physically larger body from a number of smaller ones (particles) while retaining the original particles in identifiable form. In older dryers agglomeration was a separate process which involved moistening the powder with steam and re-drying in a fluidized bed dryer. It is now possible to agglomerize powder at the atomization stage. A multiple pressure nozzle system is most suitable, the nozzles being arranged in angled clusters to produce cross-spraying, and fines are returned to the atomization area. The degree of cross-spraying can be varied to produce non-, low- or highly-agglomerated powders. Heating air is drawn from the outside atmosphere through a series of filters. Indirect steam heating *via* a heat exchanger is the most commonly used method of air heating and may be used in combination with direct electrical heating. Indirect gas or fuel oil heating is occasionally used and interest has been shown in direct gas heating using low nitrogen oxides (NO_x) combustion systems. Heat exchangers operating in the exhaust air after removal of fines recover waste heat which is used to pre-heat incoming air.

Air flow within the drying chamber is of major importance and a number of requirements must be fulfilled:

1. Air and product must be thoroughly mixed and evenly distributed to avoid formation of pockets of moisture.
2. Uncontrolled back flow of dried product into hot areas of the dryer must be prevented to avoid the risk of localized overheating.
3. The surface of droplets should be dry before contact is made with the dryer walls to prevent deposit formation.
4. The temperature of dryer walls in contact with product must be below the product melting point to avoid burning on.
5. The air flow must not permit the concentration of product to enter the explosion danger zone. Self-ignition of whole milk powder occurs at 144°C and skim milk powder at 185°C. Spark ignition of powder can occur at a concentration of 50 g/m^3, but not at 175 g/m^3.

Two basic design options exist for air flow, cocurrent or countercurrent. Countercurrent flow is considered advantageous where a low residual moisture is required since the incoming air has a low partial vapour pressure and is in mass exchange with the emerging powder. The thermal efficiency is also high due to the lower outgoing air temperature, but the temperature of the product leaving the dryer is high and the development of two and three stage dryers fitted with integrated fluidized beds (see below) has favoured the use of cocurrent flows.

Heavy particles are removed from the base of the drying chamber through a lock, while lighter particles (the fines) remain in the air flow until removal by cyclones and cloth filters. An alternative procedure is to remove all of the product with the air flow and effect separation entirely by use of cyclones and filters. The latter system can result in a high level of dust due to mechanical damage to the powder granules. Wet air scrubbers have been fitted to some plant, but while highly efficient, pose a hygiene threat which many experts consider unacceptable. Fines are returned either to the main drying chamber or to the secondary fluidized bed drying stage. Dryers are fitted with a variety of devices to assist powder removal. These include pneumatic hammers and vibrators to remove powder from the dryer wall, mechanical rakes and rotating air sweeps.

Original spray dryer designs were for single stage drying in which all of the moisture is removed in the main drying chamber. Such designs are now obsolescent and have been replaced by two and three stage dryers in which the drying process is completed in fluidized bed dryers. This has considerable advantages in terms of a reduced specific energy consumption and improved powder properties and also increased flexibility in terms of powder types produced. In a typical three stage dryer the moisture content of powder leaving the spray chamber (first stage) is 10–15%, this is reduced to 5–6% in the first fluidized bed (second stage) before final drying to the desired level in the second fluidized bed (third stage).

Fluidized bed dryers consist of a layer of product supported on a porous support plate through which air is blown. The bed may either be

---

* It has been calculated that some particles may remain in well-mixed dryers, exposed to temperatures of 70–100°C, for periods in excess of 20 min and that undesirable changes including an increase in insolubility index, protein instability and lactose crystallization can result. The temperature during the drying of moist powders should be as low as practical (Kudo, N. *et al.* 1990. *Netherlands Milk and Dairy Journal*, 44, 89–98).

stationary or vibratory. The velocity of the air is controlled between that needed to expand the product bed and that at which individual particles float. Drying (and cooling) times are influenced by the diameter of the particles, the depth of the bed and the state of the air.

Fluidized bed dryers are of two types with respect to product behaviour, the well-mixed and plug-flow types. Particle mixing is perfect in the well-mixed type and the high air flow rate results in rapid dispersal of feed and prevents lump formation. Well-mixed dryers are therefore able to handle moist and sticky powders and are used as second stage dryers. Disadvantages are the non-uniform moisture content in particles leaving the dryer and a wide distribution of particle residence time. Plug flow fluidized beds operate on a 'first in–first out' principle and give a flat particle moisture profile. For this reason plug flow dryers must be used for the final stage of drying and also for cooling.

Fluidized bed dryers may be separate from the spray dryer (external) or, more usually in current designs, incorporated into the drying chamber (integrated). In some cases the air flow is common with that of the spray dryer, but improved control and flexibility are obtained where air flows are separate. Modern spray drying plant may be arranged in various configurations including the cocurrent integrated fluid bed design and the mixed flow integrated fluid bed design. The latter offers optimal handling of high moisture content powders and considerable flexibility in terms of powder properties. Design and operation are illustrated and discussed in Figure 3.7.

An alternative means of spray drying, foam spray drying, is occasionally used where powders of very low bulk density are required. In this process air (skim milk) or nitrogen (whole milk) is injected at high pressure before atomization and dispersed as fine bubbles. The resulting powder has a bulk density *ca.* half that of conventional spray dried powder.

The Filtermat™ is a special type of spray dryer originally developed by the University of Minnesota and the Pillsbury Company for the drying of non-dairy products such as fruit juices. The drying chamber is rectangular with vertical pressure spray nozzles in the ceiling. The height of the chamber is insufficient to permit drying to the final moisture content, moist, partially dried milk building up as mat on a porous belt moving through the base of the dryer. Drying air moves down through the mat which thus acts as a filter. The mat is carried by the belt to final drying and cooling sections and is finally broken off the belt in the form of a

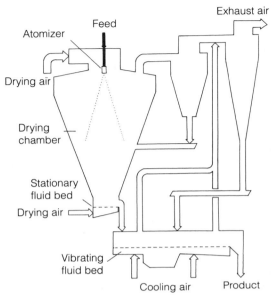

Figure 3.7 Mixed flow integrated fluid bed spray dryer. Redrawn by permission of Niro Atomizer A/S, Copenhagen, Denmark. Air enters the drying chamber, which is characterized by the short cylindrical upper part, in the centre of the ceiling and flows straight down. Atomization may either be by rotary, or pressure nozzle atomizers. A stationary fluid bed dryer is built into the base of the chamber and exhaust air leaves *via* roof vents. Powder is dried to 15% moisture in the primary drying stage and this is reduced to 5% in the stationary fluid bed. Final drying is in a vibrating fluid bed placed immediately after the outlet to the drying chamber. Fines are recirculated and fill the entire chamber with powder. Changing the residual moisture content of the airborne particles changes the degree of agglomeration and thus particle size distribution of the end product.

porous cake. Milling is required to produce a free-flowing powder and the belt is cleaned before re-entering the drying chamber. The capacity of this type of dryer is inherently limited by the size of the belt and very large throughputs are not possible. The Filtermat is, however, suitable for drying powders of high fat (up to 80%) and sugar content and is used in manufacture of such products as coffee creamers and infant foods. The thermal efficiency is very high.

3.2.5 Production of dried milk products

(a) Skim milk powder

The manufacture of skim milk powder is summarized in Figure 3.8. Manufacturing grade milk of good bacteriological quality is used. The milk is skimmed and standardized, where necessary, before heat treat-

CONTROL POINT: DRYER OPERATION CCP 1

Control

Operation of the dryer and ancillary equipment under optimal conditions to assure product quality and process economy.

Ensure that the dryer, ancillary equipment and plant as a whole do not serve as a source of pathogenic micro-organisms.

Monitoring

Dryer must be equipped with suitable instrumentation to monitor operating parameters such as temperature of entry and exit air.

Precautions against contamination from raw milk to heated milk, concentrate, or powder must be monitored on a continuing basis.

Precautions against growth of bacteria in concentrate before drying must be monitored on a continuous basis.

Precautions against cross-contamination between the wet side of the plant and the powder side must be monitored on a continuous basis.

Between production runs the dryer should be checked for efficiency of cleaning procedures and physical condition.

Employment of experienced staff is necessary at all times.

Verification

Examination of process records.

Environmental sampling to give early warning of plant contamination.

Periodic specialist examination and maintenance of plant by manufacturer, to include use of equipment to detect cracks in inner wall.

---

ment and may also be clarified at this stage. Raw milk should be subject to a heat treatment which is at least equivalent to pasteurization. For

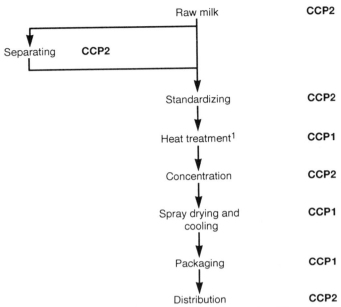

Figure 3.8 The manufacture of spray dried milk powder. [1]Heat treatment must be at least equivalent to pasteurization.

operating reasons the use of flow-diverters is usually on a 'last-resort' basis and the process must contain a safety margin sufficient to allow for normal variations.

The heat treatment applied at this stage affects the functional properties of milk proteins and the end-use of the powder (see pages 136–7). Three heat treatments producing low-, medium- and high-heat powder are used. These produce differing degrees of protein denaturation and the resulting powders may be distinguished chemically by determining the level of soluble whey protein nitrogen. Heat treatments commonly applied are 74°C for 30 s for low-heat powder, 85–105°C for 1–2 min for medium-heat and 120–135°C for high-heat. A further class of powder, high-heat, heat stable skim milk powder, used in the manufacture of recombined evaporated milk, receives a heat treatment of a minimum of 120°C for not less than 40 min. Heat treatment is also required for production of lipase-free skim milk powder used in

---

* Low-calcium milk powder is also produced for use in meat products and this requires the use of ion exchange columns to reduce the calcium content of the skim milk by 75–90%. Although low-calcium powder has very good water binding and emulsification properties, the additional processing is expensive and the product is not widely used.

chocolate manufacture to prevent lipolysis during storage. However, microbial lipases are likely to survive the commonly used treatment at 85–95°C for 1–2 min and such powder should be referred to as 'lipase-reduced' rather than 'lipase-free'.

In modern plant the heat treatment equipment is incorporated into the evaporator and offers a very high degree of control and operating efficiency. Various configurations of tubular heaters are used, although in some circumstances high-heat treatment is most conveniently applied by direct steam injection. After heating to the desired temperature, the milk is held either in vessels or holding tubes for the time period required. The milk is then cooled either in a conventional heat exchanger or, preferably, by flash evaporation. Flash evaporation is particularly efficient since flash vapour can be used for heating incoming milk. Further as much as 8% of the total water evaporated can be removed by flash evaporation thus reducing the steam consumption of the entire system.

Following heat treatment the milk passes to the concentration stage. Concentration by evaporation using, in modern plants, falling film evaporators, is most common although increasing use is made of ultrafiltration and, to a lesser extent, reverse osmosis. Milk is concentrated to 45–55% total solids, the current trend being to higher levels of concentration to meet the demand for powders of high bulk density and to reduce the energy requirements of the dryer.

Concentrated milk may pass direct to the atomizer of the spray dryer, but a small balance tank is often employed to ensure a constant flow of milk to the dryer. Microbial growth can occur in the balance tank and the residence time of the concentrate should be as short as possible. The use of dual balance tanks is common and permits regular cleaning during long, continuous production runs. Problems due to thickening and gelation are also minimized. Further preheating from *ca.* 55°C to *ca.* 80°C is advantageous and should take place directly before atomization. Specially designed spiral tube heaters are available and can be operated for long periods without fouling. The concentrate may also be seeded with α-lactose monohydrate crystals before atomizing to increase the percentage of crystalline milk sugar and reduce hygroscopicity.

Spray dryers are operated in accordance with the principles discussed in Section 3.2.4. Inlet air temperatures are in the range 180–230°C and outlet temperatures 70–95°C and the final powder has a residual moisture content of 2–5% (95–98% total solids). During operation,

Table 3.1 Examples of the relationship between spray dryer operating parameters and the quality of properties of milk powder

Operating parameter	Powder property
Outlet air temperature high	Reduced moisture content
	Reduced solubility
	Increased free fat
Inlet air temperature high	Reduced bulk density
	Reduced free fat
Degree of atomization high	Increased solubility
	Reduced moisture content
	Slightly reduced free fat
Feed total solids high	Increased bulk density

Note: Only one property, moisture content, is entirely controlled by dryer operating parameters and in all other cases there is interaction with the processing system as a whole.

control of a number of parameters is required both to ensure consistent quality and to produce a powder of the desired physical characteristics (Table 3.1; see also Figure 3.7).

Variants of basic skim milk powder may be produced using essentially the same production techniques. Powders containing casein and/or whey protein, for example, may be produced either by co-drying concentrated skim milk and a concentrated protein solution, or by introducing the dry protein at atomization. The introduction of any material after heat treatment carries the risk of microbiological contamination and a scraped surface heat exchanger should be used to pasteurize the dryer feed unless the microbiological status of the additive can be assured.

(b) Whole milk (full cream) and high fat milk powder

Whole milk powder has a legally defined minimum fat content which, in most countries, is 26%. The basic method of production is similar to that of skim milk powder, whole milk being standardized to 3.6% fat before

---

[*] A novel means of producing both high protein and low protein low heat powder using ultrafiltration in combination with evaporation concentration has been described. This involves the addition of retentate to skim milk to produce a high protein product and the addition of permeate to produce a complementary low protein product. The technology may be used to tailor powders to specific applications, in this example for making white sauce for use in frozen meals. (Muir, D.D. *et al.* 1991. *Journal of the Society of Dairy Technology,* **44**, 20–3).

heat treatment and concentration to 45–50% total solids. The concentrate is then homogenized and spray dried. The dryer is usually operated at inlet and outlet temperatures at the lower end of the range used for skim milk to minimize the amount of free fat in the powder. Whole milk powder is prone to oxidative deterioration and the preliminary heat treatment applied is designed to delay the onset, and reduce the extent, of deterioration by production of –SH groups in the milk fat globule membrane. Heating also produces –SH groups in the whey proteins and this results in cooked flavours considered undesirable in many products. Solubility may also be reduced by some heat treatments. Heating in the range 85–95°C for several minutes is widely used, but deleterious effects may be minimized by heating at higher temperatures such as 125°C for *ca.* 20 s. The situation is complicated by the fact that cooked flavour and associated flavour perceptions such as creaminess are desirable in some applications and for this reason longer holding times may be applied. In such cases it is necessary to adjust dryer operating parameters to obtain maximum powder quality. Anti-oxidants such as L-ascorbic acid and ascorbyl palmitate are permitted in most countries.

High fat milk powder is generally accepted as being a product with a fat content in excess of 35%. Manufacture involves combining concentrated skim milk, sodium caseinate and a fat source. The fat source may be anhydrous milk fat or a vegetable fat, usually hydrogenated.

Lecithinization is a process applied to whole milk and high fat milk powders to create improved instant qualities. The process involves spraying a lecithin/fat solution onto the powder granules either during drying in the well-mixed fluidized bed or during flow between the well-mixed and plug flow beds. 'Conditioning' takes place in the fluidized beds during which coating of the granules is completed.

In contrast to skim milk powder, which may be packaged for bulk sale in simple multi-wall paper sacks with a loose plastic liner, whole milk and high fat milk powders require sophisticated packaging to minimize autoxidation during storage. Gas packaging in an inert atmosphere of nitrogen is highly effective, although expensive and requires the use of

---

* An alternative process for manufacture of whole milk powder involves separating the milk and applying a heat treatment of 90°C for 12 s to the cream. The skim milk receives a low-heat treatment prior to concentration. Concentrated skim milk and cream are recombined before drying, the resultant powder having good stability against fat autoxidation and very little cooked flavour (Hols, G. and van Mil, P.J.J.M. 1991. *Netherlands Milk and Dairy Journal*, **44**, 49–52).

packaging with barrier properties. Multi-layer laminates incorporating metal foil and high density plastic films are most widely used.

(c) End-product testing

End-product testing is required for all types of skim milk. Chemical analysis is employed to ensure compositional standards are met and that functional properties are such that the product is of satisfactory quality. Additional tests may be required to ensure that the powder performs adequately in its end-use. The prime role of microbiological testing is to ensure safety with respect to pathogens, particularly *Salmonella*. It is also required to verify general standards of manufacturing hygiene and to ensure the suitability of the powder for its end-use. This is of particular importance in the case of powder for use in infant foods.

3.2.6 Relation between functional attributes of milk powders and end-use

Milk powder is multi-functional performing a variety of different functions in a variety of different applications. The requirements for use in different products vary, however, and cannot be met by a single milk powder. For this reason a range of milk powders are produced with properties matched to the end-use.

Proteins are of major importance in determining the functional attributes of milk powders. Some applications, such as the manufacture of white sauce for use in frozen products, require a powder of high protein content, but for other applications, such as chocolate manufacture, protein content is not critical. In virtually all applications, however, the properties of the proteins are significant in determining performance. The functional attributes of skim milk proteins include water absorption and water binding, foaming, emulsification, solubility, viscosity, gelation and colloidal and heat stability.

End-usage of low-, medium- and high-heat powders is summarized in Table 3.2. Ultrafiltration is currently of greatest importance with respect to cheese manufacture, for which low-heat powders are required. In this context ultrafiltration has the additional advantage of permitting the composition of the powder to be modified. The use of ultrafiltration to produce a low-heat powder of low calcium content, for example, is advantageous in the manufacture of halloumi cheese with good stretch and melt characteristics.

Table 3.2 The utilization of low-, medium- and high-heat milk powder

Product	Powder type	Attributes
Reconstituted milk and milk drinks	Low/medium heat	High solubility, minimal cooked flavour
Recombined evaporated milk	High heat	Heat stability, high viscosity
Cheese	Low heat	Rennetability
Ice cream	Medium heat	Emulsification, foaming, water absorption
Confectionery	High heat	Texture modification, water absorption
Comminuted meat	High heat	Emulsification, gelation, water absorption
Baked goods	High heat	Water binding, texture modification

Data from Augustin, M. A. 1991. *CSIRO Research Quarterly*, **51**, 16–22.

Whole milk and high fat milk powders are used where total milk solids are required for organoleptic, nutritional or functional purposes. The properties of the fat are of importance in some applications such as the manufacture of milk chocolate where a high level of free fat is required to maximize the reduction in viscosity of the chocolate mass. This may be achieved by co-drying skim milk and cream. The use of vegetable fats in high fat milk powder offers advantages over milk fat in terms of improved shelf life and better functional properties in products such as shortening.

The high rate of product innovation in the food industry has created the demand for a wide range of dried milk products. Improved dryer technology means that it has been largely possible to meet this demand, but diversification has not been supported by equivalent advances in knowledge of the relationship between the properties of milk powders and performance. A three-stage approach has been suggested for developing skim milk powder for specific ingredient applications which is applicable to all types of powder:

1. Identify the most important functional attributes required in specific food applications.
2. Apply appropriate modification procedures to achieve the required functionality.
3. Evaluate the functional performance of the powder when actually incorporated into the specific food product.

3.3 CHEMISTRY

3.3.1 Nutrient status of concentrated and dried milks

(a) Concentrated milks

The removal of water increases the dry matter content of milk in proportion to the degree of concentration. In most countries, the minimum fat and solids contents are legally defined, although there is considerable variation.

Loss of nutrients during processing reflects the extent of heat treatment. Evaporation leads to little change in the biological value of proteins and the amino acid content is similar to that of the starting milk. There is, however, a loss of *ca.* 20% of available lysine through the Maillard reaction. Feeding experiments with rats showed no detectable change in protein digestibility.

Levels of vitamins after evaporation are generally similar to those of pasteurized milk. At equivalent dilutions, however, the level of vitamin B_{12} is 17% that of pasteurized milk; B_6, 48%, thiamine 65% and folic acid 77%. Sterilization increases losses further and the vitamin levels of in-can processed evaporated milk are similar to those of in-bottle sterilized (see pages 77–9).

Enrichment with vitamin C and/or A and D is permitted in some countries.

Loss of vitamins continues during storage of canned concentrated milks. The extent of loss depends on the storage temperature and can be very significant in countries where high ambient temperatures are the norm. Sterilization results in a *ca.* 20% loss of vitamin C, which is compounded by a further loss of *ca.* 20% during storage at 21°C and as much as 60% at 36°C for 12 months. Losses of other vitamins occur, vitamin B_1 and B_2 concentration was reported to fall by *ca.* 30% during storage for 1 year, although storage temperature was not specified.

(b) Dried milks

Nutrient levels in dried milks are affected by losses during production of the concentrate as well as by those which occur in the drying process. The amino acid composition of proteins is relatively unchanged by drying and instantizing. Losses of lysine are small in comparison with those which occur during preparation of the concentrate, and usually amount to no more than 5% during spray drying.

Spray drying has relatively little effect on the vitamin content of milk, effects being most marked with vitamin B_{12} (20 to 30%), vitamin C (*ca.* 20%) and vitamin B_1 (*ca.* 10%). Loss of other vitamins is minimal. The level of fat soluble vitamins in skim milk powder is, inevitably, low and supplementation is practised where such powder is used for infant nutrition.

Nutrient losses during storage are small providing that the relative humidity is low, the powders are not exposed to very high ambient temperatures and the length of storage is not excessive. Available lysine was reduced by *ca.* 8% during storage of powder at 25 and 37°C for 1 year, the nutritive value of the proteins being reduced by the same order. Losses due to the Maillard reaction become significant when storage is prolonged even when storage temperatures are low.

Losses of vitamins B_1 and C are typically in the order of 10% during storage for 2 years, although losses of up to 33% have been reported in spray dried milk powder over 40 months' storage. Vitamin C losses are reduced by minimizing the oxygen and water vapour permeability of the packaging. Light impermeable packaging is also required to prevent loss of light sensitive vitamins, especially riboflavin.

3.3.2 Changes affecting structure and quality

(a) Influence of preheating

Thermal load during preheating plays a major role in determining the structure and quality of concentrated and dried milks. A large number of reactions occur, although the extent to which they proceed depends on the severity of the process. The major effects of preheating and the consequences for the properties of the subsequent concentrated milk are summarized in Table 3.3.

Whey proteins are sensitive to temperatures in excess of 60°C and are converted from the native globular conformation to a random conformation which readily permits protein/protein and ion/protein interactions. Denatured whey proteins, including β-lactoglobulin and other whey proteins preferentially complex with casein micelles by disulphide interchange with the κ-casein components and by calcium-mediated bonding. This is important in stabilizing whey proteins against aggregation and precipitation in heated milk systems. At temperatures of

Table 3.3 Changes resulting from preheating

Change	Consequences
Increase in titratable acidity and decrease in pH value due to lactose degradation and organic acid formation	Possible protein destabilization
Denaturation of whey protein	Change in functional properties of dried milk powder
Interaction of denatured whey proteins with casein micelles	*Stabilization against further heating*
Aggregation of casein micelles	*Stabilization against further heating*
Casein dephosphorylation	*Stabilization against further heating* Destabilization if excessive
Formation of colloidal phosphates	*Stabilization against further heating*
Maillard reaction between lactose and proteins	Loss of available lysine Production of brown pigments Production of undesirable flavours
Production of activated sulphydryl groups	*Delay in oxidation of fats* Development of 'cooked' flavour

Notes: Italic lettering indicates desirable consequences. In dried milk change in functional properties and development of cooked flavour may be either desirable or undesirable depending on end-use.

ca. 90°C the casein micelle : whey protein complex forms aggregates with molecular weights in excess of 100 000. Evaporated milk, to be in-can sterilized, requires more rigorous pretreatment for stability, the heat applied being approximately equivalent to 100°C for 15–20 min. Under such conditions casein micelle : whey protein complexes are simultaneously aggregated and disaggregated.

(b) Effect of concentration

The degree of concentration, the concentration factor (R) may be defined as the ratio of the dry matter content (by weight) of the concentrated milk to that of the unconcentrated. Concentration has a number of effects (Table 3.4) which may be modified by other factors such as preheat temperature. The consequence of these changes is to accelerate most of the reactions initiated during preheating. The effect is limited in evaporated milk where the value of R is rarely greater than 2.5 (a_w level *ca.* 0.98), but is more pronounced in sweetened condensed

Table 3.4 Changes resulting from concentration

1. Lowering of a_w level
2. Alteration to physical properties (increase in density and refractive index, decrease in thermal conductivity)
3. Increase in viscosity, liquid becoming non-Newtonian (shear thinning and viscoelastic) and then fairly solid as R exceeds 9
4. Diffusion coefficient decreases
5. Hygroscopicity increases
6. Osmotic pressure, freezing point depression and boiling point elevation (colligative properties) increase as does electrical conductivity
7. Thermodynamic activity of solutes increases
8. Ionic strength increases with a corresponding decrease in the activity coefficient of ionic species. This in turn causes an increase in ionization and solubility of salts
9. Dissolved substances become supersaturated and may precipitate
10. Conformation of proteins may change. Tendency towards formation of a compact structure and association increases
11. Casein micelles increase in size. This is due to coalescence rather than swelling and the increase is less if the milk is preheated sufficiently to form casein:whey protein complexes

milk where the value of R (before sucrose addition) usually approximates to 7.0 (a_w level ca. 0.83).

(c) Other changes during evaporation

Considerable damage occurs to the fat globule during evaporation. Globules are split to form a number of small globules with changed surface layers and are highly prone to lipolysis unless inactivation of lipases is complete. Evaporation also removes dissolved gases and reduces off-flavours by removal of volatiles.

(d) Heat stability of concentrated milk

Heating of concentrated (and unconcentrated) milk ultimately leads to coagulation and stability is usually defined in terms of the time needed to cause visible coagulation of milk at a given temperature; the heat coagulation time (HCT). Coagulation of evaporated milk during sterilization has been recognized as a problem for many years but, until recently, the underlying cause has been only poorly understood.

Early work was largely restricted to studies of the effects of calcium/magnesium and citrate/phosphate on the HCT and led to the 'salt balance theory'. This 'theory' permitted the development of the procedures still used industrially to control coagulation (see page XXX), but did little to explain the mechanisms involved. Subsequently, a 'working hypothesis' was developed, based on the stability of colloidal caseinate-phosphate and the changes induced by heating, reaction between β-lactoglobulin and the κ-casein of the casein micelles being considered the most important factor in altering susceptibility to coagulation. More recently, a model describing heat coagulation in unconcentrated milk has been developed and this has now been extended to concentrated milk.

In the case of milk which has not been preheated and which has an initial pH value below 6.5, coagulation occurs as the temperature increases during heating. Coagulation appears to be induced by aggregation of denatured serum proteins and depends on calcium and hydrogen ion activities. Coagulation of this type cannot occur in preheated milk (the commercial situation) since the serum proteins are denatured, although the stabilizing effect of preheating is determined by the manner and extent of aggregation of the denatured serum proteins rather than the denaturation itself. In all preheated concentrated milks coagulation is due to the aggregation of casein micelles. Behaviour during coagulation is, however, strongly modified by pH value of the milk.

At pH values below *ca.* 6.5 casein micelles flocculate into voluminous irregularly shaped clusters, while in contrast at higher pH values the aggregates tend to fuse into compact particles of approximately spherical shape (Figure 3.9). The compact particles become anisometric shortly before the HCT. The aggregation rate is determined by the colloidal stability of particles and the rate constant(s) of bond forming reactions, both of which are significantly affected by pH value. Dissociation of κ-casein from the casein micelles proceeds more rapidly at high initial pH value and has a profound effect on micellar stability. The formation of micellar calcium phosphate-like salt bridges is also highly dependent on high initial pH value, since the reaction only occurs significantly where casein micelles are κ-casein depleted. Other types of salt bridge (ionic and Ca^{2+}-mediated) are less affected by pH value. Covalent cross-linking of proteins may also be involved in heat coagulation, but requires close contact between specific protein residues. This would be promoted by a large heat-induced fall in pH value, or by depletion of κ-casein from the micelles.

Heat coagulation time is primarily determined by the rate of particle

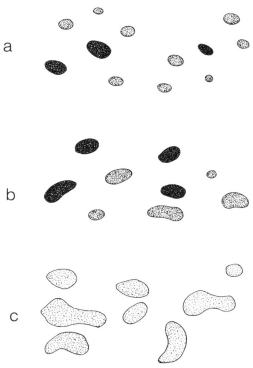

Figure 3.9 Formation of casein aggregates in concentrated milk. Heat treatments: (a) pH 6.33: heat treatment 8 min 45 s, (b) pH 6.49: heat treatment 35 min and (c) pH 6.81: heat treatment 11 min. Based on data fron Nieuewenhuijse, J.A. et al. 1991. *Netherlands Milk and Dairy Journal*, **45**, 193–224.

aggregation and the geometry of the emerging particles, the latter being strongly influenced by dissociation of κ-casein and the supersaturation of calcium phosphates. Protein content is a further major factor determining heat stability of low pH value concentrated milk, but not that of high pH value. In recently developed models, the relationship between HCT and protein content is explained by assuming that flocs formed in low pH milk are of a fractal nature, with fractal dimensionality of *ca*. 2.1. Fractal flocs need not become large to cause gelation in a concentrated dispersion and only aggregation due to encounters resulting from Brownian motion can be important. In high pH value milk, it is likely that the HCT is determined primarily by the time after which aggregates become anisometric and therefore voluminous. This time is itself determined by the ratio of the time it takes for two particles to fuse to the time elapsing between subsequent aggregation steps.

The role of phosphate in stabilizing concentrated milk has also been examined. The basic effect of phosphate may be partly explained in terms of a pH increase. The overall effect is more complicated, however, and not fully understood. Phosphate addition disturbs the existing equilibrium between calcium, phosphate and other ions and a new equilibrium is attained only slowly at 20°C. During heating at 120°C a continuing association of calcium phosphates with the micelles occurs over a period of at least 10 min, while in concentrate without added phosphate the association is completed during heating-up. The different behaviour in the presence or absence of added phosphate may have consequences affecting rates of serum protein association, κ-casein dissociation and bond formation during the sterilization process.

(e) Age-thickening and gelation of concentrated milk

Changes resulting from concentration mean that a tendency to age thickening and gelation is a property of all concentrated milk. Particular problems are encountered occasionally with canned, sterilized, evaporated milk. This is not due to protein breakdown (*cf.* UHT milk, page 84), but to the formation of a three-dimensional network in which aggregated casein micelles are connected by thread-like structures (Figure 3.10). Age-thickening and gelation is strongly promoted by cold storage of the concentrate before sterilization and this is attributed to a physico-chemical transformation of the casein micelles. Age-thickening occurs in two distinct stages, the rate being lower at higher sterilization heat loads:

1. Viscosity increases accompanied by 'spotty lump' formation, but the milk is still fluid after stirring.
2. Irreversible gelation accompanied by syneresis and flocculation after stirring.

Age-thickening is also a potential problem in sweetened condensed milk and appears also to involve aggregation of casein micelles into a three-dimensional network (Figure 3.11). Heated whey proteins are not directly involved, but may promote the development of networks by acting as links between adjacent micelles. Whey proteins may also be involved in complex formation with fats in whole sweetened condensed milk, forming a net-like structure which enmeshes casein.

A decrease in pH value occurs during age-thickening which increases casein/casein interactions and reduces ionization of seryl phosphate residues. Both reactions tend to increase gelation, although this may be reduced, but not eliminated, by addition of polyphosphates which have an affinity for casein.

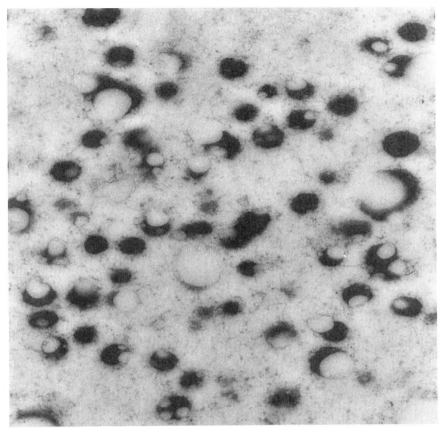

Figure 3.10 Age-thickening of canned, sterilized evaporated milk. Reproduced with permission from de Koning, P.J. *et al.* 1992. *Netherlands Milk and Dairy Journal*, 46, 3–18.

(f) Factors affecting chemical and physical properties of dried milk

In addition to its importance in determining the functional properties of dried milk, preheating has a major influence on the properties of the newly produced powder and on changes during storage. Preheat treatments in the range 110°C for 2 min to 120°C for 3 min denatured 80–91% of β-lactoglobulin and 33–45% of α-lactalbumin, but while the extent of denaturation was increased during evaporation, there was no further increase on spray drying. Incorporation of lactalbumin and lactoglobulin into casein micelles was markedly less than the extent of denaturation and not in constant ratio to it. The insolubility index of dried milk powder has been shown to be adversely affected by increases in time and length of preheat

146   Concentrated and dried milk products

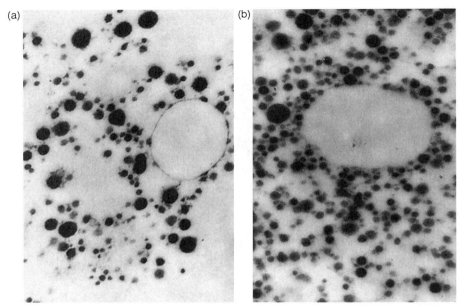

Figure 3.11 Age-thickening of sweetened condensed milk. (a) Before storage. (b) After storage at 37°C for 45 days (conditions known to favour gelation). Reproduced with permission from Alvarez de Felipe, A.I. et al. 1991. *Journal of Dairy Research*, **58**, 337–44.

treatment, although one study concerning whole milk powder suggested that only length of treatment was significant. Insoluble material from powder prepared after high temperature preheating contained the higher levels of protein and protein-lactose, while soluble material was practically depleted of whey proteins, which were utilized for complexes stabilized through disulphide bonds.

Levels of both hydroxymethylfurfural and –SH groups increase with temperature and length of preheating. The production of –SH groups, primarily from β-lactoglobulin, is of particular importance in whole milk powder and decreases the rate of autoxidation of fat by reaction with free radicals of the unsaturated fatty acids. It is not, however, clear whether –SH groups alone, –SH groups and the reductive capacity of the milk, or -SH groups in combination with the pH value are responsible for

---

* It may appear paradoxical that the *insolubility* index is used to define *solubility*. It is, in fact, a more logical approach since the parameter measured is the insoluble material present, to a greater or lesser extent, in all dried milk. The insolubility index is very widely used as an index of quality, although the value is, in some circumstances, arbitrary.

anti-oxidative activity. Preheating may produce other anti-oxidants including heated casein (which has only a few –SH groups) and lowers the redox potential.

Chemical changes continue during the storage of dried milk powder, the rates being governed by storage temperature, moisture content and, in the case of autoxidation of fat, the composition of the atmosphere and the presence of anti-oxidants. Increases in the insolubility index are most marked at temperatures above 20°C and a water content greater than 5%. Increases are also greatest when the initial insolubility index is high. The mechanism underlying the increase in insolubility is not understood, but it is not thought to be either lactose crystallization or related to the Maillard reaction. Maillard browning reactions, however, continue in dried milk with a water content in excess of 5% and play a major role in deterioration of stored powders.

The rate of fat autoxidation in air is highly temperature dependent, the reaction rate doubling with a 10°C rise in temperature. At the same time higher temperatures also favour the Maillard reaction, some products of which have anti-oxidant activity. The situation is further complicated by the influence of a_w level, water controlling fat oxidation by hydrogen bonding of hydroperoxides or hydration of metal ions and also favouring the Maillard reaction. Autoxidation may be prevented virtually entirely by reducing the oxygen content of the atmosphere to less than 2% and this is usually achieved by replacing air with nitrogen and/or carbon dioxide. Autoxidation is also reduced by foam spray drying, or by use of a steam jacketed centrifugal atomizer.

Lactose crystallization is also a common cause of quality loss. Lactose in the fresh powder is in a metastable amorphous state and highly hygroscopic. This state is maintained providing the a_w level is maintained below 0.43, the moisture content below 8.4% and the storage temperature below 20°C. An irreversible transition to the stable crystalline state occurs if these parameters are exceeded, resulting in quality loss through lumping and caking.

Reconstituted and recombined concentrated milk is subject to both heat instability and age-thickening, which may be attributed to the properties of the dried milk powder used. High-heat, heat-stable powder is required for maximum stability, but there is also an additional stabilizing effect of spray drying. This has been attributed to differences in salt 'equilibria' rather than denaturation of whey proteins, although denatured proteins may undergo a different pattern of aggregation as a result of spray drying.

Heat-resistant proteolytic enzymes derived from growth of psychrotrophic bacteria in the raw milk are partly responsible for age-thickening of recombined evaporated milk and powder used for recombining must have minimal activity of these enzymes.

3.3.3 Chemical analysis of concentrated and dried milk

(a) Concentrated milks

Chemical analyses required in production of all types of concentrated milk include fat, total solids and protein content. Ash and acidity may be required on some occasions and determination of sucrose content is necessary for sweetened condensed milk.

Methods used for liquid milk (Chapter 2, page 85) are generally suitable for use with concentrated milks although care must be taken to validate instrumental methods. For end-product testing, sucrose is determined by polarization before and after acid inversion, but determination of refractive index is adequate for process control.

(b) Dried milk

Moisture content of dried milk is of major importance, but fat and protein content are also determined on a routine basis. Determination is also routinely made of the insolubility index, incidence of scorched and burnt particles and the bulk density. Titratable acidity relates to both composition and quality and is used in some grading systems. Determination of ash, lactose, vitamins (fortified powders) and lecithin (instantized powders) are required in some cases and detection of adulteration with whey may also be necessary.

Analysis for fat in full cream powders can cause problems and the Rose–Gottlieb method is preferred, protein determination is usually by Kjeldahl, although a dye binding technique has been used successfully and moisture content is determined by drying. Insolubility index is determined by measuring the insoluble material remaining after blending of powder in warm water and centrifuging. The incidence of scorched and burnt particles is assessed by filtering blended powder and comparing the filter with photographic standards. Bulk density is determined by weighing powder into a measuring cylinder, dropping the cylinder from 150 mm onto a pad and measuring the bulk density in grams per litre.

3.4 MICROBIOLOGY

3.4.1 Concentrated and dried milk as an environment for micro-organisms

Concentrated and dried milk comprise a spectrum of a_w levels from just below that of unconcentrated milk (0.98) in unsweetened concentrates to the very low levels (below 0.60) in dried milk. Unsweetened concentrates will support the growth of all common milk spoilage organisms, although those more tolerant of reduced a_w levels will be marginally favoured. In contrast the a_w level of sweetened condensed milk (0.90–0.85) is such that considerable selective stress is applied and only osmotolerant micro-organisms, primarily moulds, are able to develop. Micro-organisms cannot grow in dried milk although survival may be prolonged.

3.4.2 Concentrated and dried milk and foodborne disease

Concentrated milks are not considered to be high-risk products provided that the milk receives a heat treatment equivalent to pasteurization. A general risk due to post-process contamination does, of course, exist with non-sterile products, while sterilized evaporated milk is subject to the same risks as all canned foods with respect to underprocessing and seam leakage.

In contrast spray dried milk powder has been implicated in a number of outbreaks of foodborne disease and the widespread use of the product in preparing infant food is a particular cause of concern.

(a) Salmonella

In recent years *Salmonella* has been recognized as the major hazard in spray dried milk powder. This follows a number of outbreaks in the US, Australia and the UK involving several different serovars. *Salmonella*

BOX 3.3 Breast is best

The practice of giving free infant formula in maternity units has been widely criticized, especially in developing countries, where breast feeding is strongly recommended. In Malaysia, the unsatisfactory microbiological status of infant formula based on dried milk has led to an embargo on free samples in maternity units (*New Straits Times*, May 30, 1992).

enters dried milk from foci of contamination within the plant where the organism is able to persist and, in some cases, multiply over extended periods. Examples of possible foci include parts of the powder handling system and cracks in the dryer walls. In the outbreak due to *S. ealing* the organism is believed to have entered the plant in raw milk, to have been disseminated through the internal environment and to have entered the insulation material of the dryer *via* cracks in the wall. A focus of contamination was established from which *Salmonella* entered the powder.

The problems with *Salmonella* have led to discussion concerning its ability to survive the spray drying processes. Results have been variable but it appears that survival is possible under at least some conditions. The debate is, however, irrelevant: spray drying is not intended to and cannot assure safety, the real problem, as noted above, is contamination of the plant and its environment. The presence of *Salmonella* in the product stream at any stage after heat treatment indicates the failure of control measures.

(b) Staphylococcus aureus

Intoxication due to contamination of spray dried milk with staphylococcal enterotoxins was identified as a major problem during the 1950s. The immediate cause is growth and enterotoxin production by *Staph. aureus* either in the raw milk before heat treatment or, more usually, in the concentrated milk before drying. The enterotoxins survive subsequent processing irrespective of the fate of the organism itself. The underlying cause of problems is poor plant design and operation permitting product to remain at favourable temperatures for microbial growth for extended periods and, in the case of growth in the concentrate, poor plant hygiene. Application of good manufacturing

---

BOX 3.4 In for a penny . . .

The most recent major outbreak of salmonellosis associated with dried milk powder arose from contamination of infant food with *S. ealing* and cost the company involved an estimated £22 million. This value represented the difference in value between the company before the outbreak and the price obtained at its sale after liquidation. The liquidation stemmed directly from loss of consumer confidence arising from the *Salmonella* outbreak.

practice has obviated problems due to *Staph. aureus* in industrialized nations, but occasional outbreaks are still reported in developing countries.

(c) Other pathogenic micro-organisms

The aerobic endospore-former *B. cereus* is common in spray dried milk powder, although normally present only in small numbers. The significance of this is debatable, some microbiologists believing that there is no major health risk, while others believe that a health risk exists but only if the reconstituted milk is stored at room temperature for an excessive time. Dried milk has been implicated as the source of *B. cereus* in outbreaks involving made-up products such as vanilla slices and macaroni cheese, but it seems likely that improper storage of the reconstituted products was a contributory factor. The anaerobic endospore-former *Clostridium perfringens* has also been associated with food poisoning due to spray dried milk and growth in the made-up product again appears to have been a major contributory factor.

Members of the *Enterobacteriaceae* other than *Salmonella* have been implicated in foodborne infection associated with spray dried milk. These include not only recognized enteric pathogens such as *Yersinia enterocolitica*, but also *Enterobacter*, a member of the family not normally associated with foodborne disease. *Enterobacter sakazakii* has been involved in a number of cases of meningitis in which the vehicle was thought to be dried milk. This illustrates the need for special care where dried milk is to be used as a base for infant foods and suggests that the significance of 'coliforms' in dried milk may need reassessment in this circumstance.

3.4.3 Spoilage microflora of concentrated milks

(a) Non-sterile, unsweetened concentrated milk

The spoilage microflora consists of heat resistant bacteria, which have survived processing, and post-process contaminants. Heat resistant bacteria are derived from raw milk and numbers in the final product largely reflect the numbers in the raw milk. Where control and hygiene are poor, however, growth may occur during processing. *Bacillus* species are a common contaminant, while *Clostridium* may also be present, although usually in smaller numbers. Survival of thermoduric species of *Enterococcus* into the finished product depends on the severity of preheating. Spoilage of the finished product by thermoduric micro-organisms is usually a consequence of slow cooling, or storage at temperatures above 10–12°C. Spoilage patterns vary, but include acid and gas production and bitter flavours and thickening due to proteolysis.

A wide range of micro-organisms may enter the product by post-process contamination, but Gram-negative bacteria are most common. Typical genera include *Pseudomonas* and 'environmental' members of the *Enterobacteriaceae* such as *Citrobacter* and *Hafnia*. Many of these organisms are able to grow at temperatures below 4°C and are dominant in the spoilage of correctly refrigerated concentrated milk. Spoilage patterns vary, but include acid and gas production, bitterness due to proteolysis and, in full cream milks, lipolysis.

(b) Canned, evaporated milk

Canned evaporated (and UHT sterilized, aseptically packaged) milk is 'commercially sterile' and under normal conditions spoilage occurs only where the thermal processing is inadequate, or where post-process leakage has occurred. The exception is the spoilage of evaporated milk by the thermophile *B. stearothermophilus*, which is sunbsequently exported to hot climates. Under these circumstances *B. stearothermophilus* survives processing normally applied.

Species of *Bacillus* are most commonly involved in spoilage due to underprocessing. *Bacillus coagulans*, for example, causes an acid coagulation and a slight 'cheesy' odour and flavour but is normally only a problem where storage temperatures are high. *Bacillus megaterium* causes an acid formation accompanied by gas and a 'cheesy' odour, while proteolysis by *B. subtilis* results in thickening and a bitter taste followed by digestion to a brown liquid. Occasional spoilage involving gas and H_2S production by species of *Clostridium* occurs, but is very rare.

Many bacteria have been implicated in spoilage due to post-process contamination and spoilage patterns reflect the nature of contamination. In an unusual case, acid and gas were produced by a co-culture of *Enterococcus faecium*, a post-process contaminant and *B. subtilis* which had survived processing. Neither organism produced acid and gas when grown in single culture.

(c) Sweetened condensed milk

Sweetened condensed milk is subject to spoilage by osmotolerant yeasts, notably *Torulopsis*. The organism enters the milk after preheating and the incidence of spoilage has been much reduced by improved hygiene. Growth of the organism is slow, especially at lower ambient temperatures, but sufficient gas may be produced to swell the cans.

Growth of moulds, usually species of *Aspergillus* and *Penicillium*, on the upper surface of the milk leads to the formation of 'buttons', small aggregates of mycelium and coagulated casein. The presence of mould is accompanied by off-flavours due to proteolysis. The fault is associated with poor plant hygiene and exacerbated by a large head space containing sufficient oxygen for excessive growth.

Most bacteria are unable to grow in sweetened condensed milk, but bacterial spoilage occasionally occurs, especially when the sugar content is relatively low. Many genera have been implicated, but the relatively osmotolerant *Micrococcus* and *Bacillus* are most common. Typical spoilage patterns involve thickening, acid production, proteolysis and, in full cream products, lipolysis. Members of the *Enterobacteriaceae* have also been implicated in the spoilage of bulk, low sugar condensed milk and produce acid and gas.

3.4.4 Microflora of dried milk powder

The microflora of dried milk powder depends on many factors including the number and types of bacteria in the raw milk, temperature of preheating, plant hygiene and dryer operating conditions. High numbers in the raw milk tend to be reflected by high numbers in the milk powder, especially if the raw milk microflora contains large numbers of thermoduric organisms. A general relationship between numbers in the raw milk and the powder has been established, raw milk counts in excess of 10^5 cfu/ml often resulting in counts in the powder of more than 10^4 cfu/g.

Both 'coliforms' and *Escherichia coli* are of indicator significance in dried milk since none are thermoduric and their presence indicates either gross underheating, or post-process contamination. *Escherichia coli* does not, however, have index significance with respect to faecal contamination since the organism is a cause of mastitis and may be present in milk.

A wide range of thermoduric micro-organisms have been isolated from

---

* There is evidence that a high 'coliform' count in the raw milk leads to the development of off-flavours during the storage of whole milk powder, even if numbers of 'coliforms' in the powder are very small. The mechanism for this is not known.

dried milk powder including *Bacillus*, *Enterococcus* and *Alcaligenes tolerans*. Actinomycetes of the genera *Thermoactinomyces* and *Micromonospora* have also been recovered.

3.4.5 Microbiological methods

(a) Non-sterile, unsweetened, concentrated milk

Media and methods used for pasteurized milk products are suitable (See page 98), although high viscosity products require an initial gravimetric dilution rather than a volumetric.

(b) Canned, evaporated milk

Canned evaporated milk is usually incubated at 25 or 30°C or, if intended for export to hot climates, 37 or 55°C for 2–3 weeks. It is not sufficient to rely on gas formation and can swelling as an indicator of non-sterility. Cans should be opened under aseptic conditions and samples streaked onto general purpose media such as nutrient agar for aerobic bacteria (dextrose-tryptone agar is more suitable for thermophilic bacteria) and blood agar (incubated under anaerobic conditions) for anaerobic bacteria. The process of 'auto-sterilization' may mean, however, that viable organisms cannot be recovered and visual examination (but *not* tasting) and determination of the pH value are valuable supplementary examinations. 'Auto-sterilization' is usually only a problem after short-term storage when incubation has been at 55°C, but is a common phenomenon when investigating complaints of spoilage after prolonged storage.

Full identification of any bacteria recovered is not usually necessary unless a public health risk is suspected. Simple tests should be applied to distinguish between non-endospore-forming bacteria, whose presence indicates post-process contamination and endospore-forming bacteria,

---

* Actinomycetes are rare in dairy products, although isolations have been made of *Thermoactinomyces* from both raw and pasteurized milk. *Thermoactinomyces* consists of branched and septate hyphae with an aerial mycelium of variable extent. The organism may be isolated from soil and self-heating, decaying vegetable material such as mouldy hay. It has been implicated as a causative organism of 'humidifier fever', associated with growth in air-conditioning systems and may play a secondary role in 'Farmers' lung'. *Micromonospora* consists of fragmentary, branched and septate hyphae and only rarely has a sparse aerial mycelium. The organism is one of a number of actinomycetes which have become adapted to aquatic habitats and its life cycle includes a motile stage. *Micromonospora* has been isolated from stored river water and may be able to colonize water distribution systems within manufacturing plants.

whose presence indicates *probable* underprocessing.

(c) Sweetened condensed milk

For most purposes media and methods used for pasteurized milk are satisfactory for sweetened condensed milk (see page 98), although gravimetric dilution is required initially and dispersion on dilution is facilitated by the addition of 1.25% sodium citrate to the dilution blank. Yeasts and moulds should also be enumerated, high sugar content media such as osmophilic yeast agar are not required and a standard medium for foodborne yeasts and moulds, such as Rose Bengal–chloramphenicol medium, is usually satisfactory.

(d) Dried milk

Methods used for milk (see page 98) may readily be adapted for determination of 'total' viable count and 'coliforms' in dried milk. *Salmonella* may be determined using classical cultural techniques (Figure 3.12). Pre-enrichment is necessary to recover stressed cells and the inclusion of 0.01% malachite green in the buffered peptone water used as enrichment medium is recommended to avoid overgrowth of *Salmonella*. Atypical biovars of *Salmonella* which are both lactose-positive and H_2S-negative are relatively common in dried milk and both typical and atypical colonies developing on selective isolation medium should be subject to confirmatory testing.

Staphylococcus aureus in dried milk is also stressed, but collaborative studies have shown that direct plating onto Baird–Parker medium is a satisfactory method of enumeration. The absence of *Staph. aureus* does not, however, assure the safety of dried milk since the enterotoxin may be present. Routine testing for enterotoxin has been practised by some manufacturers in the past, but is neither considered necessary, nor a substitute for good manufacturing practice. The determination of thermonuclease as an indicator of the likely presence of enterotoxin is a possibility, but does not appear to have been widely used (*cf.* cheese, page 341).

Many standards applied to dried milk include thermoduric organisms. Two alternative methods may be applied, heat treatment at 74°C for 15 s or 63°C for 30 min, prior to plating and incubation at 30°C for 48 h. Results obtained by the two methods are not directly comparable and the heat treatment to be used must be specified.

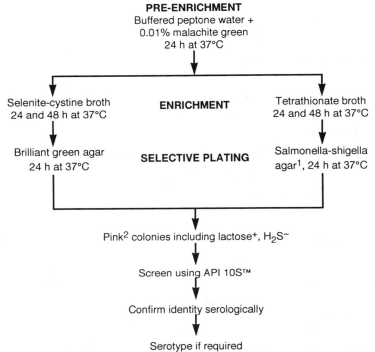

Figure 3.12 Protocol for the isolation of *Salmonella* from dried milk. Rappaport-Vassiliadis broth is the medium of choice in many foods. This medium requires incubation at 41°C and may not be suitable for salmonellas of dairy origin. [1] Media such as hektoen enteric agar may be used in place of *Salmonella-shigella* agar. [2] It is usually possible for an experienced microbiologist to differentiate lactose$^+$, H_2S^- strains of *Salmonella* from non-salmonellas.

The use of microbiological testing as a control measure in production of dried milk has created a demand for the application of rapid methodology. Electrical techniques including impedance and conductance appear to be most widely applied for assessment of the overall microbiological status and have also been applied to detection of *Salmonella*. Enzyme immunoassays for detection of *Salmonella* after a short cultural stage have been recommended for application to dried milk, but are not widely used.

Microbiology

EXERCISE 3.1.

Your company wishes to produce a low-calorie milk product resembling sweetened condensed milk in terms of taste and viscosity. Consider the possible approaches to the following:

1. Matching the organoleptic properties of the new product to that of conventional sweetened condensed milk.
2. Ensuring the microbiological safety and stability of the new product.

Design a programme to compare and evaluate the commercial viability of two new formulations, one suitable for ambient temperature storage, the other requiring refrigeration.

EXERCISE 3.2.

Some microbiologists have suggested that monitoring of producing farms is a necessary part of pathogen control procedures in production of spray dried milk powder. Discuss the difficulties implicit in establishing and operating such a monitoring system. Do you consider that monitoring would significantly increase the safety of spray dried milk powder, given that no practical system can be 100% effective?

EXERCISE 3.3.

You have recently been appointed as Development Technologist of a company producing a range of spray dried milk powders for ingredient use. A major customer is unhappy with the performance of a powder supplied as an ingredient of speciality bakery goods (breads and related products). A preliminary investigation showed that improved performance was obtained when the preheat treatment currently applied was reduced slightly. Optimal conditions of time and temperature had not, however, been established.

Draw up an outline programme for developing a suitable powder and testing its performance in use over the range of seven bakery goods the customer produces.

During the course of your investigations it becomes apparent that the temperature control device fitted to the preheater is of inadequate performance and permits an unacceptable degree of variation in the heat treatment applied. Replacement requires a significant capital expenditure which your management wishes to avoid and the managing director has stated, from the depths of his ignorance, that variations in the preheat treatment may be overcome by adjusting the operating temperatures of the evaporator. Consider the technical, chemical and financial implications and prepare a report refuting (or supporting) this statement.

EXERCISE 3.4.

Microbiological examination was undertaken of cans of full cream evaporated milk spoiled by acid formation, accompanied in a minority of cases by gas. Examination was restricted to colony counts, made using the pour plate method with nutrient agar as medium and incubation at 37°C, and a direct microscopic examination using a conventional light microscope. Colonies of a Gram-positive endospore-forming rod-shaped bacterium were isolated from two of 25 cans examined, but microscopic examination showed large numbers (more than 10^6/ml) of rod-shaped bacteria. There was disagreement over the presence of endospores.

Do you consider the examinations made, and the methods used, to be adequate in the investigation of a potentially serious spoilage problem? Consider alternative and/or additional examinations (not necessarily microbiological). What technical data concerning plant operation would be required to enable you to complete your investigations?

4
DAIRY PROTEIN PRODUCTS

OBJECTIVES

After reading this chapter you should understand
- The nature of the different types of dairy protein products
- The processing used in their preparation
- The modification and fractionation of dairy protein products
- The functional properties of dairy protein products and their use as ingredients
- The nutritional properties of dairy protein products
- Microbiological considerations associated with production of dairy protein products

4.1 INTRODUCTION

Dairy protein products are whey proteins and caseins which are generally considered as by-products. Sophisticated processing methods, however, are available to increase the value of the proteins as both nutritional and functional ingredients and many protein products have become of considerable importance in their own right.

Whey is the watery component removed after the setting of the curd in cheese manufacture. Direct use of whey as an animal feed is possible where farm and creamery are closely integrated, but in many cases this is not feasible. Many options exist for the processing of whey (Figure 4.1), but choice is, inevitably, dictated by economics.

Whey is generally classified as sweet, medium-acid, or acid depending on the titratable acidity and pH value. This in turn depends on the type of cheese making from which the whey is derived (Table 4.1). Whey contains, on average, 65 g/kg dry matter comprising 50 g

> **BOX 4.1 Sometimes I go about and poison wells**
>
> Disposal of whey has been a problem to the cheese industry for many years and disposal by dumping led to some serious incidents of pollution of rural waterways before the introduction of preventative legislation. Whey dumped into disused mineshafts has also entered water-bearing sub-strata and led to pollution of wells many miles distant. Examples of instances which had a major economic impact include pollution of a borehole supplying a poultry processing plant and of an artesian well supplying a brewery.

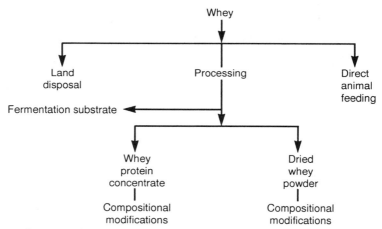

Figure 4.1 Whey processing options.

lactose, 6 g protein, 6 g ash, 2 g non-protein nitrogen and 0.5 g fat. The protein fraction contains *ca.* 50% β-lactoglobulin, *ca* 25% α-lactalbumin and *ca.* 25% of other proteins including immunoglobulins, proteose-peptone, bovine serum albumin and β-casein. Each of these has different characteristics (Table 4.2). Considerable variation in composition may occur depending on the milk supply and the nature of processing.

Casein is the principal protein in cows' milk and milk skimmed to provide cream of 40% fat, the starting point for casein manufacture, contains between 25 and 31 kg/l casein protein, representing 75–80% of skim milk nitrogen. The functional properties are summarized in Table 4.3. Casein has been extracted commercially since the

Table 4.1 Types of whey

Sweet whey	Titratable acidity 0.1–0.2%, pH value 5.8–6.6 Derived from rennet-coagulated cheese such as Cheddar and from rennet casein manufacture
Medium-acid whey	Titratable acidity 0.2–0.4%, pH value 5.0–5.8 Derived from fresh, acid cheese such as Ricotta and cottage cheese
Acid whey	Titratable acidity 0.4%, pH value <5.0 Derived from fresh, acid cheese and from acid casein manufacture

Table 4.2 Characteristics of different whey proteins

Beta-lactoglobulin	Heat labile	Dominates functional properties
Alpha-lactalbumin	Slightly heat labile	Concentrates solubility, gelation, whipping and emulsification
Proteose-peptone	Heat stable	Surface active, enhances whipping
Immunoglobulin	Very heat labile	Enhances gelation
Bovine serum albumin	Heat labile	Binds lipids
Soluble casein	Heat stable	Modifies functionality

Based on Marshall, K. R. and Hooper, W. J. 1988. *Bulletin of the International Dairy Federation*, **233**, 21–7.

Table 4.3 Functional properties of caseins

Water binding
Viscosity and gel formation
Fat emulsification
Whipping and foaming
Texturization

early part of the 20th century and is classified as rennet casein or acid casein according to the means of preparing the casein curd. Acid casein may be further classified according to the acid used; lactic acid, produced by growth of starter micro-organisms, hydrochloric or sulphuric acid. More recently acid casein has been produced using ion exchange resins to lower the pH value.

4.2 TECHNOLOGY

Technological operations in the processing of milk proteins vary in complexity from relatively simple operations, such as concentration, to sophisticated fractionation and modification techniques.

4.2.1 Whey protein products

(a). Concentrated whey and whey powder

Concentrated whey and whey powder are the oldest types of whey products. Processing is outlined in Figure 4.2 and compositional properties of the main types listed in Table 4.4.

Unprocessed whey supports microbial growth, especially when the pH value is relatively high. Care must be taken to minimize contamination and the whey is held under refrigeration for a strictly controlled period. Whey derived from cheese manufacture contains large numbers of starter micro-organisms and these must be destroyed by heat treatment. The use of anti-microbial compounds to prevent whey spoilage has been suggested, examples being benzoic acid (with pH value adjusted to 3.0–4.2), hydrogen peroxide, propionic acid and sorbate. In many

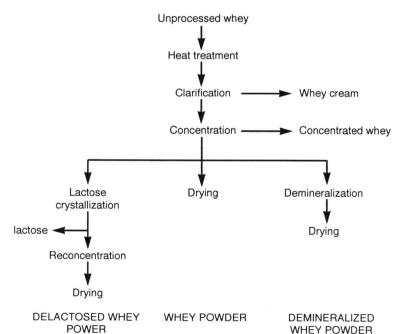

Figure 4.2 Production of whey protein products.

Table 4.4 Composition of different whey powders

	Fat (%)	Protein (%)	Lactose (%)	Ash (%)
Whey powder	1.1	12.9	74.5	8.5
Partially demineralized whey powder	2.0	15.0	78.0	
Partially delactosed whey powder	1–4	16–24	60 (maximum)	11–27

countries, however, the addition of anti-microbial compounds to whey is not permitted.

Clarification is an important part of whey processing and is of particular significance with respect to the preparation of concentrates by ultrafiltration. In this case even small amounts of lipid lead to membrane fouling and have an adverse effect on the functional properties of the finished product. The traditional means of clarification is by centrifugation, but alternative methods have been proposed, including the use of anionic precipitating agents, microfiltration and demineralization followed by sedimentation of aggregated material.

Concentration may be used either as the initial stage in the production of whey powder or for production of concentrated whey. Concentrated whey with a solids content of 30% is becoming an increasingly popular source of protein in manufacturing operations. Falling film evaporators with mechanical vapour recompression (MVR; see Chapter 3, pages 109–10) are the most common means of concentration and offer high thermal efficiency and minimal heat load. Reverse osmosis (RO) is an alternative which does not involve denaturation of whey proteins. The overall operating cost of MVR evaporators and RO is similar for large-scale operations, but in small-scale optimum efficiency is obtained by concentrating to *ca.* 25% by RO and finishing by MVR evaporation.

Roller drying has been used for the preparation of whey powder but results in a poor quality product and spray drying is now more common.

---

* Concentrated whey is not shelf-stable and normally requires refrigerated storage. Shelf-life is only 2–3 days at 5–8°C and this limits the use of the product, especially in warm countries with only limited refrigeration capacity. A method of preparing a whey stable at ambient temperatures has been developed which may be applied both to conventional and lactose-hydrolysed types. Preparation involves adjustment of the pH value to 5.0 and addition of potassium sorbate and/or calcium propionate (Leiras, M.C. *et al.* 1991. *Food Science and Technology*, 24, 12–16).

The high lactose content of whey presents a major problem since rapid drying of lactose containing solutions results in a high proportion of the α-lactose being in the highly hygroscopic anhydrous form. Moisture is rapidly absorbed as the anhydrous α-lactose converts to the monohydrate with resultant caking and lumping. Modern manufacturing processes are designed to overcome this problem by ensuring that most of the α-lactose exists as the monohydrate before drying is completed. This may be achieved either by holding the concentrate before drying to allow crystallization of α-lactose to occur, or by using a similar process to instantization of dried milk (see page 127) and thus forming the monohydrate before completion of drying. The most satisfactory system is to introduce crystallization stages both before and after drying (double crystallization). Problems are also caused by a high lactic acid content which also leads to a highly sticky and hygroscopic product.

Crystallization is also used in the traditional process for extracting lactose from whey. Whey is concentrated to 58–62% by evaporation at a maximum temperature of 70°C. Lactose nucleation and crystallization are then induced by controlled cooling and seeding. Lactose crystals are separated from the mother liquor by decanter centrifuge, washed and dried. The mother liquor is reconcentrated and spray dried as delactosed whey powder.

Whey powder has a high mineral content and for use in infant formulations demineralization is necessary. Demineralization also has advantages in general food use. The process is, however, expensive, the cost approximating to that of spray drying, and can only be justified if the product commands a proportionately high price.

Two well established methods, ion exchange and electrodialysis, are available for demineralization. The properties of whey demineralized by the two methods differ as a consequence of the different selectivities. Ion exchange is relatively non-selective and removes both mono- and polyvalent ions, while electrodialysis is more dependent on ionic mobility and preferentially removes monovalent ions. Ion exchange permits virtually 100% demineralization, while the upper limit for

---

* Electrodialysis equipment consists of sets of paired membranes, each of which has whey passing on one side and water passing on the other. A direct current is applied, one membrane in each pair functioning as an anode and attracting negatively charged ions (anions), while the other functions as a cathode and attracts positively charged ions (cations). The membranes are porous to both anions and cations, which pass into the water.

economic use of electrodialysis is *ca.* 90%, with 50–60% demineralization being more viable. Ion exchange is of low capital cost, but throughput is limited and the process is particularly suited to small plants requiring a high level of demineralization. A further disadvantage is that, while both processes produce large quantities of effluent, those produced from column regeneration are highly acid or alkaline and require special handling. Electrodialysis is most suited to high throughput plants in areas where electricity is cheap and where only a low level of demineralization is required. In other circumstances requirements may be best met by using electrodialysis for the first stage of the process and finishing by ion exchange.

Attempts to improve ion exchange techniques have largely concentrated on facilitating resin regeneration. In the SMR process, cations in the whey are exchanged for ammonium ions and anions for bicarbonate. The resulting ammonium carbonate is evaporated from the whey as ammonia, CO_2 and water, more than 80% being recovered for regenerating resins and thus reducing effluent problems. Free ammonium levels in the product are less than 0.25%. Alternatively Sirotherm™ resins may be used, which are regenerated by warm water. Difficulties may arise due to the high levels of Ca^{2+} ions in whey and to the low ion exchange capacity of the resins.

Ultraosmosis (nanofiltration) offers an alternative means of partial demineralization of whey. The process is very similar to reverse osmosis but is operated at lower pressures (0.3–0.4 MPa). The membrane is slightly porous permitting passage of water and monovalent ions.

In contrast to the US which has abundant supplies of corn syrups, a market in Australia, New Zealand and, possibly, Europe exists for lactose-hydrolysed products including whey and derivatives such as permeates from reverse osmosis. Currently used techniques involve hydrolysis of lactose by the enzyme β-galactosidase (derived from micro-organisms such as *Kluyveromyces lactis*), which may be utilized in a number of ways (Table 4.5) and acid hydrolysis at high temperature.

---

* A large ultraosmosis installation at the Murray Goulburn Co-Operative Dairy in Australia has 70–75% permeability for Na^+, K^+ and Cl^- ions, but only 5–10% permeability for Ca^{2+}, Mg^{2+} and PO_4^{2-} ions. As with reverse osmosis, effective cleaning of the membrane is an essential part of the operation, the cleaning procedure involving daily enzyme and alkali cleaning and a citric acid wash at least every 2 days. Ultraosmosis may be combined with ion exchange where high levels of demineralization are required (Sanderson, W.B. 1991. *CSIRO Food Research Quarterly*, **51**, 29–31).

Table 4.5 Manufacture of lactose-hydrolysed whey products

---
1. Single use enzymes
2. Ultrafiltration[1]
    recover enzymes by ultrafiltration
    immobilize enzymes close to ultrafiltration membrane by pressure-induced flow regimen
    use UF equipment as an enzyme reactor
3. Immobilized enzyme systems[2]
---

[1] Ultrafiltration is possible only with protein-free substrates. Whole whey must be separated by ultrafiltration, the permeate treated and recombined with the retentate. Such a process is probably too complex and expensive for commercial use.

[2] Immobilized enzyme systems have greatest potential and may be used both for whole whey and permeate.

Acid hydrolysis involves reduction of the pH value to *ca.* 1.2, either using direct addition of acid or by ion exchange, followed by heating at up to 150°C. Acid hydrolysis is only suitable for protein-free material (permeates).

After hydrolysis, the whey is concentrated to 72% by thermal evaporation. Precise control of temperature is required to minimize damage to proteins and to avoid initiation of the Maillard reaction. Although lactose-hydrolysed whey is primarily used as a source of protein, the syrup (sweet dairy concentrate) has humectant properties valuable in baked goods. These are a consequence of the colligative properties of glucose and galactose.

(b) Whey protein concentrates and isolates

Whey protein concentrates are manufactured by ultrafiltration of clarified whey. Water is removed as permeate together with some of the lactose and minerals. Ultrafiltration is capable of concentrating protein

---

* Partially demineralized, lactose-hydrolysed whey has been proposed as a cheap replacement for sweetened condensed milk in the manufacture of toffee and hard caramel confectionery. Use has been limited, however, by the lack of casein and a different sugar profile (sucrose and lactose in sweetened condensed milk, glucose and galactose in lactose-hydrolysed whey). More recently, lactose-hydrolysed whey has been tailored for toffee and caramel manufacture by addition of casein and sucrose to the same level as sweetened condensed milk. Sweetened condensed milk may be directly replaced by Novamel™, in which the lower total casein level is compensated for by the superior functionality of sodium caseinate compared with micellar casein. Use of Novamel also avoids problems of graininess due to lactose crystallization and gives a smooth texture and rich flavour. The higher water binding and humectant qualities minimize problems of water migration in two-phase systems such as wafer bars.

Table 4.6 Modification of the functional properties of whey protein concentrates

---
Preheat treatment[1]
Heating at any other stage
Replacement of calcium ions with sodium, or other, ions[1]
De-ionization
Source of whey
Protein content[1]
Lipid content

---
[1] Of greatest commercial importance.

to 65% (dry matter), but some of the lactose and minerals remain in the retentate. These may be removed by diafiltration involving dilution of the retentate with water. By this means a protein concentration of 80% (dry matter) may be obtained. Processing can be varied to produce concentrates with a very wide range of functional properties (Table 4.6). In most cases, however protein content is the most important factor determining end-usage.

Whey protein concentrates are available either as a concentrated solution, or as a dried powder. In the latter case it is necessary to raise the solids content from *ca.* 30% to *ca.* 60%. Reverse osmosis, vacuum evaporation, or a combination of the two processes may be used.

The permeate from ultrafiltration may be used in a number of ways including as fermentation substrate. Drying to produce a permeate powder is possible, but the high lactose and mineral content means that production of a high quality product is very difficult due to burning and production of sugar glasses. A more common use of permeate is as a source of lactose. Ion exchange demineralization may be used to prepare a crude lactose syrup, but demand for this type of product is low. Alternatively lactose may be extracted by crystallization.

Manufacture of whey protein isolates exploits the fact that whey proteins are amphoteric molecules, which may be considered as

---
* Isoelectric precipitation is the basis of an alternative means of recovering proteins from whey. This technique cannot normally be used because of the high solubility of whey proteins even at the isoelectric point. This problem may, however, be overcome by partial sulphitolysis (*ca.* 25–40%) of disulphide bonds using sodium sulphite and a solid state copper carbonate catalyst. The pH value is then adjusted to 5.0 when 70–80% of the proteins precipitate. Copper present in the precipitate may be removed by extraction with EDTA at pH 4.5.

charged ions. At pH values below the isoelectric point, whey proteins have a net positive charge and behave as cations, while at pH values above the isoelectric point, whey proteins have a net negative charge and behave as anions. In either case the proteins may be adsorbed onto the corresponding ion exchanger, media with suitable pore sizes and surface characteristics having been developed for recovery of proteins from dilute solutions. Two major processes are in use; the Vistec^{TM} and the Spherosil$^{(TM)}$.

The Vistec$^{(TM)}$ system uses a cellulose-based ion exchanger in a stirred tank reactor. This operates as a cation exchanger and the initial stage of the process involves adjustment of the pH value of the whey to below 4.6 before pumping into the reactor and stirring to allow adsorption of protein onto the ion exchanger. Lactose and other non-protein materials are then eluted with water and the pH value of the proteins then raised to above 5.5 with alkali to permit removal from the ion exchanger. Purification and concentration of the eluted proteins by ultrafiltration is an essential step before evaporation and spray drying to yield a whey protein isolate of 95% protein concentration.

The Spherosil$^{(TM)}$ process uses either Spherosil S$^{(TM)}$ cationic exchanger or Spherosil QMA$^{(TM)}$ anionic exchanger. In each case the ion exchange resin is situated in a fixed bed column reactor. Acidified whey (pH < 4.6) is applied to Spherosil S and sweet whey (pH 5.5) to Spherosil QMA$^{(TM)}$. The fractionation cycle in each case is similar to that of the Vistec$^{(TM)}$ system, except that the release of proteins from Spherosil QMA$^{(TM)}$ involves lowering the pH value. Ideally the Spherosil$^{(TM)}$ process produces an isolate of 85% protein concentration. Isolates have superior functional properties to whey protein concentrate, but the processes suffer from a number of serious problems including high cost (Table 4.7). Functional properties such as foaming are affected by pH value and any heat treatment applied and these

Table 4.7 Problems associated with ion exchange processes for isolation of whey proteins

1. Expense of purifying and concentrating dilute eluate fraction
2. Difficulties and expense of disposing of large volumes of effluent consisting of chemical solutions, rinse water and deproteinized whey
3. Low throughput due to long time required to complete each fractionation cycle
4. Microbial contamination of reactor and associated problems of hygiene and sanitation

parameters must be carefully controlled. Unfortunately no single set of processing conditions is ideal for all functional properties.

(c) 'Lactalbumin'

Traditional 'lactalbumin' is the protein precipitated when whey is heated at acidic pH values. The exact precipitation conditions used in commercial practice vary according to the type of whey and the properties required in the final product. Whey, which may be preconcentrated, is acidified either by addition of acid or by cation exchange, the latter process also partially demineralizing the whey. The precipitate formed on heating is recovered by various settling and decanting procedures, centrifugation or vacuum filtration. Washing may be used to reduce the mineral and lactose content before drying in fluidized bed, spray or roller dryers. Such techniques permit the recovery of 'lactalbumin' containing up to 90% protein.

(d) Whey protein fractions

Thermal fractionation of whey can be used to prepare two fractions – the alpha fraction, consisting of α-lactalbumin, phospholipoproteins, serum albumin, immunoglobulins and enzymes and the beta fraction, consisting of β-lactoglobulin and casein-derived peptides. The specific functional properties of each fraction are superior to those of whey protein concentrates or isolates (see pages 176–7).

The fractionation procedure (Figure 4.3) exploits the fact that at a pH value of ca. 4.2 and a temperature of 55–65°C, α-lactalbumin and other components of the alpha fraction have minimum solubility, while the solubility of β-lactoglobulin and casein-derived peptides remains virtually unchanged. Separation of the two fractions under these conditions is, therefore relatively straightforward.

Further fractionation to yield individual proteins is possible, although in some cases a large-scale industrial market has not yet been established. Alpha-lactalbumin differs from most proteins in that thermal denaturation is reversible and this property enables native α-lactalbumin to be recovered from the alpha-fraction. Alpha-lactalbumin, denatured during the thermal fractionation procedure used in preparation of the alpha-fraction is restored to the native form by cooling at pH 7.0. The resolubilized α-lactalbumin may then be separated from the significantly larger lipoproteins by microfiltration.

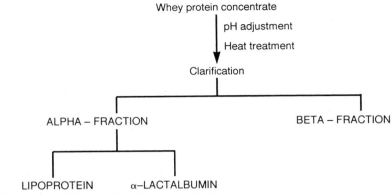
Figure 4.3 Thermal fractionation of whey proteins.

4.2.2 Casein and caseinates

(a) Casein

The basic principle of casein production is the same in all cases (Figure 4.4), but major differences occur in the means of producing the casein curd. In the case of rennet casein, the underlying mechanism is identical to that of the production of cheese curd (see pages 322–4) and depends on the unique sensitivity of the Phe_{105}–Met_{106} bond of κ-casein to hydrolysis by acid proteinases, the active components of rennet. Rennet is added to skim milk at *ca.* 29°C and held for 1 h, steam is then injected to raise the temperature to 55°C to 'cook' the clot before separation of whey. All other types of casein are produced by isoelectric precipitation. Manufacture of mineral acid casein involves mixing pasteurized skim milk with dilute (0.5N) hydrochloric or sulphuric acid to lower the pH value to *ca.* 4.6. The mixture is heated to *ca.* 50°C by steam injection and then remains for a short period in a cooking pipe and 'acidulation' vat before separation.

Adjustment of the pH value to the isoelectric point by use of ion exchange resins involves mixing pasteurized skim milk with cation exchange resin, in the hydrogen form, in a reaction column at below 10°C. This results in the pH value falling to *ca.* 2.2, untreated skim milk is then added to raise the pH value to 4.6, before heating to 50°C and separation.

Acidification of milk in the manufacture of lactic rennet involves the addition of a starter culture, usually a sub-species of *Lactococcus lactis*, and incubation at 22–26°C for *ca.* 14 h. The mixture is heated to 60°C

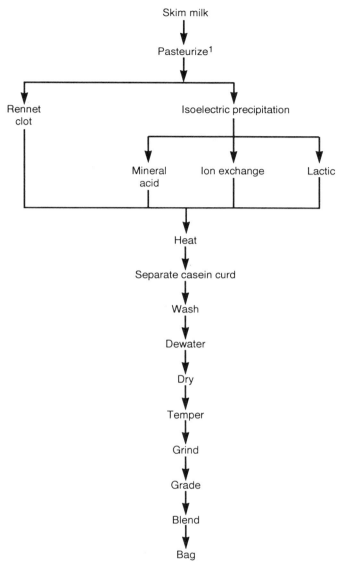

Figure 4.4 Manufacture of the various types of casein. [1]Milk for manufacture of rennet casein for non-food use is not pasteurized.

and curd particles formed, which remain in contact with the whey to permit agglomeration and the initiation of syneresis. The casein curd is separated from the whey by drainage and washed several times with water. After washing the curd is mechanically dewatered by pressing or by centrifugation. It is economically important to remove as much water as possible at this stage to minimize energy costs during the drying stage which follows dewatering. Fluidized bed dryers are popular in some countries, but roller dryers and spray dryers are also used.

Following drying the casein granules are cooled in a 'tempering' process which involves the pneumatic circulation of material through several bins to achieve equilibration of water throughout the different sized particles. The casein particles are then ground in either roller or pin-disc mills, graded in size by screening and packed into sacks. In more modern plants drying and grinding are often combined into a single operation by attrition drying.

(b) Caseinates

Caseinates can be produced from either fresh, acid casein curd or dried acid casein by reaction with one of several dilute alkalis under aqueous conditions. The caseinate solution resulting is then dried to a moisture content of 3–8%. Fresh, acid casein curd is normally used as starting material in casein-producing countries and is considered to yield a caseinate of blander flavour than dried acid casein. The use of casein curd is also economically advantageous since the additional costs associated with production of dried casein are avoided.

Sodium caseinate is the most widely produced and is usually made by reaction with 2.5 M sodium hydroxide used at 1.7–2.2% (Figure 4.5). Sodium bicarbonate and sodium phosphate may also be used but larger quantities are required and the reagents are more expensive.

Starting material must be very finely ground to permit rapid dissolution and this is commonly achieved by use of colloid mills. The dilute alkali solution is then metered into the casein slurry before passing into dissolving vats. The dissolving vats may be operated on either a continuous or a batch basis, but in either case effective agitation and recirculation is essential. Heating is also necessary at this stage and may be applied by a tubular heat exchanger in the recirculation line or by a steam-jacket fitted to the vat. Direct steam injection is possible, but may

---

* Attrition dryers consist of a fast rotating, multi-chambered rotor and a stator with a serrated surface. In operation turbulence, vortices and cavitation contribute to a highly efficient grinding which produces a very fine powder (overall average particle size 100 μm) with a large surface area. Drying occurs simultaneously in a concurrent hot air stream. Attrition dried powder is of good wettability and dispersibility due to its irregular shape and cavities formed by rapid evaporation.

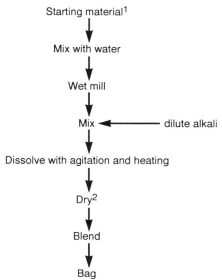

Figure 4.5 Conventional method for the manufacture of caseinates. [1] Either fresh, acid casein curd or dried casein. [2] Usually spray or roller dried.

result in a lower quality product. The solution should be held at high temperature and pH value for as short a time as possible to avoid Maillard browning and formation of lysoalanine and off-flavours. Control of pH value and viscosity is essential at this stage.

Various types of dryer can be used, the type affecting the properties of the finished caseinate. Spray drying is common, although problems can arise due to the high viscosity of the feed and the low bulk density of the finished product. Roller dryers are also widely used, especially in Europe.

A number of procedures have been developed for combining the reaction and drying stages. Roller dried caseinate may be produced by mixing casein with sodium carbonate, or bicarbonate, directly before

---

* Conventional means of producing caseinates result in a relatively dilute solution which requires the removal of large quantities of water, with associated high energy costs, during drying. Concentrated solutions of 33–47% solids may be prepared by mixing casein with water and NaOH in a steam heated reaction chamber. The reaction proceeds in two stages; at pH 5.0–5.3, when viscosity is a minimum, and pH 6.5–7.0. Solutions of 33% solids may be spray dried or, alternatively, the concentrated solution may be forced through orifices to produce jets which are dried in hot air to form sodium caseinate threads.

drying, the reaction being completed on the dryer drum. A granular casein of high bulk density may be produced by a similar process, which involves partial drying of casein curd, mixing with sodium bicarbonate and final drying of the caseinate. Either pneumatic ring or fluidized bed dryers are used. Attrition drying is used to prepare caseinate powder resembling spray dried, but of significantly higher bulk density. Dried acid casein is the starting material and sodium carbonate the alkali. Caseinates may also be prepared by extrusion cooking. In this process dried casein is the usual starting material and is mixed with water and alkali to form the extrusion mixture. Fairly high pressures and cooking temperatures are required to ensure the reaction goes to completion before the material leaves the cooker. Extrusion cooking has also been proposed as a means of preparing sodium caseinate from skim milk powder.

Ammonium, calcium and potassium caseinates may be made by reaction with the respective hydroxides using procedures similar to that for sodium caseinate production. In the case of calcium caseinate the reaction between calcium hydroxide and acid casein curd proceeds more slowly than the corresponding reaction with NaOH and is highly temperature dependent. Conditions in the dissolving vats must, therefore, be particularly closely monitored. Granular ammonium caseinate is made by exposing dried acid casein to ammonia gas, excess ammonia being removed by air-flushing in a fluidized bed degassing system. Citrated caseinate is also produced by reaction with a mixture of trisodium citrate and tripotassium citrate.

The properties of different caseinates vary considerably (Table 4.8) and are modified further by the processing received. Modern practice is to blend different types of caseinate to optimize properties for different applications. In addition modifications may be made to the properties by chemical treatment (Table 4.9).

Table 4.8 Functional properties of different caseinate products

Product	Functional property
Acid casein	Good solubility above pH 5.5. Good water absorption
Lactic casein	Moderate water absorption
Rennet casein	Insoluble below pH 9.0. Suspensions of more than 15% gel at 25°C
Calcium caseinate	Forms white colloidal dispersions
Sodium caseinate	Gels at concentrations more than 17%

Table 4.9 Modification of the functional properties of caseinates

Hydrolysis	Alkaline hydrolysis of sodium caseinate enhances foaming properties
Renneting	Renneting sensitizes caseinate to precipitation by calcium and permits formation of a gel on heating

(c) Fractionation of caseinates

A number of methods suitable for industrial use have been devised for the fractionation of casein. The method which is probably the most promising is based on the fact that β-casein exists in solution as monomers at 4°C and may therefore be separated from α_s-/κ-casein by ultrafiltration (Figure 4.6).

(d) Co-precipitates

Co-precipitates of casein and whey proteins are made from skim milk. Difficulties were encountered with early processes due to thermal denaturation of whey proteins and resulting poor solubility. This

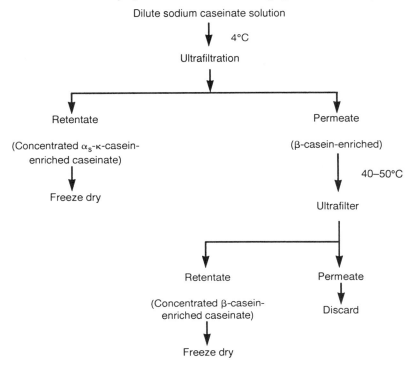

Figure 4.6 Preparation of α_s-/κ- and β-casein-enriched fractions. Based on Murphy, J.M. and Fox, P.F. 1990. *Food Chemistry*, **39**, 27–35.

problem has been overcome by current technology which involves heating at 55–75°C at a pH value between 7.5 and 10.0 to induce the casein/whey protein interaction before precipitation at pH 4.6 and resolubilization at pH 6.6–7.0.

(e) Milk protein concentrate

Milk protein concentrate may be made directly from skim milk by a combination of ultrafiltration and diafiltration. Concentrates with a protein content of up to 80% may be made using this technology, the casein is in micellar form and the whey proteins in the native configuration. The ash content is relatively high since protein-bound minerals are retained. Casein micelles may be separated from whey proteins by microfiltration using high porosity membranes.

4.2.3 Utilization of milk proteins

(a) Whey proteins

In Europe until relatively recently, whey proteins were used only for animal feeding and this remains a major market for products such as delactosed whey powder. The rising cost of skim milk powder has, however, increased demand for whey protein as a substitute. The situation is different in the US where whey protein is the most important functional protein used in human foods. Basic whey products tend to be used as fillers for improving nutritional status, while protein enriched products are usually chosen for functional products.

In a small number of cases, such as carbonated, fruit-flavoured whey-based drinks, whey is the major ingredient. More commonly, however, whey proteins are added as one of many ingredients. The use of the various whey protein products in different types of food is summarized in Table 4.10.

Fractionated whey proteins have advantages over non-fractionated in specific applications. Beta-fraction is widely used in acidic beverages, remaining soluble even after heat treatment at 90°C for 30 s. The fraction also has superior properties as an egg white replacer in

---

* Whey is also a major ingredient of the fat-replacer Simplesse™ which, in this health conscious age, is likely to find increasing application. Manufacture involves simultaneously heating and shearing egg white and whey solution to form spherical particles with a diameter of 0.1–3.0 μm (Morr, C.V. 1992. *Food Technology*, 46(4) 110–4).

Table 4.10 Examples of the use of whey proteins in different types of food

Bakery products
  improvement of baking qualities; bread (ultrafiltered only)
  egg replacer; Madeira-type cakes
  nutritional enhancer; breakfast goods

Dairy products
  nutritional enhancer; various products
  improved yield; quarg cheese
  improved quality; Ricotta cheese
  stabilizer; yoghurt
  protein base; cream cheese spreads
  flavour enhancer and texturizer; cheese-flavoured dips and spreads
  solids-non-fat source; ice cream and juice bars

Beverages
  nutritional enhancer; various products

Pasta products
  nutritional enhancer; various products
  texture improver; various products
  texture improver and structural stabilizer; frozen, microwave-reheatable pasta 'ready-meals' (undenatured protein only)

Confectionery
  egg white replacer; soft, aerated candies

Meat products
  extender; frankfurters and luncheon meat

---

confectionery such as meringues and as a binder and water-retaining agent in meat products. Alpha-fraction has fewer applications, but is widely used as a source of lactalbumin in highly humanized infant foods.

The use of whey as a fermentation substrate has already been noted (Figure 4.1). A number of fermentation processes are involved, including industrial production of citric acid, alcohol and single cell protein. Fermentation of permeate under anaerobic conditions has also been

---

* Phenylketonuria is a disorder resulting from inability to metabolize phenylalanine. The condition, which is usually diagnosed at birth, requires a special diet free from phenylalanine and other aromatic amino acids. Current diets are based on unpleasant tasting mixtures of synthetic amino acids administered as a beverage. Casein-derived peptides are devoid of aromatic amino acids, have a high content of glutamic and aspartic acid and have a bland taste. They are therefore highly suitable as the basis for a maintenance diet for phenylketonuric persons and persons with some forms of hepatic disease. Casein-derived peptides may be incorporated into a wide range of foods including infant formula, rusks, fruit jellies and teething rings.

used in production of fuel gas consisting of 50% methane with an energy of 960 kJ/l. In the long-term production of fuel gas from whey is likely to become increasingly attractive as the cost of alternative energy sources rises.

(b) Casein and caseinates

The use of casein and caseinates has been extended by the development of specialist powders including blends of sodium and calcium caseinate which maximize the attributes of each product. At the same time rennet casein, once used virtually exclusively in non-food applications, is now widely used in foods as a consequence of its flavour stability. Casein and caseinates are used as ingredients for their functional properties, but may also have a nutritional role. Usage in various types of food is summarized in Table 4.11.

Casein has a number of non-food industrial applications. Acid casein is used in the manufacture of adhesives and related materials such as paper coatings and sizes. It is also used as a pigment carrier/ dispersing agent in water-based and latex paints. Acid casein has been used for plastics manufacture, but is much inferior to rennet casein in this application. Casein-based plastics are generally used for small items such as buttons.

BOX 4.2 The Carberry Process

The first large-scale plant for production of alcohol from whey was built at the Express Dairies creamery at Carberry, Eire. Whey permeate is fermented by *Kluyveromyces lactis* in batch fermenters of over 100 000 l capacity to yield ethanol at 96.5%. At peak production the Carberry creamery produces 600 000 l of whey which is converted into 22 000 l of ethanol. Molasses is used as feedstock during the winter months. A major problem is disposal of effluent and strict precautions must be taken against contamination of the process by micro-organisms from the adjacent cheese plant. A minimum throughput of 500 000 l is required for economic operation of a Carberry-type plant, but use of immobilized yeast may make smaller plant viable. The alcohol produced is suitable for use in the production of white potable spirits such as gin and vodka, but in countries such as New Zealand it may be more attractive to use the alcohol as fuel (gasohol).

Table 4.11 Examples of the use of casein and caseinates in different types of food

Bakery products
  nutritional enhancer; breakfast cereals, high protein biscuits
  texture improver and emulsifier, frozen cakes

Dairy products
  protein base; cheese analogues
  protein base; coffee whiteners
  stabilizer; yoghurt
  flavour enhancer and emulsifier; cheese spreads

Beverages
  stabilizer and foaming agent; carbonated drinks
  stabilizer; drinking chocolate
  emulsifier; cream liqueurs and wine aperitifs
  clarifying agents; beer and wine

Desserts
  stabilizer and whipping improver, frozen desserts and instant puddings
  stabilizer, whipping improver, emulsifier and film former; whipped toppings

Confectionery
  texture improver; hard toffee and fudge

Meat products
  emulsifier, texture improver and water binder; comminuted meats

---

BOX 4.3   **A deeper shade of green**

The use of casein-based paints as artists' materials has fallen over the years as new synthetic paints, such as acrylics, have become available. Artists, however, share the ecological concerns now prevalent in the community as a whole and many would prefer to use paints based on natural materials. Casein-based paints are considered to be amongst the most 'environmentally-friendly' and this has been reflected in an upsurge in popularity. New formulations have been introduced which enable casein-based paints to emulate synthetics in terms of performance.

4.3 CHEMISTRY

4.3.1 Nutritional properties

Both whey proteins and caseins are high quality proteins. Whey proteins are of particular value due to their high cysteine content, although in non-hydrolysed products the high lactose content causes problems in lactase-deficient persons and animals. Whey also contains significant quantities of B group vitamins and minerals, although mineral content may be reduced by demineralization.

Work with pigs has suggested that whey proteins in the diet lower serum cholesterol levels and high density, lipoprotein cholesterol levels.

Losses during processing are similar to those of milk. Protein quality is also lost during storage, losses being at a maximum at a_w 0.44. The rate of loss is also greater at high temperatures, but at constant a_w, fluctuating temperatures have the greatest adverse effects.

Casein and caseinates contain a high proportion of essential amino acids, although limiting in sulphur-containing amino acids such as methionine. Losses during processing and storage are similar to those of whey proteins.

4.3.2 Functional properties

The chemistry of the functional properties of milk protein products are discussed in Chapter 1, pages 8–13.

4.4 MICROBIOLOGY

The microbiology of milk protein products has received relatively little attention, although the hygiene problems associated with membrane techniques are well known. Potential problems are the same as those encountered during the manufacture of concentrated and dried milks and similar precautions must be taken. Basically these involve heat treatment of the starting material at a level equivalent to pasteurization, prevention of the growth of micro-organisms during processing stages such as evaporation and reverse osmosis and prevention of re-contamination of either the product stream or the end product. The increasing sophistication of processing potentially creates more opportunities for microbial contamination or growth and a high level of control is necessary, especially where the product is intended for incorporation in infant foods.

The growth of species of *Bacillus* and *Enterococcus* during production of high protein whey powder by ultrafiltration at 50°C may lead not only to high bacterial numbers in the finished powder, but to an adverse effect on functionality. This may be attributed to changes to the structure of the whey proteins, which may also result in flavour defects in products containing the powders such as cheese. Microbial growth during processing of whey is also one of the causes of off-flavours during end-use in ice cream, etc.

EXERCISE 4.1

There are a number of instances where the use of preservatives in dairy foods is considered justified in developing countries but not in the western world (*cf.* page 47). Do you consider this to be morally acceptable? Are there conditions or circumstances under which the possible long-term toxicological effects of preservatives outweigh the benefits in terms of increasing availability of milk proteins?

EXERCISE 4.2

Considerable quantities of rennet casein are produced for industrial use such as plastics manufacture. An old-established manufacturer of industrial casein has decided to expand operations to the manufacture of higher value edible casein products. A plant has been erected on a separate, but adjacent site, equipped on a 'turnkey basis' by a major equipment supplier. Staff have been transferred from the existing plant and trained in operating the new equipment by the supplier. At the last minute, however, it is realized that no attention has been given to training staff with respect to the basic principles of hygiene.

You are asked to devise a hygiene training schedule for *all* production staff. This should be in two parts; an initial training programme to be attended by staff *before* production commences and a continuing programme designed to reinforce the basic training and to continually enhance hygiene awareness. Consider the most appropriate approach and use of visual aids in training staff, many of whom are elderly and considered 'set in their ways'.

5
CREAM AND CREAM-BASED PRODUCTS

OBJECTIVES

After reading this chapter you should understand
- The nature of cream and the differences between the various types
- The nature of cream substitutes
- The separation of cream from milk and type-specific processing
- The role of cream as a major ingredient in cream-based products
- The major control points
- The flavour of cream
- The nature of chemical changes associated with the processing and storage of cream
- The physico-chemical events occurring during whipping of cream
- Microbiological hazards and patterns of spoilage

5.1 INTRODUCTION

Cream consists of a concentration of milk fat in milk with the fat mainly in globules protected by the membrane. Types of cream are primarily defined by fat content (Table 5.1), secondary definitions being based on the thermal processing received (pasteurization, UHT or in-container sterilization) and on other processing designed to impart particular characteristics to cream (clotted cream, whipped cream, coffee cream, etc.). Cream substitutes are also available.

5.2 TECHNOLOGY

The basic technology of cream manufacture is relatively straightforward, the key stage – separation and standardization – being common to all types. Homogenization and other post-separation processes may then be applied to obtain creams with differing properties. The manufacture of

Table 5.1 World Health Organization standards for the fat content of creams

Type	Minimum fat content (%)
Cream/single cream	18
Half cream	10–18
Double cream	45
Whipping cream	28
Heavy whipping cream	35

Note: Fat contents quoted are recommendations and national standards may vary slightly. National standards only exist for some types of cream such as clotted cream (more than 55% fat).

pasteurized cream is summarized in Figure 5.1.

5.2.1 Incoming raw milk

The quality of the raw milk should be that of milk for the liquid market. It is particularly important that milk is free of feed taints since these partition into the fat phase and are thus of particular significance as a defect in cream. The high fat content of cream also means that problems due to the lipolytic enzymes of psychrotrophic bacteria are potentially greater and thus the refrigerated storage period for the raw milk must be carefully controlled and should not exceed 24 h. Ideally milk should be processed directly after receipt since separation becomes less efficient with increasing length of storage.

Seasonal changes in milk composition may affect the viscosity of cream. Milk fat tends to be softer during the spring months when cattle feed on fresh pasture and this leads to cream of reduced viscosity. The number of fat globules smaller than 0.8 µm (non-separable globules) increases towards the end of lactation (the autumn in the northern hemisphere) and thus fat losses in skim milk are higher.

---

* The term 'low-fat', when applied to cream in the absolute rather than the comparative sense, is a contradiction and a special descriptive terminology is required. Despite this products based on dairy ingredients, which have the properties of natural cream but a minimal fat content are under development. Technology is proprietary and details are not generally available. The most common approach, however, is to use membrane filtration to produce a concentrate high in solids-non-fat, which contains only sufficient fat to provide desirable mouthfeel and flavour. A 'coffee cream' is produced in this way from a mix of 90% skim milk, 7% cream and 3% milk protein, the finished product having a fat content of 2.5% but resembling its natural counterpart in taste and appearance (Anon 1991. *Dairy Industries International,* **56(12)**, 32–3).

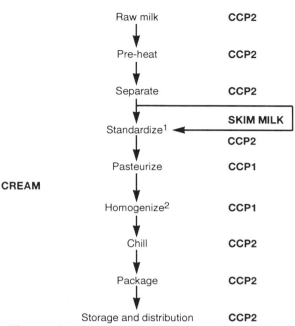

Figure 5.1 The production of pasteurized cream. *Note*: All aspects of cream handling between pasteurization and packaging have the potential for contamination by pathogenic micro-organisms and are CCP 1. [1] Separate standardization is not required where automatic control of fat content is in use. [2] Homogenizer may be placed upstream of pasteurization, in which case the process is a CCP only with respect to product quality. Homogenization is not applied to all types of cream.

Handling procedures for milk for separation must take account of the need to avoid damage to the fat globule and consequent increased loss of fat in the skim milk. The inclusion of air during pumping also reduces the efficiency of separation and must be avoided by use of suitable pumps and pipework design.

5.2.2 Separation and standardization

Separation involves the concentration of the fat globules of milk and their subsequent removal from the serum. The process is governed by Stokes' Law which may be expressed in simplified form as:

$$R = r^2 \times F$$

where R is the rate of separation, r is the radius of the fat globule and F is the applied force.

CONTROL POINT: INCOMING MILK AND HANDLING CCP 2

Control

Milk to be of good initial quality and free from feed taints.

Milk to be stored for minimum possible period.

Minimum damage to be sustained during pumping, etc.

Monitoring

Laboratory examination of incoming milk (see Chapter 2, pages 85 and 96).

Organoleptic assessment where feed taints a common problem.

Formal system to be instituted to ensure use of milk within predetermined time period.

Operation of equipment to be monitored by experienced personnel.

Verification

Quality of end-product.

Examination of plant records.

Periodic specialist examination and maintenance of equipment.

Traditional means of separation relied on gravity as the applied force but this process is slow and inefficient and, in industrialized countries, used on only a very small scale. Centrifugal separators have been available for many years and permit rapid and efficient separation. Extremely high centrifugal forces are applied but the wide variation in the radius of the fat globule means that, under practical conditions, a proportion of the smaller globules can never be separated, the milk serum phase (skimmed milk) usually having a fat content of $ca.$ 0.06%.

Modern separators are of the disc stack type which consist of a series of conical stainless steel discs placed inside the separator bowl and driven by a central supporting spindle. This design greatly increases the speed and efficiency of operation by reducing the distance the solid particle (the fat globule) has to travel before coming to rest.

In operation, milk enters the rotating bowl and is introduced to the disc stack through distributing holes. The milk is rapidly accelerated to the speed of the bowl and once the particles to be separated reach the disc surface they are effectively removed from the liquid stream. The low density fat globules move inwards along the surface of the disc and the heavier serum phase outwards to the edge, each being removed into separate discharge chambers. Solid particles of dirt, leukocytes and other debris move to the outer wall of the bowl and form 'sludge'.

The basic design of disc stack separators permits the relative quantity of skim milk leaving the discharge chamber to be controlled by varying the back pressure applied to the cream and skimmed milk outlets. This in turn provides a simple means of controlling the fat content of the cream. The fat content may be decreased, for example, by reducing the back pressure on the cream outlet and thus increasing the volume of liquid passing through this discharge chamber.

Although the principles of operation are common to all types of disc stack separator, there are three types of machine which differ markedly in design. This stems from differing approaches to the solution of a major design problem – the means of introducing whole milk to and removing the separated products from a bowl, or rotor, rotating at 6000–9000 rpm.

Open design. The open design (Figure 5.2) is the simplest and used only in small capacity machines. Milk enters through a stationary feed tube which projects at least halfway down the depth of the rotating bowl. At the point of entry the milk is moving at high speed, but with zero rotation, and acceleration to the same speed as the bowl is almost instantaneous. After separation cream and skimmed milk overflow from the top of the bowl into spouts which direct the products into containers. The fat content of the cream is usually controlled by adjusting the back pressure at the cream outlet.

Semi-enclosed (paring disc) design. The mechanism of entry of milk into semi-enclosed separators is the same as that into the open type. The means of removing cream and skimmed milk, however, differs markedly. Semi-enclosed separators are fitted with static paring discs (centripetal

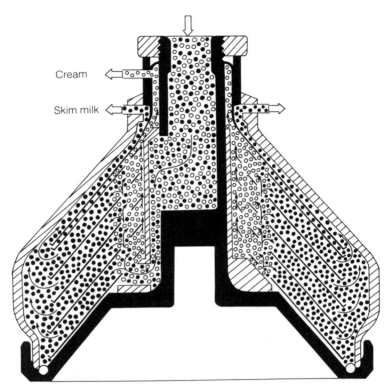

Figure 5.2 Open design cream separator. Redrawn with permission from Rothwell, J. 1989. *The Cream Processing Manual*, 2nd edn. Copyright 1989, The Society for Dairy Technology.

pumps), resembling the rotor of a centrifugal pump (Figure 5.3). These dip into separate paring chambers for cream and skimmed milk and are attached to outlet pipes. In operation each of the separated liquid products rotates at the same speed as the bowl, the surfaces covering the outer edge of the stationary paring disc. Rotational energy is converted into pressure and the products leave the separator with a pressure as high as 5 kg/cm^2. The fat content of cream may be controlled by adjusting the back pressure of both the skimmed milk and the cream, although control is simplified by fitting an automatic control valve to the skimmed milk outlet pipe. Only the highly concentrated outer layer of the fat phase is removed by the paring discs and production of cream with fat content above 45% is likely to lead to a corresponding increased fat loss in the skimmed milk.

Separators are available which combine features of open and semi-

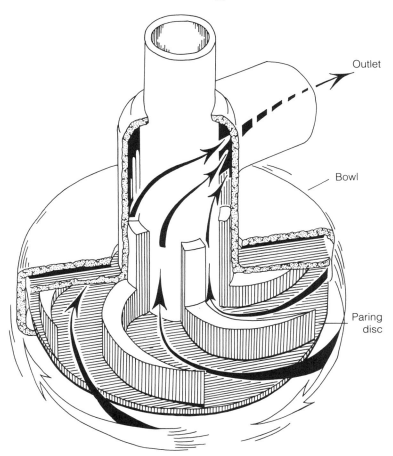

Figure 5.3 Paring chamber and disc of a semi-enclosed separator. Redrawn with permission from Rothwell, J. 1989. *The Cream Processing Manual*, 2nd edn. Copyright 1989, The Society for Dairy Technology.

enclosed designs. In one example the cream leaves the separator by overflowing into a spout as in an open design, but the skimmed milk is removed from a paring chamber by a paring disc.

Hermetic (airtight) design. Separators of the hermetic design are fitted with seals at both inlet and outlet to provide a seal between the atmosphere and the product. Whole milk is fed to the bowl through the hollow drive spindle (Figure 5.4) and accelerates gradually to the speed of the bowl. This minimizes damage to the fat globule and improves the stability and whipping properties of the cream. Cream and skimmed milk pass into separate chambers, sealed from each other and the

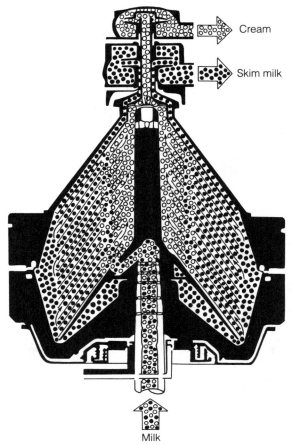

Figure 5.4 Hermetic design cream separator. Redrawn with permission from Rothwell, J. 1989. *The Cream Processing Manual*, 2nd edn. Copyright 1989, The Society for Dairy Technology.

atmosphere by axial seals, and are removed by rotating pump assemblies. The whole of the fat phase is removed from the cream chamber and creams with fat contents up to 55% may be produced without increased fat in the skim milk. Fat content of the cream is controlled by regulating the pressure on the cream outlet, sufficient back pressure on the skimmed milk outlet usually resulting from downstream equipment such as heat exchangers. A valve or orifice plate is required, however, where back pressure in the system is low.

Self-cleaning separators. Separators are conventionally manufactured with a solid bowl in which 'sludge' accumulates. Excessive accumulation reduces separation efficiency and the process stops if the 'sludge' level reaches the outer edge of the discs. This factor thus limits the length of the production run and can cause considerable disruption, especially in large creameries. In recent years self-cleaning centrifuges of semi-enclosed and hermetic design have been developed which not only permit long, continuous production but are also suitable for in-place cleaning. This considerably reduces the need for the dismantling and manual cleaning of separators, an arduous and, understandably, unpopular procedure which is all too frequently skimped. Highly effective means of hermetically enclosing this type of centrifuge have also been developed and damage to cream as a result of air inclusion has been reduced to a very low level.

The detailed design of self-cleaning separators varies somewhat but usually involves a hydraulically operated piston in the base of the bowl, the 'sliding bowl bottom', is moved downwards and opens discharge slots in the bowl body. Accumulated 'sludge' is forced through the slots by liquid pressure and collected in a receptacle incorporated into the machine cover. Hourly discharge of 'sludge' is usually sufficient and dismantling is usually required only every 2–3 months for inspection and maintenance.

The separator is, by definition, the key component in cream production but cannot be viewed in isolation from other equipment such as preheaters, pumps, etc. In commercial practice milk is usually separated at temperatures between 38 and 60°C, temperatures in this range facilitating separation and minimizing damage to fat globules and consequent greasy texture. The upper temperature limit is often defined by the greater costs involved in heating milk to higher temperatures but an organoleptic consideration is the higher viscosity of cream separated at 38–40°C than at higher temperatures. Cream separated at temperatures of less than 45°C, however, contains active milk-derived lipases

---

* Higher temperatures are used in the Alpha-Laval Scania™ system. This was originally designed to collect and process surplus cream from low fat standardization operations in Scandinavia, but is now used in cream separation plants and has sufficient flexibility to deal with the wide range of creams produced for the UK market. In this system the cream is separated at 63°C followed by deaeration. The cream is then heat-treated in a specially designed heat exchanger operating with low pressure drops to minimize damage to the fat globule. The end-product is of superior physical and organoleptic quality, having a better taste and a lower level of free fatty acids. The system also has economic advantages in that fat losses in skim milk are reduced (Rajah, K.K. and Burgess K.J. 1991. *Milk Fat: Production, Technology and Utilization*. Society for Dairy Technology, Huntingdon, UK).

which can initiate the development of rancidity during the short interval between separation and pasteurization. A minimum separation temperature of 45°C is, therefore, recommended, 55°C often being used in practice. Although the influence of other ancillary equipment is less obvious than that of preheaters, it is none the less of significance in determining the quality of the cream. It is of particular importance that all plant should be designed to ensure that the flowrate, feed and discharge pressures are constant, that air is excluded from the system and that pumping is minimized. Flowrate should be controlled either by use of a metering pump or by a mechanical flow controller.

Standardization of cream to a given fat concentration is a necessary adjunct of the separation process. Experienced operators are able to set separators to within a fat content of 0.5–1%, but higher precision is necessary and it is usual to operate separators to produce a cream of somewhat higher fat content than that required. The fat content is then determined using a rapid technique (see page 86) and whole or skimmed milk added to reduce the fat content to the required percentage. The arithmetic is not especially difficult but for convenience a method known as 'Pearson's square' is often used (Figure 5.5). Alternatively, where cream is standardized in vats of known volume using skim milk of supposedly constant fat content, tables may be provided and have the advantage of avoiding even simple calculations.

The traditional standardization process is cumbersome and, especially if delays occur in determination of fat content, can cause problems in management of production. The means of determining volumes of cream and milk are often imprecise and to avoid possible legal problems it is common practice to allow a margin of error which over a period of time leads to significant financial loss.

Potential problems resulting from standardization may be obviated by operating separators to produce cream of the exact fat content required and this is common practice in large modern creameries. The simplest means is to pre-determine the fat content of the whole milk and then to set the separator using special valves to control the flow of the cream

---

* Serious errors can occur during standardization and range from miscalculations to mishearing the fat content reported by telephone. Failure to take a representative cream sample for fat analysis is a further cause of error. The possibility of errors at the standardization stage make it desirable that a second sample is taken for analysis after standardization but it is often necessary to proceed with processing before results are available.

1. Basic square

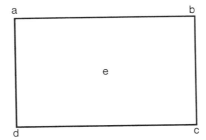

2. Problem
It is required to standardize 100 gallons of cream of fat content 23–18% using skim milk of effective fat content 0%.

3. Solution

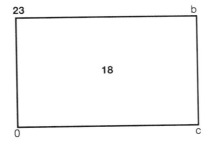

A. Substitute required cream fat content for e
B. Substitute actual cream fat content for a
C. Substitute skim milk fat content for d

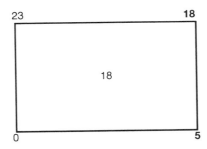

D. Substitute (e–d) for b
E. Substitute (a–e) for c
Thus 18 parts of cream must be mixed with 5 parts skim milk
Total volume of cream is 100 gallons which represents 18 parts.
Volume of skim milk = 5/18 x 100 = 27.8 gallons

Figure 5.5 Use of 'Pearson's square' in the standardization of cream.

and skimmed milk. This system can be satisfactory but depends on highly accurate calibration of flow meters and standardized operating conditions. A superior approach involves accurate measurement of the fat content of cream leaving the separator and feedback control to offset any variation due to changes in the composition of incoming milk or alterations to the operating conditions. The most sensitive means of determining the fat content of the cream is by densitometry, density being defined by the equation:

$$D = 1038.2 - 0.17T - 0.003T^2 - \theta(133.7 - 475.5/T)$$

where D is the density, T is the temperature (°C) and θ is the fractional fat content.

Density may be monitored on a continuous basis by a 'density monitor' which enables an operator to make manual adjustments as required or, preferably, forms part of a microprocessor-based automatic control system. Density monitoring is highly sensitive but is seriously affected by vibrations at the level encountered during normal plant operation. Many automatic control systems, therefore, are based on instrumental devices such as the Milko-tester™ which, although rather less sensitive is relatively unaffected by vibration. Automatic systems employing microprocessor-based feedback control loops are able to maintain a constant fat content in cream providing that the fat content of the incoming milk and the operating conditions during separation vary only within normal limits.

5.2.3 Handling of cream

All plant used for the handling of cream after leaving the separator should be designed and operated to minimize physical damage which can cause 'buttering' of high fat cream and impair whipping qualities. Pumping of cooled cream should be avoided where possible and positive displacement pumps should be used in all instances.

---

* Modern microprocessor-based control systems for cream separators are highly effective. The weak point is recognized as being the means by which the fat content is determined in real time and some systems use two complementary methods. This permits a very high level of control of separator performance and minimizes losses due to excess fat in products. Microprocessor-based control systems, however, are most effective when integrated into process control systems which regulate all of the processes and manufacturing stages of the production line. Such a system is particularly valuable with a product such as cream, where manufacture of several variants requires control of a number of variables in addition to fat content.

CONTROL POINT: SEPARATION AND STANDARDIZATION CCP 2

Control

Milk to be pre-heated to correct temperature.

Ensure cream at, or above, correct fat content.

Where required, cream to be standardized to correct fat content.

Minimize damage during cream handling.

Monitoring

Pre-heating

Equipment to be fitted with thermograph.

Fat content

Monitor performance of automatic control equipment.

When under manual control, determine actual fat content by analysis.

Standardization to be undertaken by trained and experienced personnel.

Handling

Inspection of equipment for air leaks and other malfunction.

Verification

Pre-heating

Inspection of thermograph records.

Fat content

Calibration of automatic equipment to be checked on a regular basis.

Accuracy of laboratory analyses to be checked on a regular basis.

Analysis of fat content of final cream.

Handling

Equipment to be maintained on a preventative basis.

5.2.4 Homogenization

The need, and indeed the desirability, of homogenization varies according to the nature of the final consumer product. This is discussed in more detail in Section 5.3.7. Homogenizers used are of identical design to those used for homogenization of liquid milks (see page 55) and are suitable for use with all types of cream.

From a hygiene viewpoint homogenization is preferred before pasteurization or other heat treatment. Homogenization after heat treatment, however, reduces problems of rancidity due to milk lipases and this order of processing is favoured by some manufacturers. Homogenization after heat treatment is usually considered a necessity in cream which has been UHT sterilized.

5.2.5 Heat treatment

With the exception of very small amounts sold as 'untreated cream', all cream sold in the UK must be pasteurized or otherwise processed to ensure safety. Similar regulations exist in other nations, although raw cream is commonplace in less developed countries.

(a) Pasteurization

Pasteurization may be by either the low temperature–long time (LTLT) holder method or the high temperature–short time (HTST) method. The LTLT method remains in use by small-scale producers, a minimum heat process of 63°C for 30 min being stipulated. The viscosity of cream is such that heat transfer can be a problem and heating vessels must be equipped with an adequate means of agitating the contents. Agitation should not, however, be excessive and is most effectively achieved by use of a rocking (swinging) coil vat, the heating medium being circulated both through the coil and the surrounding water jacket. To minimize the possibility of contamination, the same vessel should be used for cooling as well as heating.

CONTROL POINT: HOMOGENIZATION CCP 1 OR 2

For general control see Chapter 2, page 56

For post-UHT sterilization control see Chapter 2, pages 63–4

HTST pasteurization is in almost universal use in large creameries and is the most effective process when quantities in excess of 100 gallons are produced daily. A minimum process of 72°C for 15 s is stipulated, but higher temperatures may be used. The equipment used is essentially the same as that for milk pasteurization (see page 49) and should be fitted with a flow diversion device and suitable instrumentation for monitoring and verification of performance even where this equipment is not mandatory. Fat resistant gaskets should be fitted.

Cream should be cooled as quickly as possible after pasteurization, although in some cases (see page 201) the cooling rate is controlled as a means of imparting the required rheological properties. Packaging follows directly after cooling to avoid problems associated with a thickened product. Preformed polystyrene or polypropylene pots are used, which are usually sealed with aluminium foil. Polypropylene is preferred due to lower cost and less problems with tainting. Cream is also packaged in bulk for catering use, bag-in-box packs having largely replaced metal cans for this purpose.

(b) Extended heat treatment

There has been considerable interest, especially in Germany, in applying extended heat treatment to produce a cream containing very low numbers of bacteria. The cream is then packaged either using aseptic packing equipment, or non-aseptic equipment modified with hydrogen peroxide spray and ultraviolet irradiation. Commercially used processes use either a single heat treatment at 110°C for 30–60 s or a double treatment involving heating at 95–102°C for 15–30 s, holding for 24 h at 8°C and reheating at 120–127°C for 3 s. The underlying rationale of the double treatment is that endospores surviving the first heat treatment will germinate during holding at 8°C into heat-sensitive vegetative cells. It has been claimed that extended heat treatment permits a storage life of up to 4 weeks at 10°C. This, however, is disputed by some microbiologists (see page 94) and it is probable that in many cases the additional processing costs more than outweigh the benefits of shelf life extension.

---

* Under some circumstances the Vacreator™, a multistage steam distillation system operating under reduced pressure may be used in place of traditional means of pasteurization. The equipment removes both feed taints and butyric acid resulting from lipoprotein lipase activity. The capital and operating costs are higher than those of conventional HTST plant and can only be justified where feed taint is a continuing problem.

(c) UHT sterilization

UHT sterilization of cream involves the same principles and practices as UHT sterilization of milk (pages 58–63). A minimum $9D$ reduction in numbers of endospore-forming bacteria is recommended and the need for a more rigorous process is reflected in the UK regulations which stipulate a minimum process of 140°C for 2 s (or a combination with the same effect). The minimum permitted process does not, however, totally obviate spoilage problems due to survival and outgrowth of highly heat resistant endospores.

Either indirect or direct heating plant may be employed, direct heating giving a better quality product. High shear stresses may, however, be present at the point of mixing and this, in combination with heat, can lead to instability of the cream emulsion.

Undesirable physical changes involving formation of a cream layer (creaming) and agglomeration occur during storage of UHT cream. Homogenization is therefore necessary for all types, including whipping, to prevent creaming. Positioning the homogenizer upstream of sterilization is preferred from an operational viewpoint since aseptic operation is not necessary. Heating may, however, lead to destabilization and it is usual practice to operate the homogenizer, under aseptic conditions, downstream of the heating stage. In some plants a compromise arrangement is made by placing the high pressure homogenizer pump in the heating section and the homogenizing valve downstream in the cooling section. Separation of the milk at high temperature and addition of permitted calcium sequestrants improves the stability of UHT processed cream but impairs the whipping properties (see page 202).

In most cases packaging material for UHT cream is the same as that for milk (see Chapter 2, page 66). In addition individual portions for whitening coffee are packed into foil-sealed plastic pots and have found widespread use in catering.

(d) In-container sterilization

In-container sterilized cream is processed at 110–120°C for 10–20 min in either metal cans or glass bottles. The cream is usually of relatively low fat content, the most common types being coffee cream in continental Europe and a 23% cream in the UK. Cream of higher fat content is difficult to process due to its poor thermal conductivity and is extremely prone to separation during storage. Several types of retort are

CONTROL POINT: HEAT TREATMENT CCP 1

See Chapter 2, page 52 (pasteurization and extended heat treatment), page 64 (UHT sterilization), page 70 (in-container sterilization).

Table 5.2 Factors affecting the viscosity of cream

Milk associated
   triacylglycerol composition

Process associated
   fat content
   homogenization pressure
   heat treatment

Storage associated
   temperature
   length of storage

used including continuous and both static and rotating batch types. Retort type and can size dictate the processing applied, but the thermal load applied is high and results in a considerable degree of Maillard browning, protein denaturation and fat agglomeration which impart characteristic properties to the product.

5.2.6 Specific processing for different types of cream

With the exception of creams such as clotted and whipping most specific processing is concerned with modifying viscosity to meet consumer requirements. These often vary according to culinary end-use. Consumers may require, for example, that double cream for eating with fresh strawberries is spoonable, while cream of the same fat content for eating with trifle should be pourable.

Many factors affect the viscosity of cream (Table 5.2), indeed it may fairly be stated that every aspect of plant configuration, processing and

---

* Broken texture is a specific fault of retorted cream. Various salts, including disodium hydrogen phosphate, trisodium citrate and sodium bicarbonate, may be added to minimize the problem, quantity required varying according to the time of year. Use of stabilizing salts, however, is controlled by legislation. Problems can also be minimized by single- (*ca.* 19.5 MPa) or double-stage homogenization (*ca.* 18 and *ca.* 3.5 MPa).

handling of the cream has an effect, however slight, on viscosity. In practice factors such as milk composition are largely beyond control while others, including temperature of separation, are fixed under normal operating conditions. Modification of viscosity may, however, be achieved by homogenization together with control of cooling after heat treatment.

(a) Half and single creams

Homogenization is essential for half and single creams both to produce an acceptable viscosity and to prevent separation of the fat and serum phases. The latter factor in particular requires the use of relatively high homogenization pressures. Pressures actually used may vary according to the properties of the incoming milk.

Single cream is commonly subject to single-stage homogenization at pressures up to 25 MPa, but higher pressures of up to 30 MPa are required to produce an acceptable viscosity in half cream. In each case a homogenization temperature of *ca.* 55°C is used.

In the US a product described as 'half and half' is retailed and consists of a mixture of milk and cream with a fat content between 10.5 and 18%. 'Light' cream has a fat content of 18–30%.

(b) Double cream

Homogenization is not usually necessary for double cream with the exception of UHT sterilized. Homogenization, controlled cooling, or a combination of these processes may, however, be used to produce

---

BOX 5.1 Feed upon strawberries, sugar and cream

In the UK, sales of cream, especially single cream, increase considerably during the months when fresh strawberries are available. It is common practice for multiple retailers to promote cream at reduced prices during this period to encourage sales of the higher margin strawberries. Other outlets also seek to exploit the promotional power of the strawberries and cream combination. Milkmen offer home delivery of strawberries alongside milk, cream and other dairy goods, while many greengrocers introduce temporary sale of fresh cream. Unfortunately some fruit outlets lack refrigeration and cream may be displayed for many hours at high ambient temperatures.

'extra-thick' pasteurized creams of much higher viscosity than is normal. Single-stage homogenization is used at a pressure of 3.5 MPa, or less, at *ca.* 55°C.

A common controlled cooling protocol involves cooling cream to 20–25°C, packing into retail containers and completing the cooling in a cold store. Spoilage micro-organisms and, potentially, many pathogens are capable of rapid growth during cooling and cream produced by this system is almost invariably of inferior microbiological quality to other types.

The use of homogenization in combination with controlled cooling permits creams of widely varying viscosities to be produced. Many processes are of an *ad hoc* nature, being developed to produce a cream of desired properties within the constraints of available equipment. In all cases, however, the need to ensure microbiological safety and stability should be the over-riding factor.

(c) Coffee cream

Coffee cream is a special type of low-fat cream processed to minimize 'feathering', the coagulation and release of free fat which occurs when cream is added to hot coffee. Such cream is processed using double stage homogenization using pressures of *ca.* 17 MPa in the first stage and *ca.* 3.5 MPa in the second stage. This, however, serves to lessen the problem rather than to solve it and it should also be appreciated that the likelihood and extent of feathering is determined by properties of the coffee as well as those of the cream.

(d) Whipping and whipped cream

The quality of whipping cream is based on its ability to form, by incorporation of air, a relatively solid whip. Viscosity at the point of leaving the container is not an important attribute and a high viscosity is, in fact, undesirable due to the additional mechanical work required during whipping.

With the exception of UHT sterilized product, whipping cream should not be homogenized since even the very low pressures sometimes used to break up loose clumps of fat cause changes to the fat globule membrane which impair whipping properties. Indeed the handling of whipping cream should be even more gentle than that of other types. Homogenization of UHT sterilized whipping cream is essentially a compromise between the needs of preventing separation and impairing whipping properties. Relatively low homogenization pressures are employed, typically in the range 3.5–7 MPa, and the process may be single or double stage. Fat globules in the size range 15–20 μm are considered to give optimal whipping qualities. Whipping of UHT cream is improved by separation at low temperatures and addition of calcium ions, but this results in poor storage stability.

Higher fat creams produce more stable whips but at fat levels above *ca.* 40% the characteristic lightness of the whip is lost. Commercial practice dictates that for economic reasons the fat content should be as low as possible. A fat content of 38% is a common compromise where the whip must remain stable over a relatively long storage period, although lower fat contents may be suitable for where consumption is more or less immediate.

Whipping is most effective at low temperatures when a high proportion of the fat is in the solid form. Particular care is required during cooling of the cream to avoid supercooling and consequent damage to the fat globule membrane. This results in high viscosity and poor whipping qualities. Crystallization of the fat to a stable configuration (ageing) is not immediate but takes several hours. This is of no consequence where refrigerated distribution systems mean that ageing occurs during storage and transport.

The whipping of cream is technologically straightforward and, on a commercial scale, normally involves the use of a mechanical whisk which cuts the cream and draws in air by a planetary motion. The cream should be at 2–5°C and the equipment itself cooled to ensure maximum fat stability. From an economic viewpoint it is desirable to achieve, by incorporation of air, as great an increase as possible while maintaining stability and eating quality. The increase, the overrun is expressed as a percentage:

$$\frac{V_2 - V_1}{V_1} \times 100$$

where V_1 is the initial volume of cream and V_2 is the volume of whip.

The ideal endpoint is usually just after the maximum volume has been obtained and operator skill is required for effective operation of the whipping process.

A number of 'cream-machines' are available which operate on the principle of air-injection. The design varies but each offers, potentially, a high overrun and precise control of the degree of whip. In some cases the machines give either a continuous flow or can deposit precise quantities of whip for addition to individual pastries, etc. Very large capacity machines are based on a similar technology to ice-cream freezers (see page 409). A major problem with some air-injection machines, however, is maintaining a satisfactory standard of hygiene and while design improvements have been made, some machines have been considered 'impossible' to clean adequately.

Sugar (sucrose) is traditionally added during whipping and, in the UK, up to 13% is permitted in whipping cream sold for commercial use. Other stabilizers are permitted in whipping cream sold for commercial use, or in prewhipped cream (see below) and are considered necessary in late spring and early summer months when fat has a high proportion of low melting point triacylglycerols. In the UK permitted additives are sodium alginate (or a mixture of alginic acid, sodium bicarbonate and tetrasodium pyrophosphate), sodium carboxymethyl cellulose, carrageenan and gelatin. Of these sodium alginate and some types of carrageenan function as true stabilizers by interaction with fat globules (see page 421), but other types merely increase the viscosity. Stabilizers permit the fat content of the cream to be lowered, but the extent to which this is possible is limited by legal requirements and loss of organoleptic acceptability.

Prewhipped cream is available to bakers and confectioners, is convenient to use and offers a higher level of reproducibility than is readily available when whipping on a small scale. Aerosol-packaged whipped cream is also available to domestic consumers. The cream is UHT processed, aseptically filled and pressurized. In the UK nitrous oxide is the only suitable gas permitted as propellant, but in the US octafluorocyclobutane (Freon C-318, $\overline{CF_2-CF_2-CF_2-CF_2}$) and chloropentafluoroethane (Freon 115, $CClF_2-CF_3$) are also permitted. These propellants exist partly as a vapour in the headspace and partly as a liquid layer on top of the cream. Dispensing thus occurs at constant pressure but the contents must first be shaken to form an emulsion. Stabilizers are necessary and glycerylmonostearate may be added to compensate for the loss of whipping quality due to homogenization. On release of pressure, cream is propelled through the outlet valve. The propellant volatilizes to form a whipped structure of high overrun, a specially shaped outlet nozzle permitting the cream to be piped directly into cakes or other confections. The whip is less stable than that produced mechanically and shrinkage occurs on storage. Freons are generally considered to produce a better whip than nitrous oxide alone, but where permitted, gaseous and liquefied propellants are normally used in combination.

The aerosol packing principle has been extended to instant mousse which consists of cream, stabilizers and appropriate colouring and flavouring.

(e) Clotted (scalded) cream

Clotted cream is a regional product of south-west England, characterized by a thick and spreadable texture, nutty flavour and slightly granular texture. The colour varies from a pale to a deep yellow depending on local preference. Clotted cream was, for many years, a farmhouse product and significant quantities are still made on a small scale by traditional means. These involve the batch heating of milk in shallow pans during which the fat rises and agglomerates at the surface. Protein denaturation occurs and a crust forms which is removed, after slow cooling (ageing), by skimming spoons.

Two methods of larger-scale manufacture are used producing two types of clotted cream; float cream, which is similar to the traditional farmhouse type, and scald cream. Float cream is made in large, open-top vessels often arranged in tiers. These are part-filled with skim, or whole, milk and a layer of mechanically separated cream of fat content *ca.* 50% is poured onto the milk, the whole system having a fat content of *ca.* 15%. The milk : cream system is heated by steam, or hot water, until a crust forms (45–60 min) and is then cooled by circulating water. The cooling stage lasts *ca.* 12 h at which stage the crust temperature is 4–7°C. The crust is removed manually by a scoop, fat remaining in the milk usually being recovered for butter making.

Scald cream is produced from mechanically separated cream with a fat content of 55–60%. The cream is poured directly into trays 13–25 mm in depth and heated by hot water, steam or hot air. The body of the cream attains a temperature of 77–85°C, but the crust is cooler and care should be taken to ensure that the temperature is sufficiently high to prevent microbial growth. After heating for 45–70 min the cream is cooled to *ca.* 7°C over a 12 h period.

Scald cream manufacture is more amenable to modification than float cream and a process has been described in which hot air ovens are used at a temperature of 65.5–100°C for 45–150 min. This process permits scalding of the cream in plastic retail containers and thus minimizes handling. Small-scale use has been made of microwave heating, the reduced convection assisting clotting, and this method may well become more widely used.

In the UK the processing received is legally considered to be equivalent to pasteurization, although allowance is made for slow heating and cooling and there is no requirement for thermograph records. In practice some processes are of marginal safety and extensive growth of micro-organisms may occur during both heating and cooling stages. Flash pasteurization of milk is sometimes employed but this may affect the subsequent behaviour of the cream during cooking. Post-cooking pasteurization of scald, but not float, cream is also possible.

The manufacture and packaging of clotted cream involves considerable handling and high standards of hygiene must be maintained. Bulk containers are traditionally filled by hand, layer by layer, although this may be avoided in the scald process by cooking the cream in the final container. The practice of re-packing bulk clotted cream into retail containers in restaurants, gift shops, etc., does, however, lead to further possibilities for contamination, especially as no formal hygiene control may exist.

Clotted cream is also subject to aerial contamination during manufacture and while air flow control procedures and the use of air filtration in sensitive areas reduce problems, the addition of nisin is permitted as a means of controlling mould growth.

High-fat, spreadable creams are produced in other countries including Yugoslavia and Iran. Manufacture of some types involves passage of cream through a second, specially designed separator, but in other cases a heating process similar to that of scald cream is used.

(f) Dried cream

Dried cream may be considered to be a dried milk product with a higher fat content than dry whole milk. Depending on the initial cream the fat content is 40–70% and the moisture content less than 2% (*cf.* anhydrous milk fat, page 255). Spray drying (page 124) is used, problems arising not from the drying stage, but from handling the warm powder. The fat is in the liquid state on leaving the drying chamber and prone to membrane rupture and subsequent caking. Non-fat solids, usually sodium caseinate, and a carbohydrate carrier (lactose, sucrose or glucose) must be present to encapsulate and protect the fat globules and cyclone filtration is not suitable due to caking on the cyclone walls and blocking of filters. A satisfactory solution is to remove the powder from the dryer on a moving belt and cool to solidify the fat in a fluidized bed. Alternatively, the Filtermat™ dryer (see Chapter 3, page 129) may be used.

Dried cream powder is susceptible to oxidation and manufacture requires a high heat treatment, prior to drying, to inactivate lipases and the addition of an anti-oxidant before storage. A 'free-flow' agent such as calcium silicate should also be added and the storage temperature should be sufficiently low to maintain the fat in solid form and prevent caking.

Dried cream has only limited functionality and does not reform as the natural product unless special emulsification and homogenization procedures are used. The product does, however, have an ingredient role as a free-flowing milk fat concentrate.

(g) Frozen cream

Cream which is intended purely for ingredient use where structure is unimportant, may be frozen without any special precautions. In other cases, where it is required that the properties of the defrosted product closely match those of natural cream, fast freezing to prevent damage to the milk fat globule membrane is essential. In the UK a popular means of freezing cream is the use of modified ice lollipop freezers to cast the cream into rod-shaped moulds which are frozen by immersion in calcium chloride at −30°C. Operating efficiency is impaired by high cream viscosity and this should be controlled by separation at high temperatures. High fat cream cannot be cooled below 10°C without becoming of unacceptably high viscosity and should be stored for no more than 4 h before freezing.

Several alternative approaches to freezing are possible including blast freezing of cream in retail cartons. This is particularly suited to clotted cream which is relatively insensitive to the effects of freezing. With other types equipment is available which ensures rapid freezing by spraying a refrigerant, usually polypropylene glycol in water, onto a stainless steel belt or drum carrying a layer of cream.

Frozen cream, like dried, is prone to lipolysis during storage and prefreezing heat treatment should be of sufficient severity to destroy lipases. This usually involves flash pasteurization at 82°C.

(h) Cultured (sour) cream

Cultured cream is a fermented milk product manufactured using *Streptococcus lactis* subspecies as starter cultures. The product is discussed in detail in Chapter 8 (page 366).

(i) Substitute creams

The demand for substitutes arises primarily for economic reasons rather than the dietary concern which has stimulated the development of other dairy product substitutes. Indeed the high sugar content of many substitute creams precludes their acceptance as truly 'healthful' foods, although low calorie and reduced-fat substitutes are now increasingly popular.

Imitation cream is available in a number of consistencies corresponding to the different types of natural cream. The product resembles natural cream in being a fat in water emulsion. A fat content of 15% is common, such a formulation also containing *ca.* 7% sugar, *ca.* 3% milk solids-non-fat and 0.4% emulsifier. Vegetable fats are used, the selection of fats depending on the properties required in the end-product. Fat modification permits close matching with desirable organoleptic properties (see Chapter 6, page 246), but in all cases homogenization, usually two-stage, is an essential part of processing. The physical properties of imitation creams closely resemble those of their natural counterparts, but the physico-chemical relations of the fat globules are entirely different.

Imitation soured cream is made using acidulants such as glucono-δ-lactone and has recently found popularity as a 'party-dip'.

Some types of imitation cream are suitable for whipping, but various types of 'whipped topping' are produced. These include aerosolized products containing 24–35% fat, 6–15% sugar and 1–6% vegetable protein which, in combination with stabilizers and emulsifiers, produce a very stable whip. Frozen, ready whipped toppings and powders,

---

* Coconut oil is widely used in whipped toppings. This oil is rich in lauric acid and imparts stability to whips by its hardness at ambient and sub-ambient temperatures, but has a sharp melting point at 25°C producing an acceptable mouth-feel.

formulated for easy reconstitution, are also widely available and utilize the same basic ingredients. Flavouring and colouring may be added to produce a mousse.

Coffee whitener is prepared either as a liquid product, frozen for retail distribution or as a powder blend. The fats in such products are usually of non-dairy origin.

Low calorie cream substitutes are made in which a sweetener, usually aspartame, replaces sugar and there is an increasing demand for products in which the fat is partly or wholly replaced. Bulking agents such as carboxymethylcellulose, pectin or polydextrins may be used as the base, or the fat may be replaced directly by fat-replacers such as Stellar^{TM}, prepared from modified corn starch, or Simplesse^{TM}, prepared from whey protein concentrate. Dried cream extract may be added to improve flavour. A product with many of the properties of cream, 'cell cream', has also been developed which consists of ultrafine cellulose particles dispersed in water.

(j) End-product testing

End-product testing is required for all types of cream, whether made from dairy ingredients or substitutes. Chemical analysis is required to ensure the cream is of correct fat content with respect to legislative requirements. It is also necessary to determine the physical properties

---

BOX 5.2 **There's some are fou o' brandy**

Although cream is an important ingredient of high value products including fresh trifles and cream liqueurs, there have been few attempts to add value to the product itself. Cream has been retailed combined with chopped fruit and a few years ago 'flavoured' cream appeared briefly in south east England. A significant factor in the failure of these creams was probably the incompatibility between the flavours chosen and the natural flavour of cream, although the chances of success were not helped by the amateurish and unattractive packaging design. In contrast, cream flavoured with brandy is successfully sold on the Christmas market by at least one multiple retailer, it appearing that there is considerable synergy between the flavour of cream and that of the more mellow spirits (*cf.* Cream liqueurs, page 210).

of the cream to assure quality and adequate performance in end-use. Microbiological analysis, including sterility testing, is necessary to ensure satisfactory processing and handling.

5.2.7 Cream as a major ingredient of non-dairy products

Cream is added as an ingredient to a large number of commercial food products including canned soup, dried bakery mixes, etc. In this context the cream serves as a source of dairy fat and its structural properties are of little importance. There are, however, two major product groups, cream cakes, desserts, etc., and alcoholic cream liqueurs, where cream is both a major ingredient and a major determinant of the properties of the complete product.

(a) Cream cakes, desserts, etc.

Pastry cakes with whipped cream filling or topping are a traditional luxury product, cheaper varieties being finished with imitation creams. A highly stable whip is required for cakes sold for home consumption at a later time, collapse of the whip and leakage of the serum being serious faults. The major problem associated with cream cakes, however, is that of poor hygiene and consequent food poisoning (see page 217). Improvements have been made in recent years, particularly with respect to provision of refrigeration, but potential problems remain. These stem partly from operational factors including the amount of handling required and the difficulty of cleaning equipment such as savoy bags, used for filling the cakes, and some types of whipping equipment.

Whipped cream is also used as a topping for dairy desserts. These are based on edible starch or, less commonly, gelatin together with stabilizers, flavour, colouring and, in some cases, fruit pieces. A very

---

BOX 5.3 Skim milk masquerades as cream?

It is likely that while some decrease in the use of cream in everyday cuisine has occurred this is compensated for by increased use in luxury items such cream desserts. The dietary conscious, who routinely follow a low fat diet, often see such products as 'rewards' which may be consumed on special occasions such as expensive restaurant dining. There is, however, interest in the development of 'healthy' cream including reduced cholesterol types.

stable whip is required to withstand the long storage life of these products. Manufacture is on a large-scale minimizing manual cream handling, but a high level of hygiene control is required.

(b) Alcoholic cream liqueurs

Cream liqueurs comprise a range of high value added compound beverages containing milk fat, alcohol from various sources, sodium caseinate, sugar and, sometimes, emulsifiers, flavour and colouring. Most cream liqueurs are of similar chemical composition containing *ca.* 40% solids (15% butterfat, 20% sugar, 5% other non-fat solids) and *ca.* 14% alcohol. Low alcohol cream liqueurs containing less than 10% alcohol have been made but are prone to spoilage by species of *Lactobacillus* and pasteurization of the finished product is required.

Successful manufacture of alcoholic cream liqueurs depends on the production of a stable emulsion and the avoidance of creaming, the simplest process being the addition of cream, sugar and alcohol to a solution of sodium caseinate and homogenizing twice at 55°C and 30 MPa. The emulsion is then cooled, stabilized, flavoured and coloured.

Effective homogenization is essential to prevent creaming and formation of a 'fat-plug' in the bottle neck, but a more serious potential problem is emulsion instability, which is usually due to calcium-induced aggregation at higher storage temperatures. This phenomenon may be ameliorated by addition of citrate salts to sequester calcium and by this means a wide range of products of different solids, fat and carbohydrate content may be made. However, the use of butter oil, which is effectively calcium free, in place of cream can offer a more satisfactory solution.

---

* Most bakery operations are considered to be of low risk with respect to food poisoning and product perishability and in many cases cream cakes are the sole exception to this generalization. This inevitably leads to the danger that the nature of hazards and the corresponding precautions are not fully appreciated. A strong management commitment is required to avoid this danger and should include provision of adequate and continuing operator training as well as suitable premises and facilities. Problems are not restricted to attitudes at operative level, however, and it is considered unfortunate that some management sectors within the bakery industry are still unable to accept that the microbiological safety of cream confections, and not changes in the pastry texture, must be of over-riding importance in determining maximum permitted display temperatures.

> **BOX 5.4 The singer not the song**
>
> Cream liqueurs are generally considered to be luxury items and marketing success depends on a consumer perception of high quality. In some cases this has been obtained by brand extension, linking the cream liqueur to an established brand of spirits or, in at least one case, to a market-leading chocolate. More recent brands, however, have placed greater emphasis on the source of the cream to reinforce the quality perception. Packaging is designed to evoke images of rich cream from the pastures of the Channel Islands, Devon and Wales.

5.3 CHEMISTRY

5.3.1 Nutrients in cream

The nutrients in cream reflect those present in the unseparated milk and the degree of separation, levels of water soluble nutrients falling, while those of fat soluble increase. The concentration of vitamin A, for example, is two to three times greater than that of whole milk in 10% fat cream and 8 to 12 times greater in 40% fat cream. Losses during processing largely correspond to those of milk (Chapter 2, pages 76–80), losses of vitamin C and folic acid being of particular significance in UHT and in-container sterilized cream. Significant losses of vitamin B_{12} also occur during in-container sterilization.

5.3.2 Flavour and aroma of cream

The characteristic flavour and aroma of cream are derived primarily from constituents of the fat phase, although there is also a contribution from constituents of the aqueous phase and the milk fat globule membrane. Alkanoic acids ($C_{10, 12}$), δ-lactones ($C_{8, 10, 12}$), indole, skatole, dimethyl disulphide and hydrogen sulphide, at the levels commonly present in cream, are considered to contribute to the desired flavour, there being a marginal contribution from phenol and phenolic compounds such as o-methoxyphenol. Oxidation during whipping may improve flavour and 4-*cis*-heptanol, if present in microgram per kilogram quantities contributes to the full flavour.

Changes during processing are similar to those which occur in milk (Chapter 2, pages 80–83), the extent of change depending largely on the severity of heating. Changes in the fat phase are of greatest significance and compounds such as lactones, desirable at low levels, cause off-flavours at the higher levels associated with UHT and in-container sterilization.

Lipolysis may occur during production of double cream as a result of milk lipase activity in the separated cream before pasteurization and is enhanced by re-warming cold milk. The phenomenon is fairly unusual and unpredictable. Batch variation within a production run may be explained by factors such as the extent of mechanical damage during separation and the length and temperature of holding before pasteurization, activity being greatest at *ca.* 40°C. Day to day variation is less easy to explain and while the phenomenon may be more common in spring, seasonal change in feed plays little, or no, role.

Deteriorative changes in cream flavour during storage result primarily from fat lipolysis or oxidation. Cream which has not been heated sufficiently to inactivate casein-associated lipase is, theoretically, highly prone to lipolysis, but in practice the life of such cream is likely to be limited by bacterial spoilage. Lipolytic enzymes produced by psychrotrophic bacteria in the raw milk are of greatest significance in UHT sterilized milk, although problems can result from long refrigerated storage of pasteurized cream. The spoilage potential is high in cream due to the high fat content and to the tendency of psychrotroph lipases to partition into the fat phase. However free fatty acids are less readily detectable in cream than milk, contributing to an overall perception of deterioration. Psychrotroph-derived proteases may also cause spoilage involving thickening, gelation and bitter flavours.

Cream is highly prone to lipid oxidation especially in the light. Oxidation is accelerated by light in the wavelength range 310–490 nm, but wavelengths of 440–490 nm are most damaging. Plastics normally used for manufacture of pots for pasteurized cream offer no effective protection. Lipid oxidation is rapid in raw and pasteurized cream, but free –SH groups formed from β-lactoglobulin during UHT or in-container sterilization provide a considerable degree of protection. This permits the packaging of UHT processed cream in individual-

---

* In UHT sterilized cream saturated aldehydes formed by oxidation of lipids are an important cause of off-flavours, butanal, decanal, hexanal and nonanal being identified in 10% cream processed by direct steam injection. Levels fall during storage due, possibly, to binding to heat denatured proteins or 'scalping' by polyethylene packaging material (Hutchens, R.K. and Hansen, H.R. 1991. *Journal of Food Protection*, 54, 109–12).

serving plastic containers and some in-container sterilized cream in glass jars. Homogenized cream is more susceptible to oxidation as a greater surface area of fat is exposed to oxygen.

As with milk (page 83), the milk fat globule membrane may play either a pro- or anti-oxidant role depending on other factors. It has been proposed that the oxidation of phospholipids in the membrane may trigger the oxidation of triacylglycerols in fat globules, although it is also possible that metallo-proteins such as xanthine oxidase are primarily responsible for oxidative capacity.

Cream will readily absorb flavours from external sources including some packaging materials. In this context polyethylene is considered to be very 'clean', but may selectively remove (scalp) some desirable components.

5.3.3 Viscosity of cream

Fat content has an obvious effect on viscosity, higher fat creams tending to be most viscous. The properties of the fat are also important, cream containing a higher proportion of high melting point fats being more viscous. Homogenization is the major technological determinant at any given fat concentration. The mechanism involves breaking down existing fat globules to form a larger number of smaller globules stabilized by adsorbed protein including casein micelles and subunits. In cream the high fat content limits the extent of fat dispersion due to the competing effect of globule coalescence in the homogenizing valve and, in cream of sufficiently high fat content, the mutual obstruction of globules to their own breakup. Viscosity increases during storage due to the progressive flocculation of fat globules and to the strengthening of the structure formed by the gradual accretion of casein micelles onto globule surfaces and bridging at the points of contact. The rate of viscosity increase during storage is dependent on homogenization pressure, fat content and heat treatment.

Some double creams are unstable, especially if subject to agitation, and 'buttering' may occur due to rupture of the milk fat globule membrane and release of free fat.

5.3.4 Feathering of coffee cream

For a number of years attempts to elucidate the basis of feathering were concentrated on the properties of the cream, but it is now recognized that coffee polyphenols play an important role. Casein micelles bind to each other and to fat globules. The micelles aggregate when exposed to high temperatures, sufficient calcium and phosphate being present to permit the deposition of calcium phosphate. The aggregated micelles are destabilized by polyphenols such as chlorogenic acid (1,3,4,5-tetrahydrocyclohexane carboxylic acid-3-[(3,4) dihydroxycinnamate]) and coagulate with accompanying release of free fat. The stability of the micelles is affected by the cream pH value and total solids content and the degree of feathering is also very sensitive to small increases in fat content. For any given cream the likelihood of feathering depends on two variables, the concentration of chlorogenic acid and the ratio of the ultraviolet absorbance at 340 and 261 nm.

5.3.5 Whipping of cream

During the early stages of whipping air bubbles are formed which, initially, are stabilized by milk serum proteins. Fat globules are then adsorbed at the air interface, losing the milk fat globule membrane which accumulates in the aqueous phase. At the interface the fat globules remain discrete and protrude into the lumen of the air bubble. The original, protein-stabilized air/water interface remains between the fat globules. In the final stage of whipping a matrix of partially coalesced globules forms which traps the air bubbles (Figure 5.6). The strength and stability of the whip are determined by the strength of the air bubble and are greater in high fat cream where the matrix is more extensive and supportive. Above 40% fat, however, the bubble surface

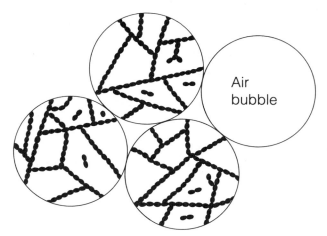

Figure 5.6 Relationship between fat globules and air bubbles in whipped cream. Redrawn with permission from Rothwell, J. 1989. *The Cream Processing Manual*, 2nd edn. Copyright 1989, The Society for Dairy Technology.

becomes more solid with coalesced fat and the organoleptic quality of the whip is reduced. The fat matrix is also more supportive at low temperatures, when a higher proportion of the fat is solid, and stability is rapidly lost as the temperature increases. Compositional changes have an important effect on the whipping quality of cream. The early stages of whip formation are affected by changes in the quantity of surface active material which in turn affects the air/milk serum interface and the subsequent adsorption of fat. This accounts for variation according to the source of the bulk milk and for seasonal variation. Cream of high somatic cell count due to mastitis is slow in whipping, producing a stiff whip with reduced overrun, but free fatty acid content, contrary to some assumptions, appears to have no direct effect on whipping qualities.

5.3.6 Chemical analysis

Chemical analysis is required both in production control (see page 192) and for end-product testing. Fat content is obviously of greatest significance and either traditional methods such as the Gerber technique, or instrumental methods such as the Milko-tester^{TM} may be used. It is necessary, however, to validate these techniques against definitive methods of fat analysis such as the Babcock or Rose–Gottlieb. Other parameters such as solids-non-fat may be determined using methods developed for liquid milk, although some modifications to protocols may be required to accommodate the high fat content.

The phosphatase test is used as an index of pasteurization, but problems due to reactivation are more common in cream than milk. The use of a technique which can differentiate residual and reactivated phosphatase may, therefore, be necessary. Alternative enzyme inactivation tests may be required where severe pasteurization is used (see Chapter 2, page 88).

As with milk, no index test exists for UHT sterilization and the turbidity test, used for in-bottle sterilized milk, is not valid for use with in-container sterilized cream. In these cases processing must be verified

---

* Simple techniques for measurements of viscosity, suitable for use in small creameries, may be devised by use of a little ingenuity. Such techniques include measurement of the time taken for cream to flow through a tube device and the pattern of cream flow from a spatula. In each case the conditions must be rigidly standardized and a common descriptive vocabulary must be used in the latter test (Rothwell, J. 1989. *The Cream Processing Manual*. Society for Dairy Technology, Huntingdon, UK).

using microbiological analysis (see pages 98–101).

Rheological and, in the case of whipping and coffee cream, in-use properties are important determinants of quality and acceptability and form an important part of end-product testing. Viscosity of cream may be accurately determined using a viscometer such as the Brookfield™. The instrument cost is, however, relatively high and skill is required in use and interpretation.

Whipping cream is subject to an in-use test in which the time taken to reach the whipping end point is measured and the overrun calculated. The stability of the whip to leakage should also be determined by weighing a standard quantity into a filter funnel containing a perforated disk. The funnel and contents are supported over a measuring cylinder and the quantity of serum leaking from the whip measured after 24 h at 4 and 18°C. Stability tests are also appropriate for prewhipped cream. Whipping tests should be carried out at low temperature and under carefully controlled conditions. The whipping endpoint can be difficult to determine and requires skilled personnel working to predetermined criteria.

Usage tests are sometimes applied to coffee cream to determine whitening power and resistance to feathering. Whitening power is determined objectively, although instruments could presumably be used. Feathering is also determined subjectively, but can give at best only a very broad indication of the degree of resistance unless the role of the coffee is taken into account. This requires sophisticated analytical equipment and is unlikely to be possible on a routine basis.

5.4 MICROBIOLOGY

5.4.1 Cream, cream-based products and foodborne disease

In recent years the safety record of pasteurized cream has been very good, although cream has been implicated, on epidemiological grounds as the source of *Listeria monocytogenes* infection. There have also been unpublished accounts of staphylococcal food poisoning most of which, although not fully investigated, appeared to involve mishandling in the home. Cream has been identified as vehicle of infection in a single case of *Bacillus cereus* food poisoning. This is one of the few cases of illness due to *B. cereus* in dairy products, despite the high incidence of the organism.

There are no known reports of foodborne disease associated with UHT or in-container sterilized cream, or with clotted cream.

Cream-based products have been associated with a number of outbreaks of staphylococcal food poisoning, despite a suggestion that growth and enterotoxin production by *Staph. aureus* in cream is poor. Cream cakes have been most commonly involved, underlying causes being poor hygiene, especially at the whipping and filling stages, the high level of 'hands-on' operations and inadequate temperature control. The situation has improved in recent years with refrigerated display cabinets in wide use, but problems do still occur. An outbreak on a cruise ship, for example, involved 215 of 715 passengers. The cream was contaminated by a carrier food handler, but temperature abuse was a major contributory factor.

Fresh cream desserts have previously been a cause of concern. Increasing popularity, however, has led to large-scale and more highly automated manufacture which has been effective in eliminating high risk operations. Trifles and similar desserts produced for catering functions are still associated with foodborne disease, usually staphylococcal intoxications. In many cases, however, the situation is confused by cross-contamination and temperature abuse meaning that a number of foods are possibly involved.

5.4.2 Spoilage of cream and cream-based products

(a) Pasteurized cream

Spoilage of pasteurized cream is similar to that of pasteurized milk (see pages 91–4). Until the advent of large-scale sales of milk through supermarkets, cream was expected to be of significantly longer storage life and greater effort was extended in obtaining a product of lower initial count and ensuring storage temperatures were adequately controlled. Some reports suggest that cream tends to be of better microbiological status than milk, but the extent to which this represents the true

---

* Elaborate trifles and desserts produced in catering establishments often require a very high level of handling during final preparation. For this reason such products (and to a lesser extent cream cakes) are considered high risk products with respect to food poisoning due to viruses such as small, round, structured viruses (SRSVs). Many outbreaks, other than those due to molluscan shellfish, have been associated with persons actually suffering SRSV food poisoning, or continuing to shed virus after infection (there is no true carrier state). The low wages and lack of job security prevalent in many sectors of the catering industry can mean that staff are unwilling to report sickness and, unlike many types of food poisoning, it is perfectly possible to work while suffering SRSV infection providing the initial explosive symptoms have subsided.

situation must be questioned. Commercial experience suggests that under refrigerated storage, microbial numbers are similar to those of pasteurized milk. There can, however, be wide variation both between different suppliers and different types of cream. The latter probably reflect differences in processing and are most marked with thick double cream subjected to a slow cooling stage. Despite this cream with high initial counts is not necessarily of shorter refrigerated storage life.

The high fat content of cream means that lipolysis tends to be more important in the spoilage pattern associated with growth of psychrotrophic bacteria, such as *Pseudomonas*, derived from post-process contamination, although proteolysis also contributes to spoilage. Heat resistant *Bacillus* species are important in spoilage at temperatures above 10°C, although comprising only 30% of the spoilage microflora. As with milk, the incidence of *Bacillus* species shows a seasonal variation and is highest in late summer and autumn. *Bacillus cereus* is very common and frequently comprises more than 50% of endospore-forming bacteria.

Yeasts are occasionally involved in spoilage of baker's whipping cream containing added sucrose. *Candida lipolyticum* and *Geotrichum candidum* are of greatest importance. Mould growth can also occur on the surface of cream stored at temperatures of 0–1°C to obtain an extension of storage life. Species of *Penicillium* appear to be most common.

(b) Extended heat treatment cream

Where post-process contamination is prevented, spoilage of extended heat treatment cream is due entirely to surviving endospores of *Bacillus* species. There is, however, doubt that the claimed extension of storage life to as long as 4 weeks at 10°C is possible in practice and at least some microbiologists consider that, without a full sterilization, it is not possible to extend the storage life beyond 14 days.

The behaviour of endospores of *Bacillus* has been studied in some detail with respect to their behaviour during heating of dairy products. In general it appears that increasing the severity of single heat treatments shortens rather than lengthens the storage life (see page 52). A double heat treatment does, however, offer some scope for a longer shelf life, by activation, germination and outgrowth of endospores during the first heat treatment and inactivating the vegetative cells during the second. The temperature combinations used commercially in Germany (see page 197) do not appear to be optimal, research having shown that

115°C is the most effective temperature for activation of endospores. This temperature should, therefore, be used for the first heat treatment, normal HTST pasteurization being adequate for the second stage. Despite this it is not possible to ensure that all endospores are activated by heating, or that activated endospores proceed to germination and loss of heat resistance.

(c) UHT and in-container sterilized cream

Spoilage of UHT cream is normally due to failure of packaging, or of packaging systems and entry of post-process contaminants. Post-process contaminants, often derived from cooling water, may also enter double seam cans, even where the seam is not overtly faulty. Post-process contamination of cream sterilized in glass jars can occur, but is less common.

Endospores of *Bacillus* species may survive both UHT and in-container sterilization, although endospores of the highly heat resistant *B. stearothermophilus* will not outgrow in temperate climates. Under some circumstances viable endospores of mesophilic *Bacillus* species will also fail to develop.

The importance of heat resistant enzymes produced by psychrotrophic bacteria in the spoilage of cream has been discussed in page 212.

(d) Clotted cream

The numbers of bacteria in clotted cream vary considerably depending on the nature of the process, the degree of control and the standard of hygiene. In most cases *Bacillus* species are dominant, although non-endospore-forming thermoduric species such as *Enterococcus* are present where lower cooking temperatures are used. Problems due to low cooking temperatures are often exacerbated by slow heating and resulting bacterial growth before the final temperature is attained.

---

* Newly formed endospores remain largely dormant, even if conditions are suitable for germination. Dormancy can be broken by a number of treatments, including heat shocking, which are collectively known as activation. Heat shocking, however, is reversible and the endospores may not proceed to germination. Germination occurs in response to a chemical trigger such as L-alanine or glucose and is a rapid process marked by loss of refractility and a burst of metabolic activity due to preformed, but inactive, enzymes in the spore protoplast. Heat resistance is lost at this stage. In the absence of a specific treatment such as heating, activation occurs as spores age, an increasing percentage of the population becoming capable of germination with increasing time.

Bacterial spoilage of clotted cream has been reported during the cooling stage, although this probably involved cream made under primitive conditions with inadequate equipment and process control. Spoilage of finished product is almost invariably by yeast or mould growth on the surface, although this can be controlled by nisin. The unusual marketing of clotted cream through souvenir shops, etc., and the practice of postal distribution with total lack of temperature control can contribute significantly to microbiological problems with clotted cream.

(e) Cream-based products

The shelf life of fresh cream cakes is usually short, although spoilage of cream in cakes by *Bacillus* species has been noted where temperature control was poor. In contrast commercially produced desserts have a long storage life and spoilage of the cream component was a relatively common problem before the general level of manufacturing hygiene was improved. Post-process contaminants, such as *Pseudomonas* and members of the *Enterobacteriaceae*, were most commonly involved, initial numbers being high due to contamination during handling.

Although the microbiological status of desserts has improved in recent years, spoilage still occurs, usually as a result of poor temperature control during extended storage. Growth of micro-organisms appears to be enhanced at the interface between the cream topping and the body of the dessert. Moisture sometimes accumulates at the interface and additional nutrients such as sugars are present. On some occasions this has led to a growth of film yeast at the interface and, on one known occasion, the development of a large colony of a *Bacillus* species.

5.4.3 Microbiological analysis

In general methods used for microbiological analysis of milk (see page 97) are suitable for cream. These include predictive methods for assessing shelf life, although there is normally no attempt to assess the effects of retail handling. The methylene blue test has been widely used as a screening test for the microbiological quality of cream, but its value must be doubted.

The high fat content of cream may interfere with some rapid methods and where used careful validation is necessary.

Microbiological examination of clotted cream should include counts for yeasts. An antibiotic-containing medium such as Rose Bengal–

chloramphenicol or oxytetracycline–glucose–yeast extract agar is preferred to low pH value media such as malt extract agar. Media used for selective recovery of yeast will also recover moulds, but the value of the colony count technique for assessing mould contamination is dubious in any circumstances. Hyphal biomass may be determined by a number of techniques but both methodology and interpretation can be difficult and beyond the resources of small-scale creameries. Incubation at *ca.* 20°C and examination for mould development may be the most suitable technique.

EXERCISE 5.1.

You are employed as all purpose scientist to a small company making a range of fresh cream products. The Managing Director considers that a niche market exists for hand-skimmed 'real' cream. Test marketing, however, is disappointing since the 'nostalgia factor' is insufficient to offset the high cost of the 'real' cream. It is therefore proposed to 'plant' articles concerning hand-skimming in up-market food magazines in which the supposed superior technical and organoleptic properties of the 'real' cream are stressed. You are asked to write a report providing a scientific basis for these claims. Are you able to find any evidence that would permit you to write such a report without compromising your professional ethics?

EXERCISE 5.2.

You are employed as a consultant to a local entrepreneur who wishes to extend his operations into re-packing clotted cream into twee ceramic containers for the tourist trade. Use the hazard analysis, critical control point technique to prepare an operational procedure which will ensure the safety of the operation from receipt of the bulk cream to display of the prepackaged product. What special precautions must be taken concerning the training and supervision of operatives?

EXERCISE 5.3.

Many attempts have been made to extend the storage life of cream by application of extended heat treatments, but little attention appears to have been given to the effect of such treatments on the chemical properties of cream. What are the likely consequences of commonly used single and double extended heat treatments (page 197) with respect to nutritional status and undesirable organoleptic changes? Consider these in relation to both pasteurized and UHT sterilized cream. Design a statistically based programme for the comparative organoleptic assessment of cream processed using three extended heat treatments.

EXERCISE 5.4.

Your employment as chemist for a medium sized manufacturing creamery involves some trouble shooting as well as routine analysis. Your company produce, for a leading multiple retailer, a brandy-flavoured double cream sold only during the Christmas period (see page 208). This is made by simply blending a double cream with a brandy syrup. After Christmas complaints were received concerning curdling of the cream. Investigations have shown that the curdling was not accompanied by any other manifestation of spoilage and examinations suggest that microorganisms were not involved. All of the affected packages were made from a batch of cream brought in from an outside supplier. This cream differed from that produced in-house by a higher fat content (50.5 versus 49%) and a lower pH value (6.55 versus 6.8). It had also been subjected to a higher temperature pasteurization (75.5 versus 72.5°C). Consider the possible interactions between the components of cream and ethanol and postulate a mechanism for the curdling. Which, if any, of the properties of the brought in cream are likely to have been predisposing factors for curdling? Suggest *simple* means by which the problem may be avoided in future years.

6
BUTTER, MARGARINE AND SPREADS

OBJECTIVES

After reading this chapter you should understand
- The nature of butter, margarine and spreads
- The ingredients used in their manufacture
- The technology of their manufacture
- The major control points
- The technology and uses of industrial dairy fats
- The basic chemistry of procedures for modifying fats
- The physico-chemical structure of butter, margarine and spreads
- Microbiological hazards and patterns of spoilage

6.1 INTRODUCTION

Butter has been produced since ancient times and was an internationally traded commodity as early as the 14th century. Butter was originally made direct from milk on a small scale and factory production dates from the 1850s when gravity separation of cream became common practice. Mass production, however, only became possible after the development of the mechanical separator in 1877.

Butter, historically, was an expensive commodity and the price remained relatively high even after the introduction of larger-scale production. Unsuccessful attempts had been made on a number of occasions to produce a cheap substitute for butter, but in 1870 the French chemist Mege-Mouries produced 'oleomargarine', which had a taste and consistency resembling that of butter. Margarine, as the product became generally known, was seen as a cheap substitute for butter and quality was judged on how close a resemblance was achieved. For many years, in the socially stratified UK society, margarine

Introduction

was seen as a food of the poor and a visible sign of an inferior life style and social status.

Improvements to the quality of margarine did lead to an increasing market share, although the status as a butter substitute remained largely unaltered. The situation was changed, however, by the desire for a product that could be spread direct from the refrigerator and by the growing perception of milk fat as unhealthy. Requirements for spreadability and healthfulness can be met by manufacturing margarine with a high level of polyunsaturated fats and such products may be chosen in preference to butter, rather than as a cheap and implicitly inferior substitute.

Both butter and margarine have a minimum fat content of 80% and for many dietary conscious persons this level is too high irrespective of the nature of the fats. This has resulted in the introduction of a wide range of spreads of lower fat content. In many countries there are currently no legal standards or definitions of spreads, although a working classification is possible based on fat content (Table 6.1). Spreads may be based on vegetable fats, a blend of vegetable and butterfat, or butterfat alone (light butter). In addition an increasing number of products are available in which the fat constituent is partially substituted by a partial fat replacer such as Nutrifat^{TM} or entirely substituted by a fat replacer such as Simplesse^{TM} or Stellar^{TM}. A probable future development is the use of a zero calorie fat such as the sucrose polyester Olestra^{TM}, which is not adsorbed by the intestine. Olestra^{TM} does not currently, however, have 'generally recognized as safe' status. A further possibility is the development of products of relatively high fat content, but containing a high level of monounsaturated fatty acids, specifically oleic acid, which is considered to have a positive protective effect against heart disease.

BOX 6.1 **Scales of justice**

In the US, the National Association of Margarine Manufacturers has claimed that the more stringent nutritional labelling required for margarine places the product at a competitive disadvantage with reduced-fat butter products. The descriptor 'no cholesterol', for example, when applied to margarine must be accompanied by a declaration of fat content in grams per serving (in addition to existing declarations). No equivalent labelling is currently required for reduced-fat butter products.

Table 6.1 Classification of yellow fat spreads

Fat content (%)	Description
72–80	Full fat
50–60	Reduced fat
39–41	Low fat
<30	Very low fat

A dairy spread is a full fat blend of dairy and vegetable fats.
Based on Charteris, W. P. and Keogh, M. K. 1991. *Journal of the Society of Dairy Technology*, 44, 3–8; Mageean, P. and Jones, S. 1989. *Food Science and Technology Today*, 3, 162–4.

The possibility also exists of incorporating n-3 fatty acids from marine oils which protect against cardiovascular disease by reducing blood platelet aggregation and clotting. At the technical level there is interest in developing spreads which can undertake the function of margarine and shortening in baking.

Although sales of butter relative to other yellow fats have fallen markedly in recent years, the western European practice of absorbing surplus milk fat by butter production means that output has not fallen proportionately. This has resulted not only in the much publicized 'butter mountains', but also in considerable efforts to utilize butter, or its derivatives, as a food ingredient. Two derivatives, anhydrous milk fat and fractionated milk fat, are of considerable importance and may be used directly as ingredients or as starting materials for a wide range of dairy fat products.

BOX 6.2 Butter will only make us fat

The use of different fat sources for spreads attracts different market sectors. In the UK, for example, dairy-based spreads are favoured by those young persons who prefer the realistic butter flavour and by persons with low income and large families who consider spreads a cheap alternative to butter. Low fat spreads based on vegetable fats are favoured by the more dietary-conscious young and by middle-income groups because of the price advantage over polyunsaturated margarines.

6.2 TECHNOLOGY

6.2.1 Butter

Butter manufacture involves four basic processes:

1. Concentration of the fat phase of milk.
2. Crystallization of the fat phase.
3. Phase separation of the oil in water emulsion.
4. Formation of a plasticized water in oil emulsion.

In conventional butter making the four processes are followed in sequence, but different process routes are followed in some types of continuous butter making. The basic process is illustrated in Figure 6.1.

(a) Cream

Cream separation effects the concentration of milk fat, although in high-fat continuous butter making further concentration is required (see page 235). Centrifugal separation is used (see page 186) and cream is usually concentrated to a fat content of 41%. The use of hermetic separators to avoid aeration is recommended. This concentration permits churning without excessive energy consumption, reduces the amount and fat content of the buttermilk and produces a butter of low base moisture content, making it possible to add salt as a slurry without exceeding a maximum moisture content of *ca.* 16%. A relatively high separation temperature of 50–55°C is used to reduce subsequent problems due to lipases.

Heat treatment of the cream, usually in a plate heat exchanger designed and operated to minimize damage to fat globules, is necessary to destroy vegetative micro-organisms and has a further important function in improving keeping quality by continuing the inactivation of lipases and, in the case of higher temperature treatments, producing anti-oxidant sulphydryl groups. The beneficial effects may, however, be partially offset by the migration of the pro-oxidant copper into the fat phase from the serum. Heat treatment involves temperatures in the range 85–112°C, but 85–95°C for 10–30 s is most common. More rigorous treatments can lead to quality defects due to production of excess quantities of sulphydryl groups and consequent poor flavour. In some countries, notably New Zealand, deodorization by vacuum (vacreation) may be used, sometimes in conjunction with heat treatment by direct steam injection. This process is also effective in removing other taints such as those derived from feed, but may adversely affect the flavour of

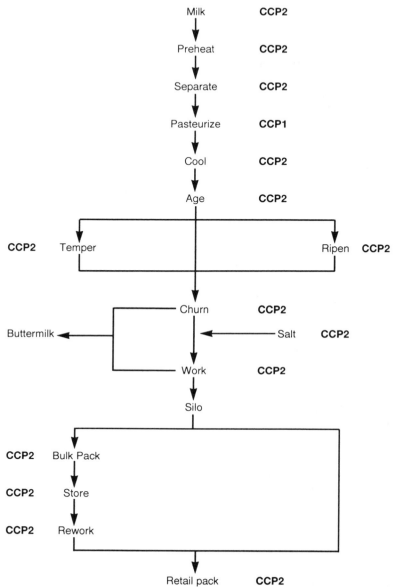

Figure 6.1 Basic process for the manufacture of butter.

butter made from good quality cream by removal of desirable flavour volatiles. When used in conjunction with heating by direct steam injection, shear effects reduce the size of the fat globule and cause high fat losses in the buttermilk. Indiscriminate use of vacuum deodorization

CONTROL POINT: SEPARATION AND HEAT TREATMENT OF CREAM
CCP 1

See Chapter 5, page 195 (separation), Chapter 2, page 52 (heat treatment).

---

should be avoided and the cause of problems such as feed taint eliminated rather than the effects.

(b) Crystallization of the fat phase

Crystallization of the fat phase occurs during the related processes of ageing and ripening. Many of the physical properties of butter are determined at this stage by the ratio of solid to liquid fat present and the shapes and sizes of the fat crystals. However, the middle and high melting point triacylglycerols which are responsible for physical properties are themselves affected by factors beyond the immediate control of the buttermaker including pasture condition, type of winter feed and breed, age and nutritional status of the cattle.

Large numbers of very small fat crystals are required in butter and other yellow fats and crystallization can be achieved by rapid cooling of the cream from above the final melting point of the fat to a temperature below the churning temperature, usually 3–7°C. Fat at pasteurization temperature is fully liquid and it is necessary to take account of the latent heat of crystallization released during cooling. The final temperature and length of holding are more important than the cooling rate and while nucleation is substantially complete within 60 min at *ca.* 5°C, a minimum of 4 h holding is necessary for the development of an extensive network of stable fat crystals.

Rapid cooling leads to the formation of impure (mixed) crystals of middle and high melting point triacylglycerols with higher levels of solid fat. This is advantageous with soft spring or summer butterfat, since it facilitates churning and minimizes the quality defect 'oiling-off' but is a

---

* For many years research in dairy cattle breeding and feeding has been concentrated on improving milk yields. Over the past 10–15 years it has become apparent that higher milk yields have been accompanied by harder winter milk fat, which has required adjustments to butter making technique.

disadvantage with hard butterfat. The extent of difficulties may be minimized by tempering which increases the relative quantity of liquid fat and improves the spreadability and consistency of the finished butter. Tempering involves the 'cold–warm–cold' cycle and the process is developed from the Alnarp process, originally used to facilitate cooling and handling of highly viscous ripened cream (see below). Cream is warmed from 6–8 to 14–21°C and then cooled again to 8–13°C. Optimum temperatures at each stage depend on the triacylglycerol content of the cream and its melting and solidification properties.

(c) Ripening

Cream is ripened by culture with a mixed starter. The composition of the starter varies, but usually comprises *Lactococcus lactis* ssp. *lactis* or *L. lactis* ssp. *cremoris* in combination with *L. lactis* ssp. *diacetylactis* or *Leuconostoc mesenteroides* ssp. *cremoris*. At its simplest, ripening involves culture at *ca.* 20°C, until the desired pH value or diacetyl level is reached, when growth of the starter is stopped by cooling to below 10°C and crystallization commences. Lower ripening temperatures are used in the summer when butterfat is soft than in the winter. The Alnarp process is now widely used both to facilitate cooling the viscous ripened

---

CONTROL POINT: CRYSTALLIZATION OF CREAM CCP 2

Control

Determine required treatment either by determination of iodine value or, preferably, by measurement of melting and crystallization properties. Personnel responsible to be trained and experienced.

Ensure optimum temperature–time combinations applied.

Fit holding equipment with thermograph.

Monitoring

Inspection of thermograph records.

Verification

Analysis of end-product.

Table 6.2 Modification of cream ripening to produce hard or soft butter

Hard butter
1. After pasteurization cool cream to *ca* 19°C, inoculate with starter, incubate at 19°C until the pH value falls to 5.2.
2. Cool to ca. 15°C, hold for 2 h.
3. Cool to churning temperature.

Soft butter
1. After pasteurization cool cream to *ca* 7°C, inoculate with starter, hold for 2–3 h.
2. Slowly warm, using water at 25°C, to *ca* 19°C, incubate until the pH value falls to 4.9.
3. Cool to *ca* 15°C, hold for 2 h.
4. Cool to churning temperature.

cream and to produce butter of the desired texture. This process may be modified to permit either hard or soft butter to be made (Table 6.2).

Traditionally used cultural processes have a number of disadvantages in addition to the high viscosity of the cream. These primarily relate to spoilage potential in that, in contrast to sweet cream butter, copper is preferentially retained in the butter leading to a shortened life due to fat oxidation. At the same time, the free fatty acid content of ripened cream butter is higher and more fat soluble leading to a greater risk of the development of rancid or soapy flavours. The NIZO process overcomes the problems associated with conventional culturing of cream by separate culture of the starter organisms and subsequent addition of culture and lactic acid to the cream.

Ageing and culturing of cream is carried out in vertical silos fitted with agitators to ensure proper mixing and jackets to provide temperature control. There is a continuing trend to larger silos to reduce variability, an important factor when using continuous buttermaking. Tempering in the silo is possible, but the use of heat exchangers, usually of the plate-type, is more efficient with cream of relatively low viscosity.

---

* The NIZO system employs specially selected starter cultures. A sub-species of *L. lactis* which produces large quantities of lactic acid is used as the acidifying organism. After fermentation the lactic acid is concentrated by reverse osmosis to avoid undue dilution of the cream. *Leuconostoc mesenteroides* ssp. *cremoris* is used to produce diacetyl. The strain used is selected for activity in production of the precursor, acetolactate and pH reduction and culture aeration are used to ensure a high level of conversion of acetolactate to diacetyl.

CONTROL POINT: CULTURING OF CREAM CCP 2

Control

Conventional

Ensure starter performance adequate (See Chapter 7, page 294).

Maintain fermentation vessel at correct temperature.

Fermentation to be stopped at correct pH value and diacetyl level.

NIZO-type

Ensure lactic acid concentrate and culture of correct composition.

Make addition to cream at correct level.

Monitoring

Conventional

Starter performance (see Chapter 7, page 294).

Fermentation vessel to be fitted with thermograph.

Progress of fermentation to be monitored by determination of pH value.

NIZO-type

Lactic acid concentration determined by analysis.

Culture concentration checked by direct microscopic examination.

Addition to be supervised by trained and experienced personnel.

Verification

All processes

Examination of plant records.

Quality of end-product.

Cream in the silos should be thoroughly mixed, but it is important to ensure that fat globules remain undamaged. A propellor-type agitator, used intermittently, offers the best solution to these potentially conflicting requirements. It is essential to avoid production of a foam and the agitator should be automatically disconnected during filling and emptying operations.

(d) Transfer of cream to churning

Transfer of cream from the silo to the churn, together with the adjustment of cream temperature to that required for churning, is an important process which may be overlooked during considerations of factors leading to butter quality. Consistency of cream is important, especially where continuous butter making is employed, key factors being:

1. Chemical composition (pH value, fat content).
2. Physical characteristics (viscosity, fat crystallization).
3. Temperature.

The optimum temperature is most conveniently obtained by passage through plate heat exchangers operated with a low pressure drop and a temperature differential of only 1–2°C between cream and the hot water heating medium to minimize burning-on. In conventional batch buttermaking the churning temperature is not dependent on the fat content of the cream, but is modified by the hardness of the butterfat, pH value of the cream and the size and design of the churn. In general terms, however, a churning temperature in the range 5–7°C is appropriate, the actual temperature used being chosen empirically on the basis of lowest fat losses into buttermilk. The situation is different in continuous buttermaking where account must be taken of the fat content of the cream as well as the hardness of the butterfat. An approximation of the ideal churning temperature may be obtained using simple formulae:

BOX 6.3 **Things weighty and solid**

Extensive release of free fat as a consequence of mis-handling can lead to cream setting solid in the silo. This is one of the ultimate disasters which can befall a butter plant. The loss of the cream and the disruption of production have significant financial consequences. From the technical viewpoint, however, the more important consideration is devising a means of emptying the silo!

---

1. Summer $T = (56-F)/2$.
2. Winter $T = (58-F)/2$.

where T is the temperature in °C (to the nearest 0.5°C) and F is the fat content.

(e) Churning and working

Churning and working are consecutive and related stages of butter manufacture during which the oil in water (o/w) emulsion is broken followed by phase separation and formation of a plasticized water in oil (w/o) emulsion. In virtually all cases these operations take place in a single piece of equipment, the butter churn in conventional batch production and the continuous buttermaker in the more modern continuous production.

Commercial-scale butter churns were originally constructed of wood and consisted of large, horizontally mounted, rotating cylinders. In most cases the churns contained pairs of rollers for working the butter. Wooden churns are difficult to clean and sanitize and have largely been replaced by churns fabricated from stainless steel. Stainless steel churns do not contain rollers but to facilitate working are irregularly shaped in, for example, cubical, single or double conical configurations. Capacities range from *ca.* 250–2500 kg. In all types of batch churn a large mass of cream is destabilized relatively slowly, the mechanical stress necessary to destabilize the emulsion being applied by rotating the part filled churn so that cream is lifted up the ascending wall and cascades to the base. Although some very large creameries continue to make extensive use of batch churns, most modern installations involve continuous buttermakers and the batch churn is increasingly restricted to use in small creameries where production of butter is on an irregular basis.

Three main types of continuous buttermaker have been developed (Table 6.3). Of these the accelerated churning, Fritz-type is generally favoured (Figure 6.2), although large numbers of the phase-inversion type are in use in the former Soviet Union and other eastern European countries. A number of manufacturers build buttermakers of the Fritz-

---

* Ideally cream handling at this stage should involve as little damage to the fat globule as possible. However a small, controlled degree of destabilization is beneficial in reducing power consumption during churning, reducing fat losses in buttermilk and minimizing the effect of any variation in the properties of the cream. Equipment is available which achieves the required degree of destabilization by injecting filtered compressed air into the cream and producing foam by passage through a static mixer.

Table 6.3 Types of continuous butter maker

Accelerated churning (Fritz-type)
Butter grains formed from cream by high speed beaters. After draining off buttermilk, the resulting grains are worked into butter.

Phase-inversion
Cream is further concentrated to ca. 80% fat. The concentrated cream is then phase inverted from an oil in water to a water in oil emulsion.

Emulsification
Cream is further concentrated and the emulsion broken. Fat, water and salt concentration are standardized, followed by re-emulsification, cooling and working.

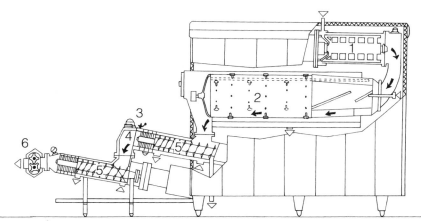

Figure 6.2 Schematic diagram: accelerated churning-type continuous buttermaker. Redrawn with permission by APV Pasilac Anhydro A/S, Copenhagen, Denmark. *Key*: 1, primary churning cylinder; 2, separating section; 3, regulating gate; 4, vacuum chamber; 5, working sections; 6, butter pump.

type and detailed design can vary considerably. The basic operating principles are, however, the same in each case.

Continuous buttermaking involves the rapid destabilization of small quantities of cream. Cream is fed to the top of the buttermaker *via* a balance tank and a variable speed positive displacement pump. The cream enters the primary churning section consisting of a chilled, horizontal cylinder where the emulsion is broken by a rotating multi-bladed dasher. The cream is aerated and the globules damaged over a period of 1-2 s. The speed of the dasher is infinitely variable to permit control of the size of the butter grains formed, a speed of *ca.* 1000 rpm being usual. Higher speeds produce larger butter grains but can also lead to retention of buttermilk and high fat losses.

The mixture of buttermilk and butter grains formed during primary churning passes into the secondary churning section which consists of a large diameter perforated drum, which rotates at a variable speed, usually 35 rpm. Butter grains undergo consolidation and aggregation during passage along the screen and the buttermilk is drained. In some designs the first part of the secondary churning cylinder is fitted with a slow-speed dasher. A portion of the buttermilk may be cooled and recycled to the secondary cylinder to maintain the butter grains at low temperature. It was also previous practice to wash the butter grains at this stage. Washing is now generally considered to be undesirable due to potential microbial contamination, increase in the water content of the butter and problems of disposing of wash water. Butter made without washing is also considered to be of better flavour.

Buttergrains pass from the secondary churning cylinder to the working section, which is normally inclined upwards to facilitate drainage of buttermilk. In the first part of the working section a pair of contra-rotating augurs consolidates the buttergrains into a heterogeneous mass and expels further buttermilk. The augurs also serve to move the mass forward through a series of perforated plates interspersed with rotating mixing vanes. The amount of work may be varied by using plates of different perforation diameters, increasing or decreasing the number of plates and mixing vanes, or by varying the angle of some of the vane blades. In this section shear forces further consolidate the mass of buttergrains and break up droplets of remaining buttermilk to form the aqueous phase of the water in oil emulsion, droplet size being ideally less than 10 μm. Working is an extremely important part of the buttermaking process and it is at this stage that colour, appearance, consistency, and spreadability are determined. Underworking produces a final butter having a crumbly consistency, while overworking results in a weak body resembling thick cream. In each case large water droplets, or free moisture are likely to be present.

Drainage of buttermilk is completed in the working section and,

---

* In addition to microbiological problems, chemical taints may be derived from wash water. In one instance 'soapiness' in butter was attributed to the presence of high levels of alkyl benzene sulphonates derived from sanitizer. The levels present were too high to have been derived from residual contamination of equipment and it was concluded that a buttermaker had substituted an alkyl benzene sulphonate-based sanitizer for the chlorine-based compound used to disinfect the wash water (Reineccius, G. 1991. *CRC Critical Reviews in Food Science and Nutrition*, 9, 197–212).

together with that draining at the end of the secondary churning section, forms a pool in the base of the machine. The pool is maintained at a constant level by a siphon device which can be controlled to alter the degree of drainage. Small fat grains entrained in the buttermilk float to the surface of the pool and may be re-incorporated into the mass passing through the working section. Recovery of fat from the butter milk is sometimes enhanced by incorporating a spinning disk clarifier into the buttermaker and by passing buttermilk over a vibrating sieve after removal from the buttermaker. Centrifugal cream separators have also been used to obtain maximum fat recovery, but some loss of fat with the buttermilk is inevitable.

Continuous buttermaking can result in the incorporation of more than 5% air and some machines are fitted with a low-pressure chamber in which the air content is reduced to less than 1%. This process results in a dense, fine textured product but, contrary to some opinion, has no significant effect on the onset of rancidity. After the low-pressure chamber the butter is reworked to complete emulsification. The two working sections can be driven independently, in which case a larger low-pressure chamber can be installed. Such an arrangement also allows greater flexibility, but the additional operational variable can make control more difficult.

Butter is salted either in the churn or in the continuous buttermaker. Salting is an important process and it is necessary to prevent loss of salt in the buttermilk. For this reason addition is made after washing (if used) and draining. Salting results in an osmotic gradient between the salt granules and the buttermilk which tends to aggregate the water droplets leading to free moisture and the defect 'mottling'. This may, however, be minimized by adequate working and the use of finely ground salt. The salt used must also be of high purity and contain not more than 1 mg/l lead, 10 mg/l iron or 2 mg/l copper.

In continuous buttermaking, salt is dosed into the first working section using a positive displacement pump. A concentrated brine may be used, but since the solubility of NaCl in water is 26% (w/w) this is practical

---

* Microbial growth in buttermilk can be rapid and under some circumstances the buttermilk pool can be a focus of contamination of the finished butter. The need for strict control of hygiene in all aspects of buttermilk handling has been recognized for a number of years, but ignorance still leads to occasional problems. Specific good manufacturing practice guidelines for handling of buttermilk and sanitization of equipment should be prepared within the master manufacturing instructions.

only when butter of a low NaCl content (<1%) is required. The butter before working must also be of low water content to allow for the water content of the brine. A salt concentration of 2% in the final butter is common in the UK and requires the addition of salt as a saturated suspension containing 50% NaCl. High salt butter containing 3–4% NaCl involves addition of a slurry containing up to 70% NaCl, although at this concentration problems with butter quality due to free water are more likely.

Salt used for preparing the slurry should have granules of a nominal 40 μm diameter, with none exceeding 50 μm. The water should be of potable quality and supplementary treatment such as ultraviolet sterilization may be required to destroy any psychrotrophic bacteria present. The slurry should be prepared at least 2 h before use to obtain saturation and to ensure that remaining crystals are very small. Agitation throughout use is necessary to maintain homogeneity.

(f) Butter handling and packing

Butter leaving the churn is usually loaded direct onto trolleys and wheeled manually to packing equipment. This process, if adequately controlled, is acceptable in small-scale manufacture, but is inefficient and involves a high risk of contamination. Continuous buttermakers may be arranged to discharge directly into the receiving hopper of packing machinery. This arrangement has the advantage of simplicity but, to ensure uninterrupted operation of the buttermaker, it is more usual to transfer the butter to a silo using a butter pump, consisting of two large interlocking rotors operated at low speed.

Butter silos are of two types, the sealed and the open. Sealed silos consist of a vertical cylinder with a telescopic feed pipe and a piston, which is acted on by an air cylinder to keep the butter under constant pressure. Butter is discharged from the base of the silo into pumps for transfer to the filling line. The sealed system minimizes contamination but is of limited capacity (*ca.* 900 kg) and not suited to large creameries. Open silos have a capacity of up to 10 tonnes and are constructed with steeply inclined, or vertical sides to prevent butter lodging. Large, contra-rotating augurs in the base are used to discharge the butter to pumps. In each case handling should either be gentle to avoid generating high shear forces leading to coalescence of water, or the butter should be re-emulsified after leaving the silo.

Butter may be packed direct into retail portions and this procedure is

CONTROL POINT: CHURNING, WORKING AND SALTING CCP 2

Control

Butter must be correctly textured and homogeneous with no free moisture.

Butter must meet legal requirements for moisture content.

Ensure butter of correct NaCl concentration and NaCl distributed evenly throughout butter (salted butter).

Prevent microbial contamination of butter from growth in buttermilk or from NaCl slurry (continuous process).

Ensure NaCl of suitable chemical purity.

Monitoring

Conventional

Ensure churn correctly filled, drained at correct time and worked for correct period.

Continuous

Ensure correct operation of buttermaker.

Determine water content of butter leaving churn or buttermaker, either by dielectric instruments fitted to buttermaker or by near-infrared analysis. Ensure NaCl crystals sufficiently fine, or slurry correctly made up. Monitor dispersal of NaCl.

Implement regular cleaning schedule for buttermilk handling in continuous buttermaker. Monitor performance of equipment used for treatment of water used in preparation of NaCl slurry.

NaCl to be obtained from a reputable supplier and of guaranteed chemical quality.

Verification

Quality of end-product.

Chemical analysis of end-product.

Calibration of automated moisture control equipment.

Microbiological analysis of water used in preparation of NaCl slurry and of end-product.

Analysis of NaCl for chemical purity.

Examination of plant records.

more economical. In many countries, however, the highly seasonal production means that initial packing into bulk containers of 25 kg capacity is common practice. Initial bulk packaging can also be advantageous with respect to the quality of the consumer pack, since the reworking required (see below) improves plasticity. For this reason it is now common practice to rework all butter before packing to ensure good spreadability.

Bulk packaging consists of cardboard boxes lined with vegetable parchment, aluminium foil or various plastic films, especially polyethylene. Polyethylene is the preferred material and is economical, virtually sterile and copper free, the most suitable type being food grade, low density and high impact. Parchment is still widely used, but supports mould growth under conditions of high humidity. Special precautions have been advised and it is recommended that butter containers should either be lined with dry parchment, or with parchment treated by immersion for 24 h in concentrated brine containing 0.5% sorbic acid.

Retail packs of butter, commonly 250 g in weight, are usually wrapped in vegetable parchment or in foil–parchment laminate. Vegetable parchment is still popular in the UK, but is permeable allowing water (and thus weight) loss during storage and tending to become greasy if the storage temperature is too high. Parchment also allows ultraviolet light to penetrate leading to an accelerated onset of oxidative rancidity, although this problem may be minimized by incorporating titanium dioxide pigment in the parchment. Foil–parchment laminate is more expensive, but this is justified by superior performance and appearance.

Transparent films provide a very good appearance and are occasionally used for 'premium' quality butter, but display life is short due to oxidative rancidity. The rigidity of cold stored butter and the cost largely precludes the use of preformed plastic packaging popular with other yellow fats. An exception is individual butter packs, widely used in catering, which are made from polyvinyl chloride in a form-fill-seal process.

Bulk packed butter is fairly stable and no significant changes occur at

---

* Phthalate esters, which are suspected carcinogens, have been detected in butter and margarine packed in foil/parchment laminates. This has been attributed to contamination of the inner parchment layer from the phthalate ester coating of the foil during storage as a roll (Page, B.D. and Lacroix, G.M. 1992. *Food Additives and Contaminants*, **9**, 197–212).

−10°C for 6 months. Within the EEC, national Intervention Boards stipulate a storage temperature of −15°C, although in practice temperatures as low as −30°C are used which enable butter to be stored for periods in excess of 1 year. On removal from the cold store butter must be tempered to 6–8°C under conditions of controlled humidity to limit condensation. The compositional and organoleptic qualities should be checked at this stage, before comminuting and reblending the butter. This is necessary to restore plasticity by breaking down the matrix of fat crystals and also provides an opportunity to increase salt or moisture levels to the maximum permitted and to correct any free moisture problems.

On a small-scale re-packing involves comminuting the butter in a 'shiver' and blending in a batch blender equipped with variable-speed, contra-rotating Z-blades. Continuous reblending equipment resembles the working stage of a continuous buttermaker, fitted with a mechanical chopper to comminute the butter.

Butter from different sources may be blended prior to packing and added-value ingredients may be introduced. These include garlic and chopped herbs. Starter culture and diacetyl concentrate may also be added at this stage.

(g) Special types of butter

A number of special types of butter have been developed. In most cases these are attempts to offset the competitive advantages of margarines and spreads in terms of spreadability and healthfulness.

One of the most effective means of producing a spreadable 'butter' involves the use of butter : vegetable oil mixtures. Such products cannot be described as butter and are effectively full fat dairy spreads, marketed under proprietary names such as Bregott^{TM} (Sweden) and Clover^{TM} (UK). The manufacturing technology, however, usually involves buttermaking processes, although margarine-based technology can also be used and avoids the production of a 'buttermilk'-like by-product. Vegetable oil may be blended with butterfat at any stage from the milk before separation to the finished butter, but the most common process is to blend the two fat types directly before buttermaking. Emulsifiers are often added to assist churning and to stabilize the final product. Batch churns or continuous buttermakers may be used, but it is necessary to take account of the softer nature of the fat. Vegetable oil must be present at levels of 15–35% to be effective,

CONTROL POINT: STORAGE AND PACKING OF BUTTER CCP 2

Control

Ensure storage temperature correct throughout storage.

Protect butter from any form of contamination during storage.

Avoid excessive shear forces during handling.

Rework butter to correct extent.

Ensure packaging used (both for bulk and retail packs) provides adequate protection and is not itself a source of taints.

Monitoring

Monitor actual product temperature as well as air temperature on a continuous basis using thermographs.

Butter should be stored away from strongly smelling foods in a properly designed constructed and maintained store.

Monitor butter handling equipment for correct operation.

Monitor operation of reworking equipment.

Obtain packaging material from reputable supplier and use as recommended. Monitor performance of machinery.

Verification

Examination of thermograph and all other plant records.

Personal knowledge of cold store and its management.

Quality assessment of end-product.

In-use assessment of performance of packaging.

unhydrogenated soya bean oil is most common, but rapeseed and sunflower oil are also used. Some products contain vegetable oils at levels considerably in excess of 35% and in this case some hydrogenated oil must be present to impart plasticity. The manufacturing technology of such products, however, is similar to that of spreads and derived from margarine making (see page 249).

Some improvement to the spreadability of conventional butter is possible using texturization. Texturization may be applied by vigorously kneading ready-churned butter. It is essential that crystallization has been completed before texturization and a 7 day resting period after churning is common. Texturization liberates liquid fat from the crystal network and spreadability at refrigerator temperature is improved. The higher spreadability decreases during use, however, due to temperature fluctuations.

Spreadability of ready-churned butter may also be improved by whipping, hardness being reduced proportionately to the quantity of gas phase whipped in. Butter is taken direct from a continuous buttermaker by butter pump and passed to a pin rotor mixer. Gas (nitrogen) is injected into the butter between the pump and mixer and the butter allowed to expand before packing into thermoformed plastic tubs. An overrun of 75% imparts satisfactory spreading qualities and the butter is relatively stable to temperature abuse. The structure, however, is coarse and spongy and the butter differs in appearance from conventional butter. Whipped butter has met some popularity in the US, but not in Europe.

The industrial application of fat modification techniques, notably fractional crystallization, to butterfat has enabled spreadable butter to be made by combining a hard fraction with a very soft fraction. The technology involved is basically that of spread manufacture and involves blending of the fractions, cooling in a scraped-surface heat exchanger and texturization. An expensive double fractionation procedure is required and the finished product, which competes with low cost spreads and margarine, must be sufficiently attractive to command a premium price. A separate market must also be found for the medium melting point fat fraction.

The fatty acid composition of milk fats, and thus the spreadability of butter, may be altered by modification of the diet of the dairy cattle (see pages 18–19). At present, however, economic considerations have prevented exploitation of the well established scientific knowledge available.

Low-fat (light) butter may be made by a number of processes. Although continuous buttermakers have been used successfully in some cases, large-scale production involves methods employed for spreads. There is, indeed, considerable overlap between low fat butters and low fat spreads containing butterfat and both types of product are discussed with spreads (pages 252–5).

Recombined butter is made in countries which have no indigenous dairy industry. The starting material is a coarse water in oil emulsion made from anhydrous milk fat, skimmed milk powder, water and NaCl, which is processed and textured using margarine-making technology. Recombined butter tends to lack flavour, but is acceptable under some market conditions.

(b) End-product testing

End-product testing is required. Chemical analysis is largely concerned with ensuring compositional standards have been met. However physical attributes are also of importance especially with speciality butters such as whipped and spreadable properties. Microbiological analysis is used primarily as an index of satisfactory hygiene during manufacture.

6.2.2 Margarine

(a) Ingredients

Margarine, like butter, is a water in oil emulsion, although a number of other ingredients may be present (Table 6.4). The fat phase is of major importance in determining the physical properties of margarine, espe-

---

BOX 6.4 Honest, truthful and decent?

The term low-fat, or light, butter is itself controversial and in some countries products containing less than 80% butterfat cannot be described as 'butter', even when the term is qualified. The use of 'butter' as is descriptor is particularly strongly protected in traditional dairy farming countries such as the UK. A low-fat spread (vegetable oils and buttermilk) with the brand name 'I can't believe it's not butter'^{TM}, was the subject of considerable controversy and television advertising was initially banned. The resulting free publicity and an aggressive newspaper marketing campaign ensured success in the market-place.

Table 6.4 Representative composition of margarine

Fat phase
Aqueous phase[1]
Emulsifier
Colour and flavour
NaCl or substitute
Vitamins
Preservative[2]
Other minor additives[2]

[1] May, or may not, contain protein
[2] Permitted only in certain countries

cially the structure and its effect on consistency and plasticity. These factors depend in turn on the melting point of component triacylglycerols, the solid fat contents at any given temperature, the distribution of these solid fats over a temperature range and the polymorphic modification or crystal habit of the fat composition.

It is desirable that the mouth-feel of margarine should resemble that of butter and the solid fat content at 35°C should be low to avoid a lingering greasy sensation during consumption. Before the widescale application of domestic refrigeration, margarines were formulated to be fairly hard and yet remain spreadable at cool room temperature, but spreadability at refrigeration temperatures is now considered to be of major importance as a physical attribute. There remains a demand for traditional hard margarine for use in baking where the quality of the baked product is of importance rather than mouth-feel and spreadability. Speciality margarines are also produced for use in making particular types of pastry. Puff-pastry, for example, requires a margarine of high melting point which has a waxy consistency at room temperature to facilitate rolling into the dough.

A wide range of fats and oils have been used in margarine manufacture including those of animal, marine and vegetable origin. The role of margarine as a cheap substitute for butter means that the choice of fats has been strongly influenced by price and availability. In Scandinavia, for example, whale oil was previously widely used as a margarine ingredient, while many Greek margarines are based on olive oil. Olive oil has not previously been widely used in other countries for margarine manufacture, but is now used elsewhere in production of a 'healthy' margarine high in the monounsaturated oleic acid. In some cases use of fats and oils is limited by technical considerations such as the phenomenon of flavour reversion (see page 266) in soybean and marine oils which contain significant quantities of linolenate. In some countries margarine may contain up to 10% of the fat as butterfat.

Vegetable oils are now very widely used, especially in soft margarines of high polyunsaturate content, the most common being coconut, cottonseed, palm oil, peanut and sunflower. Sunflower, which has a high proportion of the polyunsaturated linoleic acid, is particularly popular in Europe.

All fats and oils must be fully refined to avoid the contribution of undesirable flavours. The required properties of the margarine may be imparted either by using a blend of fats or by modification of some, or all, of the fat present. A number of means of modifying the properties of fat are available (Table 6.5), but of these hydrogenation is of greatest importance and is used to raise the melting point of the fat (see page 260). Indeed some experts consider that modification other than by hydrogenation and blending is generally unnecessary. The use of co-randomization is necessary, however, to permit the manufacture of 100% sunflower seed oil margarine, while directed interesterification of olive oil is considered preferable to hydrogenation.

From a technological viewpoint the blend of fats and oils used is determined by the required consistency of the end product, as defined by the solid fat index. Blends used for different types of margarine are illustrated in Table 6.6. Other factors must also be considered, canola oil, for example, which is widely used in Canada, requires the incorporation of 10–15% palm oil to stabilize the β' structure.

The aqueous phase of margarine was originally skim milk, although water itself could be used. The usual current practice is to make up a 'milk' consisting of water and a source of dry protein. This may be skim milk powder, but whey products are now widely used.

Table 6.5 Modification of the properties of edible fats

Non-selective hydrogenation
Selective hydrogenation
Fractionation
Interesterification
Directed interesterification
Co-randomization

Table 6.6 Fat and oil blends used in different types of margarine

1. *Blend*
   high level of high iodine value oil and a low level of low iodine value fat
   Margarine
   fluid margarines; 75–80% liquid oil stick margarines; high, polyunsaturated soft tub products

2. *Blend*
   intermediate level of high iodine value oil and an intermediate level of intermediate iodine value fat
   Margarine
   high polyunsaturated margarines, 50% liquid oil stick margarines, soft tub products

3. *Blend*
   low level high iodine value oil and high level intermediate iodine value fats
   Margarine
   low polyunsaturated, soft tub margarines; all hydrogenated stick products

4. *Blend*
   blend of intermediate iodine value fats
   Margarine
   as 3

Note: Iodine value is a measure of the degree of unsaturation (see page 267). Data from Wiederman, L. H. 1978. *Journal of the American Oil Chemists' Society*, **55**, 823–9.

An emulsifier is necessary to stabilize the emulsion. Mono-and diacylglycerols of fatty acids are used at levels of 0.1–0.3%, usually in combination with lecithin at a level of *ca.* 0.1%. Lecithin encourages the reversion of the water in oil emulsion to an oil in water emulsion under the shear forces occurring during chewing. The creation of an oil in water emulsion avoids oily mouth-feel and enhances release of flavour compounds. Sodium sulphoacetate derivatives of mono- and diacylglycerols are also used and are particularly effective in preventing separation of the fat and aqueous phases on melting and thus minimize 'spattering' during cooking.

Ideally the fats and oils used in margarine manufacture should be bland and make no contribution to the flavour of the final product. For many years the skim milk component was cultured with butter starter micro-organisms to provide flavour, but flavouring derived from butter is now used. Salt is also added both to enhance the flavour and exert some anti-microbial effect.

Most margarine produced for the consumer market is fortified with vitamin A, or vitamins A and D. Colouring is also added to enhance the natural yellow–orange colour of many oils. Beta-carotene is most widely used, although in some countries palm oil concentrate is an important colouring.

The preservative potassium sorbate is permitted in the US, but not the UK, and the need for a preservative is questionable. Rancidity is not usually a problem in vegetable oil-based margarines due to the presence of tocopherols. Rancidity can develop in margarines containing animal fat and many countries permit the addition of antioxidants. Butylated hydroxyanisole, butylated hydroxytoluene and propyl gallate have all been used, but in each case doubt has been cast concerning adverse toxicological effects. Natural extracts of tocopherols or synthetic tocopherols are therefore preferred.

(b) Processing of margarine

Detailed procedures used during margarine processing may vary, but the basic process is the same in each case (Figure 6.3). Processing is now organized on a batch-continuous, or fully continuous basis.

The two phases are prepared separately, the oils and fats being blended and emulsifiers added. A 'milk' is prepared from water and a dry protein source and other water soluble ingredients such as salt and preservatives are incorporated. The two phases are then metered into an

---

CONTROL POINT: INGREDIENTS CCP 2

Control

Ensure ingredients conform with formula requirements.

Monitoring

Obtain ingredients from reputable source.

Verification

Chemical analysis.

Determination of physical properties such as hardness.

Technology

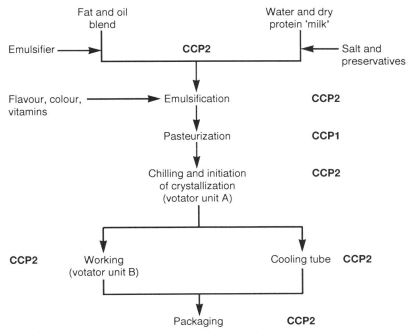

Figure 6.3 Basic process for the manufacture of margarine.

emulsifying unit at 45°C and combined under conditions of vigorous agitation. Flavouring and a vitamin/colour premix are usually added at this stage.

Emulsifying equipment is difficult to sanitize and pasteurization should be carried out downstream of the emulsification stage. Pasteurization is at 80–85°C for 2–3 s in a scraped surface heat exchanger.

The emulsion may then pass to a stirred tank (pre-crystallizing unit) where crystallization is initiated. More commonly, however, the emulsion is chilled immediately in a scraped surface heat exchanger (votator) to initiate crystallization. The heat exchanger consists of one or more steel cylinders cooled externally by a refrigerant, such as ammonia or freon, to −10 to −20°C. A shaft fitted with two or more rows of blades is mounted horizontally in the cylinder and rotates at 300–800 rpm. Only a small annular space exists between the blades and thus the chilled cylinder walls. High internal pressures and shear forces are generated which both favour the dispersion of the aqueous phase into small droplets and induce nucleation and fat crystallization during the 10–20 s residence time, the product leaving the votator as a super-

CONTROL POINT: BLENDING AND EMULSIFICATION CCP 2

Control

Ingredients must be fully blended.

A stable emulsion must be formed.

Monitoring

Monitor correct operation of equipment.

Verification

Examination of process records.

Quality of end-product.

---

cooled emulsion. The most satisfactory results are obtained by use of multiple votator units, since heat transfer and shear requirements change as the emulsion cools and becomes more viscous.

Processing after the votator varies according to the type of margarine, especially between stick margarines and soft tub products. In the case of stick margarine the emulsion passes to large diameter cooling tubes, fitted with sieve plates to produce shear conditions, in which fat crystallization continues to produce sufficient crystallization ('set') to allow moulding and subsequent processing. The extent of crystallization is a function of time and temperature and varies with different blends due to differences in the supercooling properties of the constituent fats and oils. Solid fat levels increase as crystallization proceeds giving the texture required at packing.

CONTROL POINT: PASTEURIZATION CCP 1

Control of pasteurization is similar to that of other pasteurized dairy products such as milk (Chapter 2, page 52).

In the case of soft tub margarines an extension of the plastic range is required and this is obtained by working during crystallization in a second votator unit. The working unit is usually a pin rotor texturizer consisting of a large diameter cylinder fitted with stator pins on the walls in which a horizontally mounted shaft fitted with rotor pins rotates at 20–150 rpm. The residence time in the texturizer is up to 3 min, during which α-polymorphs rearrange to the β'-form (see page 258). Rearrangement releases heat which induces melting and recrystallization, while the vigorous mixing action prevents crystals forming a rigid, interlinked network. The average size of moisture droplets is also increased slightly, preventing greasy mouth-feel.

Depending on the hardness, margarine is either moulded into sticks or filled into plastic tubs. Crystallization is completed after packaging, but before dispatch. Baking margarines may be tempered by holding the packaged product at a slightly elevated temperature to improve plasticity and creaming properties.

Most margarine manufacturers employ variants of the basic process in order to most readily obtain the desired properties in the finished

---

CONTROL POINT: CRYSTALLIZATION AND TEXTURIZATION CCP 2

Control

Correct operation of votators and associated equipment.

Control of temperature during tempering.

Monitoring

Instrumentation to be fitted to processing equipment.

Thermographs to be fitted at stages where temperature is a control parameter.

Personnel to be trained and experienced.

Verification

Quality of end-product.

product. Several methods are considered to impart superior melting and eating properties. These include the use of a double emulsion, a water in oil emulsion in which the aqueous phase contains emulsified oil droplets and precrystallization in which some of the chilled, partially crystallized emulsion, is fed back to the votator feed. It has also been claimed that similar effects to precrystallization are obtained by feeding a portion of the aqueous phase into the part crystallized mass leaving the votator. Blending occurs during subsequent operations.

A great many other variants of the basic process have appeared in patent literature. In most cases the intention is improvement of organoleptic qualities, especially with respect to texture and mouthfeel, but in many cases the improvement is too slight to justify the additional cost.

(c) End-product testing

End product testing is required for all margarines and spreads (see below). The basic purpose is the same as that of end-product testing of butter. It is necessary, however, to pay more attention to physical characteristics and, especially in the case of special purpose margarine, in-use testing in applications such as baking. In very low fat spreads, it may be necessary to test for emulsion stability during storage.

6.2.3 Spreads

As noted above high fat dairy spreads may be made using a continuous buttermaker, while margarine technology may be applied directly to high fat vegetable oil spreads. With reducing fat content, however, the aqueous phase becomes of increasing importance and the technical difficulties of maintaining a stable emulsion increase. Most current products are of the water in oil type, although some oil in water types are also available.

---

* It has been argued that it is not possible to maintain a water in oil emulsion below 34–43% fat, since there is insufficient fat to separate oil droplets. Conversely, other workers believe that oil continuous emulsions can be maintained at 10% fat or less. There may, however, be no clear distinction between the two types of emulsion and indeed in some products the two may coexist. The structure of the product has potentially important repercussions with respect to description and legal definitions (Glaesner, H. 1990. *Dairy Industries International*, **55**(9), 9–11; Moran, D.P.J. 1990. *Dairy Industries International*, **55**(5), 41–4).

(a) Ingredients

Fats and oils may be vegetable, dairy or a blend of the two. Choice of vegetable fats reflects the 'healthful' nature of the product and material high in polyunsaturates, such as sunflower seed oil, is widely used. Dietary considerations have also led to the proposed use of marine oils high in eicosapentaenoic and docosahexanoic acids and human milk fat substitutes for infant feeding. Hydrogenated fat is present to ensure the presence of fat crystals which are important in stabilizing the emulsion. Melting of the fat crystals in the mouth during consumption permits emulsion inversion, the product viscosity decreases and aqueous phase components such as salt and flavouring are released on to the palate. This process is important in determining the organoleptic quality of spreads.

It is possible to stabilize high water content spreads by using high levels of mono- and diacylglycerol emulsifiers, but this method has now been superseded by the combined use of relatively low levels of emulsifiers and aqueous phase structuring agents. This allows closer matching of the rheological properties of the aqueous and fat phases and minimizes the coalescence of water droplets during processing and spreading. Wide use has been made of milk proteins such as sodium caseinate and buttermilk powder at levels of as much as 12% and polysaccharide stabilizers such as carrageenan, alginates and pectins at lower levels. Considerable use has also been made of gelatin.

The level of use of structuring agents is critical, since too little results in an unstable product, while excess quantities lead to reduced organoleptic quality as a result of high residual viscosity after fat melting during consumption. Very high levels of structuring agents are required in very low fat water in oil-based products and in all types of oil in water-based products, where the spread-like properties depend on a gelled aqueous phase. Combinations of milk protein and modified starch, gelatin and monoacylglycerols are all suitable. Starch products such as Paselli SA2^{TM} or N-Oil^{TM} are particularly useful in very low fat spreads since they impart a fatty or creamy texture and are highly effective in preventing coalescence of water droplets. Interactions between starch and milk proteins are of considerable importance in stabilizing very low fat dairy spreads such as St Ivel Gold Lowest^{TM} which contains only 25% fat.

The high water content of spreads means a lower level of microbiological stability and the use of potassium sorbate and, in some countries, benzoate as preservative is widespread. Preservatives are not, however, permitted in some countries, including France and Luxemburg. The refrigerated storage life of very low fat products tends to be very short in comparison with equivalent products of higher fat content and the development of novel preservative systems may be required.

(b) Processing of spreads

Most water in oil spreads are processed using conventional margarine processing equipment, although there is considerable variation in the detailed processing of different spreads. In general, processing conditions are even more critical than with margarine, particular care being required when adding the aqueous phase to the fat phase to obtain the initial water in oil emulsion. Shrouded propeller mixers are widely used and it is necessary to vary the rate of addition of the aqueous phase, applied shear rates and intensity of mixing as the viscosity of the emulsion increases. Crystallization and working must also be carefully controlled to obtain the desired body and consistency. Use of two chilled votators is common, although in other cases an unrefrigerated crystallizing tube situated after the first votator unit is employed. Extensive use is made of pin rotor texturizers for working the spread during continuing crystallization. Variations include the use of precrystallization units situated before the votator to produce large crystal nuclei and a softer product, while crystallization may also be promoted by recirculating part of the crystallized mix.

An alternative means of processing involves preparation of an initial oil in water emulsion and using high speed mixing and blending to bring about phase inversion during subsequent processing. Problems may arise with the stability of the final water in oil emulsion and this method is not widely used.

At present few yellow fat, oil in water-based spreads are available, although the technology has been applied, on a limited scale, to processed cheese analogues and a 'spreadable mayonnaise'. Oil in water-based spreads have technical advantages in that processing is simpler and a wider range of fats may be used, while from the organoleptic viewpoint, there is a more rapid flavour release during consumption. Formulation, with respect to fat level and choice of hydrocolloids as water structuring agents, requires very careful control and tends to negate the advantages. The key manufacturing stage is homogenization and altering the conditions allows a range of products of different rheological properties to be made.

Spreads made with fat replacers are entirely aqueous systems. These resemble oil in water-based spreads with respect to the importance of water structuring agents and the high potential of microbial spoilage.

6.2.4 Industrial milk fat products

Milk fat and butter oil both have a fat content of not less than 99.7%, the prefix 'anhydrous' being applied where the moisture content is less than 0.2%. The terms are often used interchangeably but whole milk is the starting point for milk fat and butter for butter oil.

(a) Manufacture of anhydrous milk fat

The basic process involves disruption of the milk fat globule membrane (MFGM) in order to break the emulsion and subsequent removal of the solids-non-fat and water. The MFGM may be disrupted mechanically or by acidification.

Mechanical disruption is achieved by homogenization of a concentrated cream (70–80% fat). Specialist equipment is available, such as the Alfa-Laval centrifixator, which applies high shear forces to the fat globules. This results in rupture of the MFGM and phase inversion. After phase inversion the fat is further concentrated by centrifugation up to 99.6% and heated under vacuum to *ca*. 95°C for removal of residual moisture. Globules remaining intact after phase inversion are recycled into incoming concentrated cream.

Acidification involves the addition of an acid, usually citric, to lower the pH value to *ca*. 4.5. At this pH value the MFGM is destabilized by precipitation of the casein fraction. Destabilization is followed by rupture of the MFGM and the release of free fat.

(b) Manufacture of butter oil

Butter is softened at *ca*. 50°C before heating to 70–80°C. During heating the fat melts and the emulsion breaks. The oil phase is then concentrated by separation in a series of two to four separators and dried under

---

* Ghee is a traditional product of the Middle East and Asia, which is obtained from milk fat and/or fat-enriched milk products of various animal species. A common means of manufacture involves heating butter at high temperatures which leads to the development of a powerful 'buttery' flavour. Traditional means of manufacture and storage are unsophisticated and the product develops a rancid taste which many consumers consider typical. Within the EEC, however, ghee is produced by non-traditional means involving the addition of ethyl butyrate to anhydrous milk fat (Rajah, K.K. and Burgess, K.J. 1991. *Milk Fat: Production, Technology and Utilisation*. Society for Dairy Technology, Huntingdon, UK).

vacuum. In some cases the oil is washed before the final separation to remove impurities.

(c) Fractionated milk fat

Fractionation (see page 262) is the most commonly used means of modifying the physical properties of milk fat. At its most basic, fractionation is used to prepare high melting point (hard) and low melting point (soft) fractions, but the process may be used to tailor milk fat to individual requirements. Fractions may either be used directly or blended with unfractionated milk fat.

(d) Hydrolysed milk fat

Controlled hydrolysis of milk fat, using either microbial lipases or pregastric esterases from the mouths of calves, lambs or kids, is a means of producing butter or cheese flavours. The use of pregastric esterases has advantages in that there are fewer problems with soapy or bitter flavours. Microbial enzymes, however, are cheaper and avoid ethical issues resulting from the use of animal tissues. The extent of hydrolysis determines the flavour development, butter flavours being produced at low levels of lipolysis and cheese flavours at higher levels. Enzymes may be used either in a batch system with the fat present as an homogenized emulsion, or a continuous system with the enzyme immobilized onto a support. A novel system has also been proposed using *Candida cylindracea* lipase entrapped in the water pools of reversed micelles.

(e) Blending and plasticizing of milk fat

The blending and plasticizing of milk fat permit the manufacture of a wide range of products, each with distinctive properties. The basic technology is derived from margarine manufacture. A blend of milk fat with milk fat fractions is prepared, which is then cooled rapidly in a scraped surface heat exchanger. As the fat crystallizes, the blend passes through a pin rotor texturizer, which weakens the crystal network. Crystallization, and consequently the properties of the final product, are controlled by adjusting the configuration of equipment and processing conditions. Milk fat blends can be prepared which range in texture and consistency from soft, whippable products for use in confectioner's

---

* Reversed micelles are thermodynamically stable, nanometer scale aggregates of ampiphilic molecules solubilizing aqueous drops in a continuous hydrophobic medium.

cream and cakes to products which can be worked without significant softening, which are used in pastry making.

(f) Uses of industrial milk fat products as ingredients in non-dairy products

The functional properties of milk fat products are summarized in Table 6.7 and the ingredient uses summarized in Table 6.8. Flavours derived from partially hydrolysed milk fat may be used to enhance flavour in dairy products and their analogues as well as in snacks and bakery products.

6.3 CHEMISTRY

6.3.1 Nutritional status of butter, margarine and spreads

Butter, margarine and spreads are primarily fat sources and can be important sources of dietary energy, especially where other sources are deficient. Butter is an important source of fat soluble vitamins containing *ca*. 20 times more vitamins A and D than whole cream milk. For this reason supplementation of margarine and spreads with vitamins A and D is mandatory in some countries and general practice in others. Butter contains 10 times lower levels of water soluble vitamins than whole milk, and only traces of minerals, but this is of no significance in a normal diet.

6.3.2 Fatty acid composition of milk fat and other edible fats

The proportion of fatty acids in milk fat and in vegetable fats commonly used in margarine and spread manufacture is summarized in Table 6.9.

6.3.3 Crystalline morphology of edible fats

Edible fats of vegetable, marine or animal origin all form three main types of crystal, the α, β' and β forms. The α form is the least stable and existence in this form is usually transitory. The α form consists of fragile,

Table 6.7 The functional properties of milk fat

Air incorporation	Flavour carrier
Anti-staling (bakery products)	Glossing agent
Creaming	Lightening
Flavour	Shortening

Table 6.8 Uses of milk fat, butter oil and fractionated milk fat in non-dairy products

General culinary	
shallow frying:	ghee, cooking butter[1]
deep frying:	anhydrous milk fat (bakery, patisserie)
Pastry, cakes and biscuits	
biscuits:	anhydrous milk fat/soft fractions
ice cream cones:	soft fractions
puff pastry:	anhydrous milk fat/hard fractions
cakes:	anhydrous milk fat
Chocolate and confectionery	
milk chocolate:	anhydrous milk fat/hard fractions
'creams':	anhydrous milk fat or hard fractions
Convenience foods	
various applications including sauces (flavour), frozen ready meals and meat products	

[1] 'Butter' of low water content and containing lecithin and additional milk solids-non-fat.

translucent crystals *ca.* 5 µm in length. The β' form consists of tiny, delicate crystals *ca.* 1 µm in length and the β form of relatively large and coarse crystals, which average 25–50 µm in length, but which may grow as large as 100 µm or more. An 'intermediate' form has also been described, which comprises crystals 3–5 µm in length which tend to aggregate in coarse clumps.

Liquid fats and triacylglycerols cooled without agitation tend initially to form α crystals. Further cooling of the α form leads to tighter chain packing and a gradual transition to the β form. Cooling with agitation, however, favours formation of the β' form.

Although fats may exist in any of the three polymorphic forms, different fats tend to transform into either the β or β' form. The polymorphic behaviour is largely influenced by the composition of the fatty acids and the positional distribution in the acylglycerols. Fats that consist of relatively few triacylglycerols of similar structure tend to transform to β forms, while heterogeneous fats transform more slowly. In most cases fats which tend to transform into the β type contain low (*ca.* 10%) quantities of palmitic acid, while those which transform into the β' type contain *ca.* 20%.

Milk fat is of the β' type, as is beef tallow and whale oil. Pork lard, however, is of the β type. Of vegetable fats commonly used in

Table 6.9 Approximate percentage fatty acid composition of milk fat and vegetable fats

Coconut	Caproic	<1.0	Capryllic	7.0
	Capric	7.0	*Lauric*	*48.0*
	Myristic	17.0	Palmitic	9.0
	Stearic	2.1	Oleic	5.7
	Linoleic	2.6		
Cottonseed	Myristic	1.0	Palmitic	25.0
	Stearic	3.0	Oleic	18.0
	Linoleic	*51.0*	Linolenic	<1.5
	Arachidic	<1.0		
Milk fat	Butyric	3.2	Caproic	2.0
	Capryllic	1.2	Capric	2.8
	Lauric	3.5	Myristic	11.2
	Palmitic	26.0	Stearic	11.2
	Palmitoleic	2.7	*Oleic*	*27.8*
	Linoleic	1.4	Linolenic	1.5
Palm kernel	Caproic	<0.5	Capryllic	4.0
	Capric	5.0	*Lauric*	*47.0*
	Myristic	16.0	Palmitic	9.0
	Stearic	2.0	Oleic	18.0
	Linoleic	1.0		
Rapeseed[1]	Palmitic	4.0	Stearic	2.0
	Oleic	*64.0*	Linolenic	9.0
	Arachidic	2.0	Erucic	<5.0
Soybean	Myristic	0.5	Palmitic	11.0
	Stearic	4.0	Oleic	22.0
	Linoleic	*53.0*	Linolenic	8.0
	Arachidic	<1.0		
Sunflower	Myristic	0.5	Palmitic	6.0
	Stearic	3.0	Oleic	23.0
	Linoleic	*64.0*	Linolenic	<1.0
	Arachidic	<1.0	Erucic	0.5

[1] Canadian low-erucic acid cultivars
Note: Major fatty acids in each fat are indicated in italic.

margarines and spreads, coconut, olive, peanut, soy bean and sunflower oil are of the β type, while rapeseed, cottonseed and palm oil are β′

Beta-prime crystals are desirable for butter, margarine and spreads and, in the latter two cases, a serious defect 'graininess' may result from the formation of large coarse β-type crystals. Potential problems of 'graininess' may be overcome by using β-type fats as the 'soft' component of the fat blend, although other conditions may be manipulated to favour formation of β' crystals. For this reason co-randomization is used to enable margarine to be made from unblended sunflower oil without problems associated with β crystal formation.

6.3.4 Physical and chemical modification of fat

A number of industrial-scale methods are available for the modification of edible fats (Table 6.5). Modification of fats for margarine manufacture is well established, but is a more recent development in the case of milk fat. Only the more important processes, however, are discussed in this section.

(a) Hydrogenation

Hydrogenation involves a reduction of the degree of unsaturation by addition of hydrogen to double bonds in the fatty acid chains. The mechanism is thought to involve the reaction between unsaturated liquid oil and atomic hydrogen adsorbed onto the surface of a metal catalyst. The process increases the melting point and also the oxidative stability. Hydrogenation is of major importance in the preparation of oils for margarine manufacture, but is not generally applied to milk fat because of its existing low degree of unsaturation and high cost as a feedstock.

Fat is usually hydrogenated by a batch process which involves mixing the oil with nickel or another suitable catalyst and heating to 140–225°C under hydrogen at pressures up to 60 psi. Agitation is necessary to completely dissolve the oxygen, to ensure uniform contact between oil and catalyst, and to dissipate the heat of reaction. The course of hydrogenation is monitored by refractive index and when the desired degree is obtained, the oil is cooled and filtered to remove the catalyst.

In addition to saturation of some double bonds, hydrogenation may also lead to relocation and/or transformation from *cis* to *trans* configuration, the isomers being commonly referred to as iso acids. A complex mix of reaction products may, therefore, result from partial hydrogenation, the

---

* The influence of constituent fatty acids on crystalline morphology may be illustrated by comparing high- and low-erucic acid content rapeseed oil. Erucic acid is toxic and its presence in large quantities in rapeseed oil restricted its application in foods. The development of low-erucic acid cultivars (canola) was of considerable economic importance and canola oil is widely used for margarine manufacture in Canada. High-erucic acid rapeseed oil crystallizes into the stable β' form, but the removal of erucic acid increases the oleic acid content leading to an unstable β' form which transforms to the β form with resultant development of 'graininess'.

composition depending on which of the double bonds are hydrogenated, the type and degree of isomerization, and the relative rates of the various reactions.

The relative rate of hydrogenation of the more unsaturated fatty acids to that of the less unsaturated acids is referred to as selectivity. A quantitative selectivity ratio (SR) is defined as the rate of hydrogenation of linoleic to oleic acid/rate of hydrogenation of oleic to stearic acid. Selectivity is affected both by operating parameters and the nature of the catalyst. Higher SR values, for example, are higher at high catalyst concentration, low hydrogen pressure, high operating temperatures and under mild agitation. Manipulation of selectivity offers a means of imparting specific characteristics to an oil and thus meeting specific requirements in end-use.

(b) Inter-esterification

Inter-esterification is a chemical process by which the distribution of fatty acids among the triacylglycerols is altered. This results in a change from the unique fatty acid distribution patterns of many fats to a random distribution with a resultant change in the physical properties. Two mechanisms have been proposed; enolate ion formation and carbonyl addition. Inter-esterification has a number of applications in the food industry and inter-esterified fats are used in the manufacture of high stability margarines. Inter-esterification has also been proposed as a means of modifying milk fat, milk fat fractions or mixtures of milk and vegetable fats, either as a means of improving the nutritional status or spreading properties. Inter-esterification of milk fat, however, results in an increase in the solid fat content at 35°C, which produces a 'tallowy' mouth-feel. There is also a loss of the characteristic butter flavour during processing supplementary to inter-esterification.

Industrial-scale inter-esterification involves heating fats in the presence

---

* Although it is unlikely that milk fat modification by hydrogenation will be applied on a significant scale, it appears that there is considerable potential for the reverse procedure dehydrogenation, in which the degree of unsaturation is increased. Current thinking tends to favour the use of desaturase enzymes to increase the proportion of unsaturated fatty acids by treatment of isolated triacylglycerols. Such a procedure, however, would involve partial hydrolysis of the triacylglycerol to release the free acid specifically required by all known desaturases, enzymic desaturation of the 18:0 component of these acids and, finally, re-esterification. The relationship of the final product to milk fat would be tenuous as not all of the re-esterified fatty acids would return to their original positions in the triacylglycerol moiety (IDF 1991. *Bulletin of the International Dairy Federation*, No. 260).

of alkali metal or alkali metal alkylate catalysts. Sodium methoxide is most commonly used as catalyst and permits interesterification to be completed in *ca.* 30 min, at temperatures as low as 50°C. Conventional inter-esterification produces a random distribution of fatty acids which is not always desirable. Directed inter-esterification involves maintaining the fat at a temperature below its melting point, which results in selective crystallization of the trisaturated acylglycerols and their removal from the reaction mixture. This, in turn, changes the fatty acid equilibrium in the liquid phase and enhances the formation of further trisaturated acylglycerols. The newly formed trisaturated acylglycerols also crystallize and precipitate out of the reaction mixture and the process thus continues until virtually all of the saturated fatty acids present have precipitated. Directed inter-esterification may be used to stiffen the consistency of oils such as olive oil without hydrogenation (see page 246).

(c) Fractionation

Fractionation provides a means of producing fats and oils with sharply defined melting characteristics. Fractionation is of importance as a means of modifying milk fat and finds some application in the manufacture of margarines.

Several means of fractionating fat are possible (Table 6.10), but dry fractionation has been preferred for milk fat to preserve the characteristic flavour. Dry fractionation involves heating the fat to *ca.* 65°C to destroy crystal nuclei. The fat is then cooled under controlled conditions to allow the growth of crystals. The higher melting point crystals are then separated from the lower melting point liquid phase by filtration, centrifugation or by a combination of these methods.

6.3.5 Changes in fat globules during the ageing and churning of cream

In recent years, the use of freeze-fracture techniques for preparing specimens for electron microscopy has been of major importance in determining changes in milk fat globules during the transition from

Table 6.10 Methods of fat fractionation

Dry fractionation (fractionation from the melt)
Fractionation by short-path distillation
Fractionation by supercritical fluids
Fractionation by crystallization from solvents
---

cream to butter. Four distinct types of globule have been identified, differing in the presence, or absence, of a shell structure and in the quantity of crystalline fat present.

Shell-free globules appear to be present only in soft summer milk fat, all other types having a shell. Shells consist of a monolayer of microcrystalline fat and are found only in oil in water emulsions. During the warm stage of cold–warm–cold cream treatment (see page 230), thick-shelled globules with a liquid centre are present. During the second cold stage, some of the liquid fat crystallizes. This stresses the globule, causing deformation and surface fracture, releasing liquid fat. Globules with very thick shell, however, often remain intact through this stage and subsequent churning.

6.3.6 Structure of butter, margarine and spreads

(a) Structure of the emulsion

Butter, margarine and high-fat spreads exist as stable water in oil emulsions, the aqueous phase is relatively evenly distributed through the product and the water droplets are, ideally, in the size range 2–4 µm (Figure 6.4). The higher water content of lower fat spreads results in a general tendency to larger water droplets and coalescence (Figure 6.4). In such spreads the water droplets are usually in the size range 4–20 µm, although some may be as large as 80–90 µm. The situation becomes more complex as the water content increases and several possible structures exist. The average size of water droplets increases and a single spread may have the properties of both a water in oil and an oil in water emulsion.

(b) Structure of the fat phase of butter, margarine and spreads

The fat phase of butter, margarine and spreads consists of a crystal matrix of solid fat which retains the liquid oil in suspension. In addition churned products such as butter and some types of spread contain some globular fat which limits the tendency of the higher melting point crystals to form an over-rigid network resulting in a brittle texture. The globules are of the very thick shelled type (see Section 6.3.5), and also contribute to mouth-feel and spreadability by behaving like microscopic ball bearings.

The presence of both a liquid and solid fat phase is necessary to attain a state of plasticity. It is also important that the correct ratio is

Figure 6.4 General structure of butter, margarine and spreads. Full fat (80%) butter and margarine have water droplets of approximately equal size, distributed evenly throughout the fat phase (lower part of illustration), with little tendency to coalescence. In products of higher water content, there is a tendency to form larger water droplets, which readily coalesce (upper part of illustration).

obtained between the two phases. Sufficient crystals must be present to retain the liquid phase, but an excess leads to a very rigid structure, lack of plasticity and brittleness. As discussed above, many of the processing operations involved in the manufacture of margarines and spreads are concerned with obtaining a crystal network of the correct rigidity.

In the case of butter, crystallization continues for a significant period after manufacture, maximum firmness developing after three weeks' storage. This is due to the formation of two types of secondary crystal structure:

1. A weak, thixotropic association of a large number of small (less than 1 μm) crystals to form a three-dimensional structure.
2. Formation of strong bonds between large fat globules following realignment.

Reworking causes a disruption of both types of secondary crystal structure. The disruption is irreversible in the case of the strong bonds, but only temporary in the case of the weak association. Under most conditions, the weak association re-forms after *ca.* 8 weeks.

6.3.7 Flavour of butter, margarine and spreads, and changes during storage

The flavour of sweet cream butter is derived from the milk fat itself (see Chapter 1, page 27) and from odourants dissolved in the milk fat. There is also a contribution from compounds, such as lactones, formed during heating and mild oxidation. The perceived flavour is the result of the balance between many components, but differs from that of the parent cream, or pure butterfat, due to the emulsion structure. In ripened cream butter the distinctive flavour of diacetyl overlies the more subtle flavours of milk fat and is the dominant flavour impact compound.

Dissolved odourants are usually undesirable and may be derived from animal feed, medication, etc., or be acquired during storage and handling of milk, cream or the finished butter. Animal feed is the major source of odourants acquired *in vivo* and contaminants are often associated with particular pasture plants. Not all odourants, however, are considered to be contaminants, the γ-lactones bovoliode and dihydrobovoliode, for example, which contribute to butter flavour are natural constituents of grass.

Lipids are good solvents for most odourants and the high fat concentration in butter means that taints are readily acquired from many sources including packaging and other foods stored in close proximity. The low vapour pressure of odourants in solution in lipids does, however, mean that relatively high quantities must be present to achieve sensory impact.

Fat oxidation can lead to serious organoleptic faults, generally perceived as rancidity, during storage. Autoxidation, involving reaction with molecular oxygen is most important, but photosensitized oxidation is also involved. Metals such as copper play an important role as catalysts.

Oxidative reactions are involved in two faults specific to soya and marine oils. The most common of these faults, flavour reversion, results in beany or grassy flavours. The underlying cause is thought to be the autoxidation of linolenate and production of flavour-active compounds such as 2-*n*-pentylfuran and *cis*- and *trans*-2-(1-pentenyl) furans. Other compounds such as 3-*cis*- and 3-*trans*-hexenals and phosphatides may also be involved.

The second fault, hardening flavour, is believed to arise from autoxidation of isomeric dienes (isolinoleates) which are formed during hydrogenation. A number of compounds may contribute to the defect and include 6-*cis*- and 6-*trans*-nonenal, 2-*trans*-6-*trans*-octadecadienal, ketones, lactones and aldehydes.

6.3.8 Chemical analysis

(a) Butter

Chemical analysis is required both for control purposes during manufacture and for end-product testing. Determination of fat content is of obvious importance with respect to the cream composition and can be determined by methods such as the Gerber, or by instruments such as the Milko-tester™. Such methods are not suitable for the determination of fat in butter and conventionally the fat content is determined by subtracting values for total solids-non-fat (see below) and moisture from one hundred. The use of instrumental methods such as Fourier transform infrared spectroscopy may be anticipated.

Moisture content is of prime importance as a production control parameter (see page 239) and is also required for end-product testing. Instrumental methods such as near-infrared spectroscopy are suitable, but are not definitive and a standard oven drying method should be used as reference. The washed residue is weighed to determine solids-non-fat ('curds' plus NaCl).

Salt content of butter, margarine and spreads may be determined by conventional methodology using silver nitrate titration. An instrumental method (Oxford Instruments) is now available which involves use of a small X-ray source to excite fluorescent X-rays from chlorine. The fluorescent X-rays are detected and related, by computer software, to calibration values and thus the chlorine (and NaCl) content of the sample.

The high cost of butter fat and some vegetable fats used in margarine and spreads means that adulteration with cheaper fats is always a possibility. In the past detection depended on determination of the Reichert Meissel number which indicates the quantity of water soluble volatile fatty acids. The Reichert Meissel number of butter is usually 24 to 34 times that of other fats. More sophisticated methods for detecting 'foreign' fats, usually based on gas chromatography, have been available for some time, but are extremely time consuming. Near-infrared spectroscopic methods are now available which are relatively simple to use and which can detect as little as 3% adulterating fat. Determination of diacetyl levels is also an important indicator of adulteration since levels greater than 4 mg/l are indicative of supplementation to disguise the presence of 'foreign' fats or to mask defects. A method based on reverse phase liquid chromatography coupled with a fluorescence detector is available which is of significantly greater sensitivity and specificity than earlier colorimetric methods.

The sensitivity of high fat products to oxidation means that methods are required for measuring the extent to which the reaction has proceeded. Many methods are used, but none is fully satisfactory and results must always be interpreted with care. Peroxide value is most widely used despite doubts concerning its reliability and a poor correlation with organoleptic changes. In some cases an accelerated oxidation test is performed in which air is bubbled through molten fat at a controlled temperature before determination of the peroxide value. The test is widely used with margarine and non-dairy spreads, but appears to have poor predictive value when applied to butter or milk fat. Determination of free fatty acids by gas chromatography is preferred in some cases and is widely used in quality control of milk fats.

(b) Margarines and spreads

Non-dairy fats for use in margarines and spreads are subject to chemical analysis to ensure that any modification has proceeded satisfactorily and that specifications are met. The most important of these is the iodine number which determines the degree of unsaturation by formation of addition compounds between the double bonds of unsaturated fatty acids and iodine. Fat, dissolved, in chloroform or carbon tetrachloride, is reacted with an excess of iodine with iodine bromide or iodine chloride to act as an accelerator of the reaction. At the end of the reaction period excess iodine is titrated against thiosulphate.

The use of iodine number and other chemical analyses in process control of margarine manufacture is limited. Melting point determination has been widely used in the past as a means of determining fat hardness, but is too empirical for good control and cannot distinguish between fats and fat systems. Dilatometry, the change in specific volume with temperature, is more satisfactory and is plotted in terms of solid fat index. Nuclear magnetic resonance spectroscopy, however, is now increasingly used on a routine basis.

Performance testing of the end product is of considerable importance, especially with spreads. Performance testing is of two types; physical and sensory.

Physical testing is primarily concerned with structure and texture. Characterizing the solids–temperature relationship is of major importance, solids level being determined by sophisticated techniques such as pulsed nuclear magnetic resonance or, at lower cost, ultrasonics. Structure can be compared using either penetrometry or extrusion tests, while empirical tests for parameters such as oil loss on compression and spreadability can also be used.

Sensory testing may involve either individual trained tasters or various configurations of panels. Taste and mouth-feel are the most important characteristics and are often related to butter. 'In-use' tests are applied, where appropriate, with margarine to assess suitability for use in baking. Such tests involve preparing pastry to standard recipes.

6.4 MICROBIOLOGY

6.4.1 Butter, margarine and spreads as an environment for growth

Three main product-related factors determine the ability of microorganisms to grow in butter, etc.; the water content, the physical distribution of the aqueous phase and its nutrient content and the presence, in the aqueous phase, of inhibitors. In the case of butter made with ripened cream by traditional methods the pH value may also be sufficiently low to contribute to inhibition.

In butter and traditional margarine, which have no more than 20% water, the extent of microbial growth is severely restricted by the physical size of the droplets and extensive multiplication is not possible. This compartmentalization also means that nutrients may be limiting. Low fat spreads are of higher moisture content and the size of the water droplets is greater, many spreads containing droplets up to 80 µm diameter. In such cases physical restrictions on growth are much less and the presence of inhibitors in the aqueous phase becomes of much greater significance.

Sodium chloride is an important inhibitor of microbial growth in some types of butter and margarine. The NaCl content of butter varies considerably and in many countries, especially continental Europe, unsalted butter is preferred. Elsewhere butter usually contains 1.5–2.0% NaCl, although there is a general trend to lower levels as a result of concern over dietary sodium intake. Butter containing 2% NaCl will contain 12.5% in the aqueous phase, a concentration strongly inhibitory to most micro-organisms. Distribution of NaCl through the aqueous phase is, however, irregular and thus the preservative effect, while still significant, is less than that predicted on theoretical grounds.

Lactic acid levels in cultured cream butters may be sufficiently high to exert an inhibitory effect, especially if the pH value is below 6.0.

Where permitted, sodium benzoate and sorbate have an inhibitory effect on the major spoilage organisms, yeasts and moulds. Preservatives are most commonly added to spreads, but where permitted in margarine are considered to be unnecessary additives. Despite the presence of preservatives, spreads are less stable than butter or traditional margarine and refrigeration is required over even a short period.

Oil in water-based spreads and spreads made with fat replacers are entirely dependent on preservatives for intrinsic stability. The presence of water structuring agents does result in some lowering of the a_w level, but this is not significant with respect to the spoilage microflora.

6.4.2 Butter, margarine and spreads, and foodborne disease

Providing that butter is made from pasteurized cream under good hygienic conditions the risk from foodborne pathogens is very low. Complaints of food poisoning are occasionally received by retailers and

---

* *Listeria monocytogenes* is considered to present a greater hazard than *Yersinia enterocolitica* in low-fat dairy spreads. This results from its higher level of thermotolerance and faster growth in the products during storage at 4°C. The faster growth rate has been attributed to *L. monocytogenes*, being less affected by constraints of space and nutrient limitation than *Y. enterocolitica*. *Listeria monocytogenes* has been isolated from creameries producing a variety of dairy products and it is considered that safety assurance procedures in plants producing low-fat spreads should include specific precautions against this organism (Lanciotti, R. *et al.* 1992. *Letters in Applied Microbiology*, **15**, 256–8).

manufacturers but, where substantiated, are usually associated with oxidative rancidity and not of a microbiological origin. The only published accounts of microbial food poisoning concern growth of, and enterotoxin production by, *Staphylococcus aureus* in whipped butter. At least two outbreaks of staphylococcal intoxication have resulted from consumption of whipped butter, although temperature abuse during storage was a contributory factor. Other pathogens may persist in butter even when numbers are low, due to removal of the majority of cells with the buttermilk. In the case of *Listeria monocytogenes* for example, numbers were reduced by *ca.* 95% following removal of buttermilk but the organism remained recoverable for more than 70 days. Limited growth may occur during the initial stages of storage.

The safety record of margarine is also very good although, like butter, rancidity can lead to revulsion and sickness immediately after consumption. There have been no published accounts of food poisoning associated with spreads, although concern has been expressed that some products of this type may support the growth of pathogens. Inoculation experiments with a spread made from milk fat and vegetable oil showed multiplication of *Staph. aureus*, considered the major hazard, to be at most limited. In contrast, an inoculation study made with two brands of 'light' butter showed both *L. monocytogenes* and *Yersinia enterocolitica* to be capable of growth during storage at 4°C. Each of these pathogens was capable of faster growth on the 'light' butter than the indigenous microflora.

6.4.3 Spoilage of butter, margarine and spreads

In the past a number of micro-organisms have been implicated in the spoilage of butter. In many cases, however, problems related to butter made with unpasteurized cream, under relatively poor conditions of hygiene and with a lack of refrigeration at retail and domestic level. The relevance to modern practice is therefore questionable.

Micro-organisms are unable to grow at the low temperatures (below −10°C) used for bulk storage of butter and some decline in numbers may be expected. The lethal effect of low temperatures is selective, survival of *Micrococcus* ssp. and yeasts generally being greater than that

---

* Low temperature storage *must* not be seen as a means of eliminating pathogenic micro-organisms. Recent work on survival mechanisms in Gram-negative bacteria suggests that the extent of death at low temperatures may be exaggerated and practical difficulties arise due to the lack of suitable resuscitation techniques.

of Gram-negative, rod-shaped bacteria such as the *Enterobacteriaceae*.

Surviving micro-organisms, however, and any entering the butter during reworking and packaging, are potentially capable of growth during retail and domestic storage at temperatures greater than 0°C. Species of *Micrococcus*, many of which are lipolytic, have been important spoilage organisms in the past, especially in sweet cream butter. *Micrococcus*, however, grows only very slowly, if at all, at temperatures below 5°C and the organism is now rarely involved in the spoilage of butter unless temperature abuse occurs. Gram-negative bacteria such as *Pseudomonas* have been involved in lipolytic spoilage of butter at refrigerator temperatures and spoilage of butter by lipolysis may also result from the presence of heat-stable enzymes present in the butter as a result of the growth of *Pseudomonas* ssp. and other psychrotrophs in the raw milk. Spoilage due to preformed lipases can be very rapid, as little as 2 days at 5°C, although in such cases an unacceptable level of growth must have occurred during raw milk storage. Wash water is often considered to be the major source of *Pseudomonas*, but now washing is less common the plant contamination is probably most important.

Yeasts are also important butter spoilage organisms, capable of growth and lipolysis at low temperatures. A number of yeasts have been implicated including *Candida lipolyticum*, *Torulopsis*, *Cryptococcus* and *Rhodotorula*. Yeasts are favoured at the low pH value of some cultured cream butter, but are potential spoilage organisms in all types.

Mould spoilage normally involves the lipolytic genera, *Aspergillus*, *Cladosporium*, *Geotrichum* and *Penicillium*. Spoilage is lipolytic and the butter may also be discoloured. Problems associated with moulds have become less common with improved control of aerial contamination and a higher hygienic standard for packaging materials such as parchment. The practice adopted by some retailers of extended storage at −2 to 0°C, usually in order to build up stocks for special promotions, etc., can, however, result in mould spoilage as many strains grow relatively rapidly at such temperatures.

---

* Choice of method for determination of lipolysis caused difficulty in the past. There has been much discussion over the relative merits of single and double agar layer methods but it appears that while double layer methods give highest counts, the results are difficult to interpret and use of a single layer of lipid-containing medium is preferred. It is necessary to match the lipid source with the product being examined, although clarified butter is generally suitable. The incorporation of Nile blue sulphate dye simplifies interpretation and the base medium should contain 1% peptone to enhance production of lipolytic enzymes.

Micro-organisms responsible for spoilage of butter have also been implicated in the spoilage of margarine. In general vegetable fats are more resistant to lipolytic enzyme activity than milk fat and the spoilage potential of many margarines is lower than that of butter.

There are few published reports of the spoilage pattern of spreads, but yeasts appear to be most commonly involved. *Yarrowia lipolytica* was found to be an important spoilage organism in a low-fat dairy spread, but *Bacillus polymyxa* and *Enterococcus faecium* were also identified as being of potential spoilage significance.

6.4.4 Microbiological analysis

Analysis for specific pathogens is not considered justified and testing is restricted to potential spoilage micro-organisms together with *Escherichia coli* and 'coliform' bacteria. Methods used for liquid milk are generally acceptable (see page 97–8), but the high fat contents means that it is necessary to use a warm diluent containing a surfactant to disperse the sample properly when making the initial dilution. A selective plating medium for yeasts should be included (see page 155) and the inclusion of a colony count to determine the numbers of lipolytic micro-organisms is usual practice.

EXERCISE 6.1.

You have recently been employed as a technologist at a dairy manufacturing ripened cream butter using conventional starter technology. Over a period of years the factory has suffered intermittent problems due to insufficient diacetyl production by *Leuc. mesenteroides* ssp. *cremoris*. The problem, which occurs more frequently in the spring and after Sunday shutdown, has been investigated by a consultant. Beyond concluding that the cause is *not* starter failure due to phage infection, however, no solution has been found.

Consider other possible causes of starter failure and design a plan to investigate the underlying nature of the problem.

Your preliminary investigations show that failure of the starter to grow at an adequate rate is not the cause of the problem. Further it appears that all the available *Leuconostoc* strains produce satisfactory levels of diacetyl when grown under laboratory (i.e. ideal) conditions. Extend your investigations to determine the underlying causes into failure to produce diacetyl in cream. What short-term measures could be tried to remedy the problem before your investigation is complete?

Further information concerning the formation of diacetyl may be obtained from Cogan, T.M. and Accolas, J.-P. 1990. In *Dairy Microbiology, vol. 1. The Microbiology of Milk*, 2nd edn (ed. Robinson, R. K.). Elsevier Applied Sciences, London, pp. 77–114.

EXERCISE 6.2.

What measures are available by which the microbiological stability of a low fat (40% w/v) spread of parameters; NaCl 1.5% (w/v), pH value 6.7, a_w level 0.97, may be enhanced without using preservatives such as benzoate and sorbate. Particular attention should be paid to the potential problems posed by *Staphylococcus aureus* and psychrotrophic pathogens such as *Listeria monocytogenes*. What are the likely effects of the chosen measures on

1. The physico-chemical properties of the spread?
2. The organoleptic properties of the spread?

It is recommended that consideration should be given to use of 'hurdle technology'. Further discussion is available in Leistner, L. 1992. *Food Research International*, **25**, 151–8 (meat products are used as the examples in this paper, but the principles are applicable to all foods).

EXERCISE 6.3.

You are employed as chemist/technologist by a company manufacturing a spreadable butter (80% total fat content), the fat phase of which consists of 75% milk fat and 25% unhydrogenated soya oil. Manufacture involves buttermaking technology. The price of soya oil is rising and your employers wish to consider cheaper alternatives. You are asked to prepare a report summarizing, from a technical viewpoint only, which oils could be used as to replace soya oil. Indicate any changes in the product formulation and processing which may be necessary and any chemical modification required to the oils.

7

CHEESE

OBJECTIVES

After reading this chapter you should understand
- The different types of cheese
- The role of micro-organisms in their manufacture
- The importance of starter micro-organisms
- The technology of cheesemaking
- The relationship between technology and the properties of the finished cheese
- The application of novel technology to cheesemaking
- The major control points
- Chemical changes during the making of the young cheese
- The chemistry of ripening
- Microbiological hazards and patterns of spoilage

7.1 INTRODUCTION

The manufacture of cheese is an extremely ancient activity, it being postulated that the process derived from the eastern Mediterranean practice of carrying milk in animal skin sacks, or in stomachs and bladders. Over the centuries, cheese making has been modified and refined although the dried curd cheese *Kishik*, produced by nomadic tribes in North Africa, may be considered the direct descendent of the primitive product. Cheese making remained an essentially small-scale activity until the application of scientific principles, commencing around the early 20th century, permitted large-scale manufacture. Today manufacture of the more popular varieties is on a very large-scale and cheese is an important export commodity in the economies of the major producing countries such as Eire, France, Australia and New Zealand. Over the years a very large and apparently bewildering number of cheese varieties have evolved, although in some cases differences are very small and involve no more than shape or type of packaging. All varieties of cheese, however, share a common basic technology in which starter cultures of lactic acid bacteria usually play a key role. Further, all varieties may be classified according to moisture content (Table 7.1) and according to the means by which any ripening is achieved.

7.2 TECHNOLOGY

The basic technology for the manufacture of all types of cheese is similar (Figure 7.1), relatively small changes in procedures during manufacture resulting in large perceived differences in the final cheese. The technology is well established but in recent years has been subject to a considerable degree of refinement and automation.

7.2.1 The general role of starter micro-organisms in the manufacture of cheese

In modern practice bacteria of the group commonly referred to as lactic acid bacteria (LAB) are added to milk as starter cultures, the key role being the production of lactic acid by fermentation of lactose. Lactic

Table 7.1 A scheme for the classification of cheese

Hard (26–50% moisture)
 internally ripened, no added ripening micro-organisms
 e.g. Parmesan, Cheddar, Double Gloucester
 internally ripened, added ripening bacteria
 e.g. Emmental
 internally ripened, secondary surface ripening by mould
 e.g. Blue Cheshire

Semi-hard (42–52% moisture)
 internally ripened, no added ripening micro-organisms
 e.g. Lancashire, Edam
 internally ripened, ripening mould added
 e.g. Stilton, Roquefort

Semi-soft (45–55% moisture)
 surface ripened, ripening bacteria added
 e.g. Limburger, Port du Salut

Soft (48–80% moisture)
 surface ripened, ripening mould added
 e.g. Brie, Camembert
 unripened
 e.g. Cottage, Coulommier

Others
 e.g. brined varieties, Whey cheese

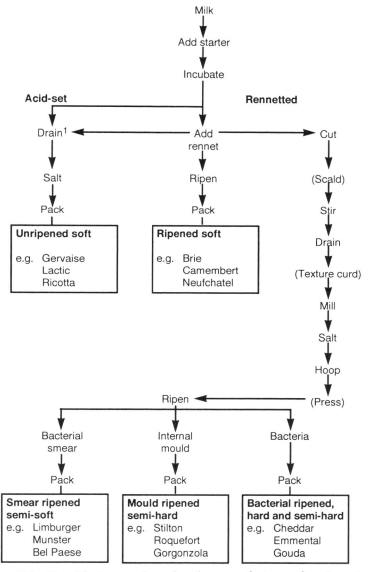

Figure 7.1 Simplified procedure for the manufacture of various types of cheese. *Note*: Stages in parentheses are not involved in the manufacture of some varieties. [1] Manufacture may involve some light cutting and scalding.

acid is responsible for the fresh acidic flavour of unripened cheese and is of importance in the formation and texturizing of the curd. In addition, starters play other essential roles: the production of volatile flavour

Table 7.2 Lactic acid bateria used as cheese starter micro-organisms

Lactobacillus	rod shaped, grow at 45°C but not 10°C; do not metabolize citrate
Lb. delbrueckii	
ssp. *bulgaricus*	does not produce ammonia from arginine; does not ferment galactose
ssp. *lactis*	some strains produce ammonia from arginine; some ferment galactose
Lb. helveticus	does not produce ammonia from arginine; ferments galactose
Lactococcus	coccal shaped, grow at 10°C but not at 45°C; ferment galactose
L. lactis	
ssp. *cremoris*	does not produce ammonia from arginine or grow at 40°C. Does not metabolize citrate.
ssp. *lactis*	produces ammonia from arginine and grows at 40°C. Does not metabolize citrate.
ssp. *lactis* biovar *diacetylactis*	some strains produce ammonia from arginine and some grow at 40°C. Metabolizes citrate
Leuconostoc mesenteroides ssp. *cremoris*	coccal shaped; does not produce ammonia from arginine; grows at 10°C but not at 40°C; Metabolizes citrate and ferments galactose
Streptococcus salivarius ssp. *thermophilus*	Coccal shaped, grows at 45°C but not at 10°C; does not produce ammonia from arginine or metabolize citrate; does not ferment galactose

Note: Bacteria other than those described may occasionally be used as starter cultures. Non-starter members of the genera described may have different properties.

compounds such as diacetyl and aldehydes, and the synthesis of proteolytic and lipolytic enzymes involved in the ripening of cheese and the suppression of pathogenic and some spoilage micro-organisms. Species of four genera, *Lactobacillus*, *Lactococcus*, *Leuconostoc* and *Streptococcus* (Table 7.2) are most widely used, the use of a fifth genus, *Pediococcus*, having been proposed more recently.

(a) Lactose metabolism

There are major differences in lactose transport and metabolism between the different starter bacteria. In *Lactococcus*, transport into the

---

* Micro-organisms other than starter species are involved in the ripening of cheese. These micro-organisms can be of major importance in determining the final properties of the cheese and may either be of adventitious origin or added as a part of the manufacturing process. The role of these micro-organisms should not be confused, however, with that of starter.

cell is by the phosphoenol–pyruvate phosphotransferase system and involves the simultaneous formation of lactose-phosphate. Lactose-phosphate is hydrolysed by phospho-β-galactosidase to glucose and galactose-6-phosphate which are metabolized by the glycolytic and tagatose pathways, respectively.

Lactose enters the cell intact *via* a permease system in starter species of *Lactobacillus, Leuconostoc* and *Str. salivarius* ssp. *thermophilus* and is subsequently hydrolysed to glucose and galactose by a β-galactosidase. Glucose is metabolized by the glycolytic pathway in *Lactobacillus* and *S. salivarius* ssp. *thermophilus* and by the phosphoketolase pathway in *Leuconostoc*. Galactose is usually metabolized *via* glucose-1-phosphate in the Leloir pathway. *Lactobacillus delbrueckii* ssp, most strains of *Str salivarius* ssp. *thermophilus*, and some strains of *Lb. delbrueckii* ssp. *lactis* are unable to metabolize galactose, which is excreted.

(b) Other technologically important characteristics

Protein metabolism. The amino acid content of milk is insufficient to support extensive growth of starter LAB and the culture must possess proteinase activity. Proteinases are situated in the cell wall and differ in number, location and specificity according to strain. Proteinase activity results in hydrolysis of proteins to oligopeptides, which in turn are hydrolysed to low molecular weight peptides and amino acids by cell wall-associated aminopeptidases. Proteolytic enzymes derived from starter micro-organisms also contribute to the ripening of cheese (see page 327).

Citrate metabolism. *Lactococcus lactis* ssp. *lactis* biovar *diacetylactis* and species of *Leuconostoc* are the only starter bacteria capable of metabolizing citrate. It is an important property, however, and the products diacetyl and acetate are responsible for flavour development in unripened and, to a lesser extent, ripened cheese. In Dutch-type cheese CO_2 produced during citrate metabolism is responsible for the desirable eye formation. The basic metabolic pathway *via* oxalacetate and pyruvate is the same in all citrate utilizing bacteria but whereas acetyl-CoA is the precursor in *L. lactis* ssp. *lactis* biovar *diacetylactis*, acetolactate may be involved in *Leuconostoc* ssp.

(c) Genetics of starter micro-organisms

Most strains of lactic acid bacteria employed as starter cultures contain plasmid as well as chromosomal genes. Plasmid-borne genes are of particular importance both because of their accessibility for molecular

Table 7.3 Plasmid-borne, technologically important properties of starter bacteria

Lactose fermentation	Lactose plasmids identified in most starter bacteria; possible that lactose genes may be either plasmid or chromosomal in *L. lactis* ssp. *cremoris*.
Proteinase	Cell-wall proteinases recognized as plasmid mediated; probable that both lactose and proteinase genes are encoded by same plasmid in *L. lactis* ssp. *lactis* but genetic linkage often broken by deletion
Citrate metabolism	Citrate transport by *L. lactis* ssp. *lactis* biovar. *diacetylactis* plasmid mediated.
Phage resistance (see Figure 7.1)	At least three resistance mechanisms in *Lactococcus* plasmid mediated; some may be encoded on same plasmid

genetic analysis and because they code for many properties essential for successful starter activity (Table 7.3). There is considerable interest in applying knowledge of the genetics of lactic acid bacteria to improve existing, or produce new strains. Both conventional techniques such as conjugation and recombinant DNA technology are used and directed integration of genetic material into the bacterial chromosome, is also being investigated with the intention of improving the stability of genetically manipulated cultures.

Mesophilic strains of *Lactococcus* have received most attention although progress is also being made with lactobacilli and *Str. salivarius* ssp. *thermophilus*. Work is currently concentrated in areas such as bacteriophage resistance, proteolysis and flavour generation and production of antagonistic compounds. Work is also progressing to stabilize properties of fundamental importance such as lactose metabolism.

(d) Antagonistic properties of starter bacteria

Lactic acid, at low pH value, is known to play an important role in the suppression of undesirable bacteria by starter cultures, while diacetyl has activity against Gram-negative, but not Gram-positive pathogens. It

---

* Special conditions must be fulfilled when undertaking the genetic manipulation of cultures for use in food fermentations. It is necessary to ensure that selectable markers do not compromise human drug therapy and that vector DNA is natural to lactic acid bacteria approved for food use. A suitable vector has been developed which uses nisin resistance as selectable marker.

is also known that at least some starter strains produce specific antagonists. Strains of *L. lactis* ssp. *lactis*, for example, produce the polypeptide antibiotic nisin, which has a broad spectrum of activity against Gram-positive bacteria and which is a permitted preservative in specified foods including processed cheese. Genetic modification has been used to introduce the gene mediating nisin production into a wider range of starter bacteria.

There has also been considerable recent interest in the use of starter micro-organisms which produce antagonists during fermentation and which remain active in the finished product. The use of antagonists to control specific pathogens such as *Listeria* is particularly attractive but there is also interest in the control of spoilage micro-organisms in more perishable cheeses. Doubts, however, persist over effectivity of antagonists in cheese and in at least some cases, inhibition of undesirable micro-organisms is primarily due to acid production. In a systematic study inhibitor-producing strains of lactic acid bacteria were effective against *Listeria* in Camembert cheese providing that the level of contamination was low and the organism used was responsible for the fermentation. There was no inhibitory effect if contamination occurred during ripening when the pH value of the cheese was rising. Spoilage of cottage cheese, however, is minimized by the use of inhibitor producing strains of *L. lactis* ssp. *lactis* biovar *diacetylactis* to 'cream' the dressing, a procedure which is now established practice in the US.

(e) Classification of starter cultures

Cheese starter cultures may be classified in a number of ways. The micro-organisms themselves, for example, may be classified according to optimal growth temperature. Mesophilic starters comprise *Lactococcus* and *Leuconostoc* and have an optimal growth temperature of *ca.* 30°C, while thermophilic starters comprise the more widely used *Lactobacillus* species and *Str. salivarius* ssp. *thermophilus* and have an optimal growth temperature of 40–45°C.

---

* Commercial bacteriocin preparations such as pediocin AcH™ are available but their usage is controversial. A non-bacteriocin preparation, Microgard™, produced by starter micro-organisms growing in skim milk is also available and has been shown to be effective when used as an additive to control the spoilage microflora of cottage cheese and yoghurt. No activity, however, could be demonstrated against either Gram-negative or Gram-positive pathogens in pure culture experiments (Salih, M.A. *et al.* 1990. *Journal of Dairy Science*, 73, 887–97; Motlagh, A.M. *et al.* 1992. *Journal of Food Protection*, 54, 873–8).

In either case, starters may be mixed, in which the number of strains is unknown or defined in which a known number of strains are present. The traditional mesophilic mixed cultures of northern Europe, are commonly further classified on the basis of the presence, or absence, of the 'flavour-producing' organisms *Leuconostoc* species and *L. lactis* ssp. *lactis* biovar *diacetylactis*.

Mesophilic defined cultures may consist of single, paired or multiple strains. In recent years it has been found possible to reduce the number of strains in multiple cultures from the six originally used to two or three without any adverse effect on performance.

Single or multiple strain defined thermophilic starters have also largely replaced the traditional mixed strain cultures in large-scale usage.

(f) Production of starter cultures

Traditionally, cheese factories maintained their starter cultures as liquid stocks which were sub-cultured through mother and feeder stages to provide a sufficiently large bulk inoculum. This procedure is not fully satisfactory due to a relatively high risk of contamination and of changes in starter properties resulting from the continuous sub-culturing required. It is now more common to use a culture supplied by a commercial laboratory either dried, liquid, or as a frozen concentrate. Although these cultures may be supplied as stocks and propagated through the mother and feeder stages, it is more convenient to use cultures which can be inoculated direct into the bulk tank, or direct into the cheese milk (direct-in-vat inoculation).

Problems with phage infection of starter cultures means that stringent precautions must be taken during preparation of the bulk inoculum. Specific precautions are of two main types: the use of mechanically protected systems, the common approach in the UK, Australia and New Zealand, and cultivation in phage-inhibitory media (PIM), widespread in the US. Specific precautions, however, can be effective only in the context of good practice in all aspects of starter handling and in the factory as a whole. The starter room should be situated upwind of whey handling operations and movement of personnel between the two areas prevented. Air filtration, ultraviolet light or 'fogging' with hypochlorite solution may be used as an additional means of controlling airborne phage. Bulk media for *Lactococcus* containing starter cultures must be heated at 90°C for a minimum of 30 min to inactivate lactococcal phages which are of markedly greater heat resistance than their host cells.

Table 7.4 Mechanically protected systems for production of bulk starter cultures

Lewis system	
starter tank	pressurized
media	heat treated in tank
culture transfer	from polythene bottles using two way hypodermic needles[1]
sterile barrier	hypochlorite solution
Jones system	
starter tank	unpressurized, air entering/leaving is sterilized by heat and cotton wool filtration
media	heat treated in tank
culture transfer	poured through point of entry sterilized by flame or steam

[1] In the otherwise similar Alfa Laval system filter-sterilized air is used to transfer culture.

A number of mechanically protected systems have been developed (Table 7.4) but all are variants based on the same principles. The bulk starter medium, antibiotic-free milk or reconstituted skim milk powder, is inoculated under semi- or fully aseptic conditions and protected within the bulk tank by air filtration or sterilization and by maintenance of a slight positive pressure.

The underlying rationale behind the development of PIM is the restriction of available calcium ions and the consequent limitation of phage multiplication. Phage-inhibitory media, in current use, are whey-based media in which pH control is applied internally using buffer salts, or externally using NH_4OH. Whey-based, pH controlled PIM are cheap and effective and are suitable for use with both mesophilic and thermophilic cultures. Further technological advantages are the limitation of daily variation in acid development in the bulk culture, and the production of

---

* Particular care must be taken when preparing starter cultures containing two species to ensure that the ratio between the two remains correct. Failure to maintain the correct ratio increases the likelihood of starter failure due to phage infection and also leads to specific defects in different cheese varieties. Bitterness and over-acidification in Cheddar cheese, for example has been attributed to dominance of *L. lactis* ssp. *lactis* over *L. lactis* ssp. *cremoris*. Pink discolouration of ripened Brie or Camembert has been attributed to dominance of *L. lactis* ssp. *lactis* biovar *diacetylactis* over *L. lactis* ssp. *cremoris*.

Misbalance between starter species has several possible causes including continued subculturing of a frozen, concentrated culture, a too high milk solids content in the bulk starter milk and over-heating bulk starter milk. Problems are avoided by use of direct-in-vat inoculation (Champagne, C.P. *et al.* 1992. *Food Research International*, 25, 309–16).

a starter of high cell count and prolonged activity.

Direct-in-vat inoculation reduces culture handling to a minimum, offering a convenient system with a low risk of phage infection and highly consistent starter performance. Cultures for direct-in-vat use are concentrated, separated from their metabolites and preserved. Both freeze dried cultures and cultures preserved in liquid nitrogen are available on a commercial basis, the former being more convenient and finding wider application, especially in small-scale plants. Freeze dried cultures may also be 'tailored' to produce a cheese with characteristics matched to a specific market niche.

(g) Starter failure

Starter failure occurs when growth of, and acid production by, the starter micro-organisms is significantly slower than the accepted norm. The resulting cheese is of low quality requiring downgrading and the need to 'nurse' slow vats causes considerable disruption to factory operations. Starter failure also has serious public health implications and can result in growth of enterotoxigenic strains of *Staphylococcus aureus* during fermentation and enhanced survival of *Salmonella* during maturation and storage (see pages 333–4).

Intrinsic failure resulting from the use of insufficient or inactive starter culture has been effectively eliminated from modern production, although such problems still occur occasionally where cheese is made by enthusiastic amateurs. Bacteriophage activity is now of overwhelming importance as the cause of starter failure although other extrinsic factors can be of importance under some circumstances. Bacteriophage have been isolated from all starter genera of LAB although studies have tended to concentrate on *Lactococcus*.

The phage–host relationships of LAB are the same as those of other bacteria, phages being either of the virulent (lytic) or temperate type. The two types of phage have different relationships with their hosts, the lytic cycle and lysogeny, respectively.

The lytic cycle can be the cause of total starter failure and is thus of greater industrial significance. Following infection, multiplication takes place in four sequential stages – adsorption, DNA injection, intracellular multiplication and phage release following bursting of the host cell. As many as 200 phages per cell are released after a short latent period, between infection and release, of *ca.* 60 min and infected cells may thus lose activity very quickly.

In most cases temperate phages behave like virulent and cause host cell lysis, but in a small number the phage is either integrated into the host chromosome (prophage) or is maintained as a plasmid. In these two states the phage replicates with the host cell without causing lysis. Bacterial cells containing a prophage are also immune from infection with virulent phage of the same, or closely related type (lysogenic immunity). Most lysogenic cultures contain free phage and temperate phage present in earlier starter cultures were the probable origin of many of today's virulent phage.

A starter culture may achieve equilibrium with phage present in its environment leading to a third condition, the carrier state. In this state only a proportion of host cells are infected, despite the presence of free phage and the failure to incorporate phage DNA into the host chromosome.

The presence of virulent phage is necessary for the promotion of resistance and exposure of sensitive strains to phage provides a means of selecting bacteriophage-insensitive mutants (BIMs), which are of major importance in many modern cheese making systems. Three main resistance mechanisms exist – prevention of phage adsorption, abortive infection and restriction/modification, although lysogenic immunity is a further important control mechanism.

Although phage infection is of major importance as a cause of starter failure a number of other factors can be involved. Agglutinins are present in nearly all milk and cause the aggregation of starter micro-organisms with casein micelles leading to uneven starter distribution and resulting slow acid production and a poor quality, 'grainy' curd. Problems resulting from agglutinin activity were previously only recognized in cottage cheese manufacture, but it is possible that their importance in manufacture of other cheese types has been underestimated.

Starter failure due to antibiotic residues, formerly a major problem, has been largely eliminated in many countries by milk testing schemes and financial penalties levied on producers of antibiotic-containing milk. Continuing problems result from inadequate rinsing of equipment after cleaning and the consequent presence of sanitizer residues. Sensitivity

---

* The phage carrier state appears to be common in the traditional mixed strain starters used in the Netherlands. Such cultures are still capable of fast acid production but may themselves be a source of virulent phage.

of starters to sanitizers varies greatly according to strain and species, but iodine and chlorine have little effect on starter performance under practical conditions, problems being potentially greatest with quaternary ammonium compounds. Spring or late lactation milk has enhanced activity of the lactoperoxidase system which can lead to starter failure. There is considerable variation in sensitivity and only resistant strains should be selected as starters.

7.2.2 Cheese milk

Cheese may be made from the milk of any species and while cows' milk is most commonly used in the US and Western Europe, there is increasing interest in manufacture of goats' and, to a lesser extent, sheep milk cheese. In regions where fresh milk is scarce, cheese has been successfully made from recombined anhydrous milk fat and reconstituted skim milk powder.

The chemical composition of the milk affects the nature of the final cheese and in the past seasonal variation has restricted manufacture of some types to particular months of the year. In modern large-scale manufacture, however, it is general practice to standardize milk composition to ensure consistency and maximum yield. In whole milk cheese, such as Cheddar, the sum of the casein and fat content is the principle factor affecting yield. The two main factors influencing cheddar quality, the fat on dry matter content and the moisture in non-fat solids content are largely determined by casein:fat ratio.

The fat content of the milk is important in determining the characteristics of the different varieties and cream, whole, reduced fat, or skim milk is used (Table 7.5). Homogenization increases yield by reducing fat loss in whey. It may also be applied to promote the development of lipolysis in mould-ripened varieties and to improve the texture of soft cheese. Homogenization is not used to treat milk for hard varieties such as Cheddar. The handling and storage of cheese milk prior to processing requires care. Proteolysis affects both quality of the final cheese and yield and must be minimized by storage below 5°C for a strictly limited time period.

The heat treatment of cheese milk is a contentious, indeed emotive, issue. A heat treatment of equivalent lethality to pasteurization is considered necessary to ensure the destruction of milkborne vegetative pathogens such as *Campylobacter* and *Salmonella* and has the additional advantage of favouring the growth of starter micro-organisms by

Table 7.5 Fat content of milk used for making various cheese varieties

High fat	
Stilton[1] (cream added)	up to 4.5%
Gorgonzola	*ca* 4%
Whole milk	
Cheddar and related varieties	3.5–4.0%
Edam (whole milk)	2.9–3.1%
Low fat	
Parmesan	2.7–3.0%
Edam (low fat)	1.5–1.6 or 2.5–2.6%
Skim milk	
Cottage cheese	trace

[1] Addition of cream is rare in current practice.

reducing the competitive microflora. Heat treatment also increases yield slightly due to heat-induced casein/whey protein interactions.

Heat treatment of cheese milk is, however, opposed by some manufacturers and consumers on the grounds that the organoleptic quality is reduced. The evidence for this remains somewhat anecdotal, but there is a consensus that cheese made from pasteurized milk tends to have a blander, if more consistent, flavour. The development of ripening cultures based on beneficial constituents of the raw milk microflora is a relatively new development, which goes at least some way to resolving the conflict between organoleptic quality and safety. It must be stressed, however, that in cases of doubt, safety considerations are of paramount importance.

The use of sub-pasteurization heat treatment of cheese, thermization, has been suggested as a means of limiting heat-induced changes in milk without compromising microbiological safety. Various time–temperature combinations have been proposed, but all are likely to permit the survival of *Listeria monocytogenes* and *Salmonella*.

Milk containing endospores of lactate-fermenting clostridia is associated with 'late blowing' defect in the finished cheese (see page 339). The

---

* The lactoperoxidase/thiocyanate/hydrogen peroxide system (LPS) has also been investigated for treatment of cheese milk and has been shown to be acceptable for manufacture of either hard or soft cheese. The use of LPS is of particular interest in developing countries, where refrigeration may not be available, and is also permitted in Sweden under 'emergency' conditions.

treatment of such milk presents particular problems since neither heat treatment, at an acceptable level, nor the LPS method provide effective control. Endospores may be removed by centrifugation (bactofugation), although use of this process in isolation involves the loss of *ca.* 3% of the milk. This problem may be overcome by sterilizing the centrifugate and recombining it with with the liquid milk. Centrifugation requires considerable capital expenditure and reduces plant throughput, it can only therefore be justified in large plants for treating milk with particularly high levels of clostridial endospores.

In recent years there has been a considerable amount of interest in modifying cheese milk to increase the incorporation of whey proteins into the curd. This increases yield and reduces the problems associated with handling and disposal of whey. Heating milk at temperatures above 100°C permits incorporation of whey proteins into the curd with a resulting yield increase but under normal conditions the heated milk loses its ability to be coagulated by rennet. This loss may, however, be reversed by addition of $CaCl_2$ or by acidification. Bitterness associated with acidification of milk may be overcome by reducing the quantity of rennet added but further modifications to the cheesemaking process are required. The resulting cheese is considered to be of acceptable, but not high, quality.

Concentration of milk by reverse osmosis or ultrafiltration permits undenatured whey proteins to be incorporated in the curd by entrapment. Less rennet is required to form a coagulum and this leads to economic benefit and a high potential for automation and continuous processing. Ultrafiltration is the more promising technology due to fewer problems with membrane fouling and the use of ultrafiltered milk is now well established in the manufacture of soft unripened cheese such as Quarg. Ultrafiltered milk has also been used on a small-scale in the manufacture of semi-hard cheese such as Gouda. Texture and flavour may, however, be affected and modifications to established procedures are needed to obtain optimal quality (see page 314). Attempts to make a hard cheese of satisfactory quality from ultrafiltered milk have, however, been unsuccessful.

---

* Vacuum concentrated milk has also been used for cheese making with an increase in yield and efficiency. Partial homogenization of fat occurs during concentration which leads to improved fat retention, although the major benefit lies in improved utilization of equipment. The resulting cheese is 'curdy' when young but improves as ripening proceeds.

CONTROL POINT: CHEESE MILK CCP 1

Control

Milk of good initial quality and free from antibiotics.

Milk must be correctly heat treated (see Chapter 2, page 50).

Ensure milk correctly standardized.

Additional processes, such as bactofugation and homogenization must be controlled to meet technological objectives.

Monitoring

Monitor temperature of milk throughout storage and minimize storage period.

Test incoming milk for antibiotics **before** acceptance.

Monitor time and temperature of heat treatment (see Chapter 2, page 52).

Determine fat and crude protein (see note) values by analysis and standardize on basis of crude protein:fat ratio or, preferably, casein:fat ratio (Cheddar 0.69–0.71). Standardization to be supervised by trained and experienced personnel.

Verification

Cheesemaking properties.

Examination of plant records.

Re-analysis of fat and crude protein content.

Note: The casein:fat ratio provides a better criterion for standardization, than crude protein:fat, but analysis for casein is currently difficult.

7.2.3 Manufacture of hard, semi-hard and semi-soft cheese

Hard, semi-hard and semi-soft cheese form a spectrum ranging from the very hard grating cheeses such as Parmesan, which have a moisture content as low as 26%, to semi-soft varieties such as Bleu d'Auvergne, which have a moisture content as high as 50% and which are organoleptically similar to soft varieties. A number of varieties of each type is produced but all differ from soft cheese in the degree of cutting and/or scalding the curd receives. Cutting and scalding together with acid production, nature of curd handling and ripening pattern dictate varietal differences within the hard, semi-hard and semi-soft cheeses (Table 7.6).

Cheddar is by far the most important variety of hard cheese and is produced on an extremely large-scale and on a world-wide basis (Figure 7.2). Whole milk, standardized to a casein: fat ratio of *ca.* 0.7 is normally used, although low-fat 'Cheddar-type' cheese is now available (see page 316). Starter cultures are mesophilic, usually containing *L. lactis* ssp. *cremoris* alone or in combination with *L. lactis* ssp. *lactis*. Large-scale manufacture is particularly susceptible to problems with bacteriophage and special management systems have been developed to overcome this. Initially pairs of phage-unrelated strains were used in a four day rotation, a different pair being used each day, but this system was unable to cope with the much greater problem of phage associated with the development of automated cheese making. The subsequent development of defined multiple strain starters has, however, permitted the control, but not the elimination of phage-related problems in large-scale manufacture. Development of such starters took place in New Zealand but similar cultures are now used in Australia, the US and Eire. The key to the development of multiple strain starters is the use of a simple test for predicting phage resistance. Potential starter strains are then chal-

Table 7.6 Relation between cheese variety and manufacturing technology of hard, semi-hard and semi-soft cheese

Cheddar
  moderate starter activity, scald 39–40°C, curd 'cheddared' (see text)
Caerphilly
  low, then high starter activity, scald 31–33°C, coarsely cut curd
Cheshire
  enhanced starter activity, scald 31°C, mill for free-flowing curd
Dunlop
  moderate starter activity, scald not more than 36.5°C, mill for friable curd
Double Gloucester
  low starter activity, scald 35–37°C, rubbery curd, lightly 'cheddared'

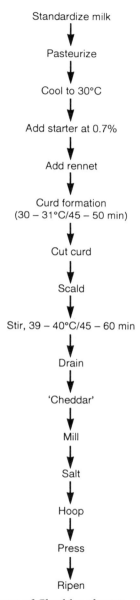

Figure 7.2 The manufacture of Cheddar cheese.

lenged, using as large a number of phage as possible. Whey from previous production cycles is included and the testing is continued over several growth cycles. Phage-resistant strains of satisfactory technical performance are placed in a panel from which those for starter use are selected. Multiple strain starters consisting of six strains used without rotation have largely been replaced by two or three strain cultures used with, or without rotation. The starter must be monitored for phage-sensitivity on a daily basis. Phage-sensitive strains are removed from use immediately and replaced with a resistant strain from the panel. The number of available resistant strains can be rapidly depleted by this practice and it is now usual, where possible, to isolate BIMs from sensitive strains and to re-introduce these to the panel. Difficulties may, however, arise in isolating BIMs, or from slow acid production and reversion to phage sensitivity.

The use of proteinase-negative strains (prt^-) of starter microorganisms offers a number of potential advantages and has been of considerable interest in recent years. Such strains exhibit a high level of phage and antibiotic resistance, improved cheese yield due to reduced casein hydrolysis, and improved flavour due to reduction in the extent of bitter peptide formation. The performance of prt^- strains has been variable in practice, Cheddar cheese, but not Gouda, having been successfully made using prt^- strains only. Use of prt^- strains requires a high level of starter culture or supplementation of milk with yeast extract. The production cycle is also lengthened and some experts consider that 10–20% of the starter population should be prt^+ to provide sufficient proteinase activity. It is now appreciated that prt^- strains differ in their proteinase profiles, some offering no advantage in terms of improved yield. Analysis before use is therefore considered essential.

Pasteurized milk is delivered to the cheesemaking vats at a temperature of 30°C. Delivery of milk at lower temperatures and heating in the vat can lead to a 'dark seam' defect. Cheese vats vary in design from small, rectangular, open vats stirred by hand, or by sweeping agitators to large, cylindrical, enclosed vats fitted with rotating agitators and cutters. Starter culture is added at a rate of 1–2% followed after 20 min by addition of rennet to produce a coagulum. In production of coloured varieties, such as Red Leicester, annatto is added to the cheese vat before the coagulum forms.

The major active ingredient of rennet, chymosin, is secreted in the

---

* An alternative system of starter management has evolved in the Netherlands and other north European countries in which phage resistant strains are selected by the continuous challenge of mixed strain starter cultures. Strains are not rotated but are used day after day.

bovine abomasum (fourth stomach), calves being traditionally the main source of supply. A shortage and the corresponding expense of calf rennet has led to the use of substitutes, including proteases from other animals and proteolytic enzymes produced by micro-organisms. Moulds are the most common source of coagulating enzymes, *Mucor miehei*, *M. pusillus* and *Endiothia parasitica* being the most common sources. Recently chymosin produced by genetically modified *Kluyveromyces lactis* has been introduced.

Calf rennet contains variable quantities of pepsin A in addition to chymosin. Rennet quality may be expressed in terms of the ratio between milk clotting and general proteolytic activity. Residual rennet proteolytic activity plays a role in ripening (see page 327), but a high level results in loss of yield and production of bitter peptides.

Substitution of fungal enzymes for rennet leads to an economically significant reduction in yield, although this may be at least partially offset by the lower cost. Yield with genetically modified chymosin is similar to that with calf rennet although there may be textural differences.

In cheeses such as Cheddar rennet, or a substitute, is responsible for coagulum formation, since at this stage acid formation is insufficient to promote gel formation. The degree of acidity is, however, of considerable importance in determining the properties of the coagulum and hence the nature and the quality of the finished cheese. Variations in the rate of acidification resulting from differences in starter activity or milk composition can lead to inconsistency of quality. Acidification of process milk with glucono-δ-lactone is permitted in France to minimize this problem.

BOX 7.1 **In his mother's milk**

One of the advantages of milk clotting proteases produced by moulds is their acceptance for use in production of cheese acceptable to vegetarians. The position is unclear with respect to chymosin produced by genetically modified micro-organisms, since the gene controlling chymosin synthesis is of animal origin. There has been much discussion of this issue amongst vegetarian groups, but the situation currently remains unresolved.

CONTROL POINT: ACIDIFICATION AND CURD FORMATION CCP 2

Control

Acidification must proceed at correct rate.

Curd formation must follow correct pattern.

Monitoring

Determine activity of starter culture before use or, for DVI cultures, obtain culture from reputable supplier.

Monitor temperature of milk in cheese vat.

Monitor development of acidity (pH value).

Obtain rennet from reputable supplier.

Addition of rennet to be supervised by experienced personnel.

Coagulum formation to be monitored by experienced personnel.

Verification

Properties of coagulum.

Inspection of plant records.

Incubation at 30°C continues until the acidity reaches 0.10–0.14% lactic acid when the coagulum is cut into cubes of *ca.* 5 mm and stirred into the whey. Cutting initiates and 'encourages' syneresis and the extent is of importance in determining varietal characteristics. At the same time care must be taken to avoid damaging the coagulum with consequent fat loss.

Following cutting, curds for Cheddar cheese are scalded (cooked) at 39–40°C, stirring continuing to ensure uniform heat distribution. Scalding enhances syneresis and whey expulsion, higher scald temperatures leading to cheese of low moisture content and firm body and lower scald temperatures to a cheese of high moisture content and soft body. For this reason, scald temperatures are also important in determining varietal characteristics.

Curds and whey are separated when acidity and curd firmness have developed sufficiently. In modern practice, curds and whey are removed from the vat at this stage, the vat being cleaned ready for immediate re-filling. The curd is then 'cheddared', a procedure which involves two basic principles, stretching and matting. Individual curd particles coalesce, drainage of whey continues and finally the curd attains the characteristic 'chicken breast texture'. Traditional cheddaring procedures involve manually piling and turning slabs of curd and in large, modern plants the procedure is mechanized to avoid this hard and expensive labour. Two approaches to automation exist; the cheddaring tower, a tall tower holding *ca.* 2 h production in which the curd mats under gravity, stretching taking place in the square sectioned lower end, and the use of flexible belts running at different speeds which alternately squeeze and stretch the curd. Such machinery is often combined with equipment for curd milling and salting. Curd texturing procedures for related varieties such as Double Gloucester and Derby differ from that of Cheddar and this, together with differences in scald temperature and acidity, accounts for differences in the final cheeses (Table 7.7).

Acidification by starter micro-organisms continues throughout scalding and cheddaring, the rate of acidification, but not growth, being highest during cheddaring. The degree of syneresis and thus the moisture content of the final cheese is affected by the rate of acidification, while the extent of acidification affects the calcium content of the curd and cheese texture, high acid, low-calcium curds such as those of Cheddar-

Table 7.7 Characteristics of Cheddar and related cheese varieties

Cheddar
  buttery but firm body with close texture; clean 'nutty' flavour
Caerphilly
  firm, springy body, flaky but short texture; mildly acid when young, becoming sharp and strong with age
Cheshire
  open textured, firm body with mild, lactic flavour
Dunlop
  close and waxy texture, mild and creamy flavour sharpening with age
Double Gloucester
  intermediate character between Cheddar and Cheshire (hybrid cheese)

type cheeses having a crumbly texture. The quantity of chymosin, but not of fungal pepsins or rennins, retained by the curd is also affected by acidity, larger quantities being retained at low pH value with a resulting higher rate of proteolysis during ripening. Control of acidification throughout the process is necessary to ensure a high quality cheese, optimal levels for cheddar being 0.15 to 0.19% lactic acid at the point of draining, 0.20–0.22% lactic acid at commencement of cheddaring and 0.60–0.80% lactic acid at the end of cheddaring. Harsh bitter flavours result if acid production is excessive and it is common practice, where paired starter strains are used, to use a 'fast' acid-producer which continues to grow during cooking with a 'slow' producer which does not grow but continues to produce acid at a slower rate. Multiple strain starters containing BIMs are highly efficient and over acidification may occur unless manufacturing procedures are modified. In varieties such as the related Cheshire, however, the use of specially selected, highly active strains of *L. lactis* ssp. *lactis* to produce a fast rate of acid production is an essential part of manufacture.

After cheddaring the curd is milled, dry salted, moulded and pressed. Traditionally cheddar cheese was pressed into circular or rectangular blocks weighing up to 20 kg. Paraffin wax was used to protect the rind and minimize moisture loss. The traditional cheese press is a manually operated screw press, but this has been largely replaced by hydraulically powered equipment operating multiple pressing heads. Such presses are usually operated in conjunction with automated moulding equipment. In very large-scale production as much as one tonne may be pressed in a single operation, the pressed cheese being cut into blocks and vacuum packed. The cheese is packed before the curd has 'set' and cooling is applied to prevent deformation of the blocks. To some extent moulds and presses are being replaced by block makers. Several types are available, the widely used tower system holding milled and salted curd under vacuum in a tower. The curd compacts under the weight of its own mass, blocks are cut at the base of the tower, and vacuum packed in a barrier film. Care must be taken to avoid distorting the blocks during handling. Under some circumstances the curd compacts relatively loosely and some block makers incorporate a pressurized section to overcome this problem.

The conversion of lactose to lactic acid should be completed, as far as possible, while starter micro-organisms remain active. A low level of residual lactose minimizes heterofermentative metabolism by non-starter lactobacilli and pediococci leading to over-production of formic acid, ethanol and acetic acid and resulting loss of cheese quality. The conversion

CONTROL POINT: DRAINING AND TEXTURIZATION CCP 2

Control

Curd to be cut at correct time.

Curd to be cut to correct extent.

Damage to curd to be avoided.

Correct heating to be applied.

Acidity development to continue at optimal rate.

Texturization processes (e.g. cheddaring) to be correctly applied.

Monitoring

Monitor acidity (pH value) to determine correct time of cutting.

Manual cutting and texturization to be carried out by experienced personnel.

Automated cutting and texturization to be monitored by experienced personnel.

Monitor heat applied on a continuous basis.

Monitor development of acidity on a continuous basis.

Verification

Properties of curd.

Inspection of plant records.

Automated equipment calibration and maintenance to be on a regular basis.

of lactose to lactic acid is completed within 24 h in the freshly pressed curd if starters remain metabolically active. This may be achieved by salting the curd to a salt-in moisture level of 4%. This procedure may, however, result in the under-production of flavour compounds and an alternative method involves rapid cooling to 10°C, a temperature at which non-starter LAB metabolize lactose homofermentatively.

Mesophilic starter cultures are also used in the manufacture of semi-hard cheese varieties. These are a diverse group, although all are made from coarsely cut and lightly scalded curds of high moisture content. Varietal differences arise largely from the extent to which acid development is permitted.

Edam and Gouda are produced on a large-scale in Holland, a considerable degree of mechanization being employed in the larger factories. These are low acid, sweet curd varieties which are prone to microbial spoilage and process milk must be of high microbiological quality. Pasteurization is invariably employed and bactofugation widely used. In addition up to 0.02% $NaNO_3$ may be added, the reduction product, $NaNO_2$ preventing the germination of bacterial endospores. Calcium chloride, to enhance coagulum formation, and colouring may also be added prior to inoculation. Starters include *L. lactis* ssp. *lactis* biovar *diacetylactis* or *Leuconostoc*.

The extent of acid production in the curds is controlled by replacing *ca.* 25% of the whey with water at a temperature of 50 to 60°C. This has the effect of both lightly scalding the curds by raising the whey temperature to *ca.* 37°C and restricting further starter activity by reducing the lactose content. This results in a high moisture content curd of pH value 5.3 to 5.4 which develops a rubbery, elastic texture in the final cheese as a consequence of the high calcium content. After whey drainage the curds are filled into moulds and lightly pressed, placed in a concentrated NaCl brine for 24–48 h and dried.

In contrast to the Dutch-type cheese, acid production in the curd is encouraged in other semi-hard varieties such as Caerphilly and Lancashire. Caerphilly is salted by brine immersion and eaten without maturation, whereas Lancashire is unusual in being made by blending an older, high acid curd (1.5% lactic acid) with a young low acid curd (0.18–0.2% lactic acid) to produce a process curd with *ca.* 0.9% lactic acid. This results in a characteristic soft, crumbly texture.

Semi-soft cheeses include the bacterial surface-ripened type such as

CONTROL POINT: MILLING, SALTING AND PRESSING CCP 2

Control

Curd to be milled to correct size.

Ensure NaCl of suitable grade and evenly distributed.

Apply correct pressure during pressing.

Ensure sufficient vacuum after packing.

Ensure seal integrity.

Monitoring

Milling and salting to be supervised by experienced personnel.

NaCl to be obtained from reputable supplier.

Pressing to be supervised by experienced personnel and operating parameters monitored on a continuous basis.

Verification

Quality of young cheese.

Inspection of plant records.

Periodic determination of vacuum within blocks.

Visual inspection of seam integrity.

Automated equipment calibration and maintenance to be on a regular basis.

Limburger and Port du Salut and the mould-ripened, blue-veined cheese such as Stilton and Roquefort. Surface-ripened cheese all share a similar technology and are usually made with mesophilic starters, although in some cases *Str. salivarius* ssp. *thermophilus* is present. Starter growth during manufacture is rapid and continues while the curd is draining in moulds. The rate of acid development determines the rate of whey drainage and thus the final moisture content of the cheese. This in turn influences the pattern of ripening and the organoleptic properties of the final cheese (see page 328).

Young surface-ripened cheeses are salted either by brine immersion or by dry surface salting followed by wiping with a brine soaked cloth. The inoculum of ripening bacteria is derived from the ripening cellar and smeared over the surface.

Lipolysis is important in flavour development during ripening of blue-vein cheese (see page 330) and for this reason high fat milk is used if possible. Roquefort cheese must be made from sheep milk, which is also widely used for other varieties in France and Spain. The fat content of cows' milk may be increased by addition of cream in the manufacture of Stilton and Gorgonzola and homogenization may be used to aid lipolysis during ripening by increasing the surface area of the fat globules. The smaller size of the globules is also important in promoting a smooth curd texture.

Mesophilic starters are used in manufacture of blue-vein cheese and include *L. lactis* ssp. *lactis* biovar *diacetylactis* or *Leuconostoc* which produce CO_2 and impart an open texture to the cheese, encouraging subsequent mould growth. The curd is of the high acid type, acidity developing slowly over a long draining period. The curd is neither cooked nor pressed but allowed to consolidate under its own weight.

Ripening involves the mould *Penicillium roquefortii* which may be added either to the cheese milk or to the curds. The use of moulds inevitably raises the question of possible mycotoxin formation and, in the past, consumer anxiety has been raised by poorly informed press reports. While there is no evidence that mycotoxins cannot be produced in mould-ripened cheese, the solution, as in all situations where mould is used in food production, lies in ensuring that commercially used strains are atoxigenic.

Thermophilic starters are used where manufacture involves the use of

---

* When manufacturing cheeses of different size it is important to obtain an optimal ratio between depth of cheese and surface diameter to permit ripening to proceed correctly. *Brevibacterium linens* is usually of greatest importance in determining flavour, but yeasts and other bacteria are also of importance.

> **BOX 7.2 Half as old as time**
>
> Mention should be made of the Norwegian blue-vein cheese Gammelost (literally 'old cheese'), which is of ancient origin and predates established cheesemaking technology. Gammelost is a skim milk, acid-set cheese and no rennet is used in manufacture, the coagulum forming as a result of acid formation by starters resembling *L. lactis*. Two main regional manufacturing methods are used, but in each case the curds are heated to 100°C. The cheese is not salted. Spores of *Penicillium* are either added to the curd, or inoculated by piercing with coated needles, but adventitiously derived moulds such as *Mucor* and *Rhizopus* are important as surface-ripening agents.

high cooking temperatures. The main groups are the very hard Italian cheeses such as Parmesan and hard Swiss-types such as Emmental. Parmesan and similar cheeses are made from low-fat milk standardized at *ca.* 2%. *Streptococcus salivarius* ssp. *thermophilus* and *Lb. delbrueckii* ssp. *bulgaricus* are used as starters and are added to milk in a 'kettle' containing sufficient milk (*ca.* 150 gallons) to make a single cheese. After a short incubation at 32–36°C rennet is added, in traditional practice a special high-lipase preparation or a rennet paste being used. The coagulum is fine cut into 3 mm pieces using a special tool, the *spino*, the curds and whey being stirred and heated in a two-stage process. During the first stage the temperature is raised to 43°C over a 15 min period and held for a further 15 min, in the second stage the temperature is raised to *ca.* 55°C over a 15–30 min period. This temperature is maintained until the curds are judged to be of sufficient firmness, the curd particles are then allowed to settle before removal from the whey in a cloth. After a short draining period in the cloth, the curds are placed in moulds, pressed overnight and then held for 3 days at 15°C during which time starter micro-organisms ferment residual lactose. The cheeses are then immersed in brine for 14 days and dried for several days at 15–20°C before maturation.

Whole milk is used in the manufacture of Swiss-type cheese such as Emmental and bactofugation may be desirable to reduce the number of clostridial spores. Starters should include *Lb. helveticus* or gal$^+$ strains of *Lb. delbrueckii* ssp. *lactis* to ferment galactose excreted by *Str. salivarius* ssp. *thermophilus*. This ensures that the cheese, after brining, is effectively sugar-free and minimizes growth of spoilage micro-organisms. *Propionibacterium* is added with the starters but is of no importance in the pre-ripening stages of manufacture. After cutting the coagulum into 3 mm pieces, the curd temperature is raised from 35 to 45°C at the rate of 1°C every 2 min and then to *ca.* 55°C at the rate of 1°C every minute. This protocol produces a cheese which is both sufficiently elastic to permit eye formation by the non-starter *Propionibacterium* and sufficiently firm to maintain shape. It is also essential that the starter micro-organisms survive cooking and are able to produce sufficient acidity in the cooling curd to mat the curd particles into a dense mass.

The more rapid cooling of the periphery of the cheese to the growth range of the starter bacteria means higher post-cooking production of lactic acid and the establishment of a concentration gradient between the edge and the centre of the cheese. After pressing the cheese undergoes a complex and lengthy brining procedure which establishes a NaCl concentration gradient. Development of *Propionibacterium* is greatest in the low NaCl and lactic acid concentrations and anaerobic environment at the centre of the cheese and this is an important factor in producing the characteristic properties of Swiss-type cheese.

The Italian Pasta filata cheeses are a miscellaneous group comprising the hard varieties Cacciocavallo and Provolone and the soft Mozzarella and whole milk Ricotta. Provolone is unusual in that hexamine, acting through its decomposition product formaldehyde, is permitted in the process milk as a preservative.

Cacciocavallo and Provolone are both made using *Lb. delbrueckii* ssp. *bulgaricus* as starter. A cooking temperature of *ca.* 48°C is used and acidity development in the curd encouraged. The pH value of the curd falls to 5.1–5.2, at which point the curd is softened in water at a temperature of 85°C. The softened curd then undergoes a kneading and stretching process and develops a very smooth and elastic texture which permits moulding by hand before hardening in cold water and brining.

Mozzarella and whole milk Ricotta cheeses are traditionally made with a mixed starter culture comprising the mesophilic *L. lactis* ssp. *lactis* and the thermophilic *Str. salivarius* ssp. *thermophilus* and *Lb. delbrueckii* ssp. *bulgaricus*. Replacement of *Lb. delbrueckii* ssp. *bulgaricus* by *Lb. helveticus* has been recommended to improve cooking qualities of Mozzarella. The curd undergoes a similar treatment to that of the hard

Pasta filata varieties, but the water content is significantly higher and the cheeses are eaten fresh without ripening.

7.2.4 Ripening of hard, semi-hard and semi-soft cheese

All types of hard, semi-hard and semi-soft cheese are ripened before consumption, the period of ripening varying from up to 2 years for the very hard grating cheese Parmesan to *ca.* 2 weeks for the semi-hard Caerphilly and semi-soft Romadour and Monterey varieties. The ripening process is summarized in Figure 7.3 and the biochemical and chemical reactions discussed in detail in pages 326–31.

With the exception of Emmental and other Swiss-type cheese and mould-ripened versions of hard cheeses such as Cheshire, hard and semi-hard varieties are ripened by enzymes derived from rennin, starter micro-organisms, non-starter micro-organisms and, in raw milk cheese, the milk itself.

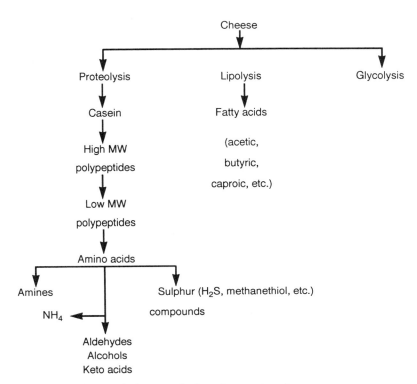

Figure 7.3 Chemical changes during cheese ripening.

For a number of years it was assumed that the non-starter LAB *Lactobacillus* and *Pediococcus*, were involved in the changes taking place during ripening, and at least equalled the starter organisms in importance. Subsequently, however, it was shown that cheese of normal organoleptic properties could be made with starter organisms alone. Non-starter LAB may play a secondary role under commercial manufacturing conditions, but this may be insignificant compared with the production of defects such as calcium lactate crystals (see page 339) or, in some varieties, production of the biogenic amines histamine and tyramine (see page 336).

Ripening of Swiss-type cheese such as Emmental follows the same basic pattern as that of other hard cheese, but is strongly influenced and modified by the presence of *Propionibacterium freudenreichii*. *Pr. freudenreichii* metabolizes lactic acid to propionic and acetic acids and CO_2, the latter being responsible for the characteristic eye formation.

Ripening of the body of smear-ripened varieties follows a similar pattern to that of Cheddar-type cheese, but at the surface non-starter organisms play a major role. *Brevibacterium linens* is generally considered to be of greatest importance and is responsible for the brownish-red coloration of the mature cheese. Salt-tolerant yeasts, *Geotrichum candidum*, *Caseobacter* and micrococci may also be present and contribute to the final properties of the cheese. *Geotrichum candidum* is of particular importance in initiating the growth of *Br. linens* by metabolizing lactate and raising the pH value.

The extent to which ripening by *Br. linens* proceeds depends on the cheese variety. In Limburger proteolysis proceeds to the extent that the consistency may resemble that of warm butter and the cheese is strongly flavoured. In contrast, growth of *Br. linens* on the Bel Paesa and Monterey varieties is limited by the relatively dry curd and the mature cheeses have a firm texture and a low flavour intensity. Over-ripening is, however, a potential problem with any surface-ripened cheese and may be controlled by lowering the temperature, physically removing surface growth or restricting the area of the cheese exposed during ripening.

Ripening of blue vein cheeses is initiated by enzymes derived from rennet and, probably, from starter bacteria, but the major changes result from the growth of the mould *P. roquefortii*. Air is admitted to the interior of the cheeses during ripening by piercing with needles and mould growth follows the lines of piercing. Roquefort and Gorgonzola are pierced after 2–3 weeks maturation and Stilton at 5–6 weeks, growth usually reaching a maximum after 1 to 3 months. Mould growth in the surface-salted varieties (Roquefort and Gorgonzola) is greatest in the mid-zone of the cheese where the NaCl concentration is an optimal 1–3%. Excessive growth of *P. roquefortii* during ripening is undesirable and with many blue vein varieties it is usual, once the mould has grown to the desired extent, to wrap the cheese in tin foil. This limits further mould growth but permits continuing enzyme activity.

Formation of a smooth, impervious rind which protects the interior from airborne contamination is an important aspect of blue-vein cheese manufacture. The surface microflora is important in the development of the rind, the composition depending on the moisture content of the cheese. Salt-tolerant, lactate-metabolizing yeasts usually appear first and by raising the pH value permit growth initiation by other microorganisms. *Brevibacterium linens* is usually dominant in cheeses of higher moisture content such as Bleu d'Auvergne and Wensleydale, but in drier varieties such as Stilton a varied microflora containing yeasts, *Lactobacillus*, *Caseobacter*, *Micrococcus* and, occasionally, *Bacillus* species develops.

A traditional blue-vein variant of the hard English variety Cheshire is produced and attempts have been made to develop similar variants of other hard varieties. In such cheese *P. roquefortii* develops to a lesser extent and has less involvement in ripening than in semi-soft blue vein cheese.

Ripening of cheese involves storage under conditions of controlled temperature and, unless the cheese is sealed into an impermeable film, humidity. Mechanical refrigeration is normally required although some cheese is still ripened in naturally cooled cellars and in limestone caverns. The temperature used is often a compromise between the need to ripen the cheese as rapidly as possible and the need to prevent development of atypical flavours and the growth of undesirable microorganisms. Ripening temperatures are usually in the range 10–15°C according to variety and the relative humidity controlled at 85–90%. The first stage of ripening of Emmental, however, is at 18–22°C to encourage initiation of growth of *Pr. freudenreichii*, while the relative humidity during the growth of *P. roquefortii* on blue-vein cheese is as high as 96%.

The length and extent of ripening depends primarily on the cheese variety. With some varieties, including Cheddar, Cheshire and Parmesan, the extent of ripening is varied according to the preferences of different groups of consumers and the end use. Parmesan, for example, may be sold as a natural cheese after *ca*. 1 year ripening, or matured for a further 2–4 years and used as a grating cheese.

The ripening of cheese is an expensive procedure and there is continuing commercial pressure towards reducing costs by accelerating the ripening process. Interest in accelerated ripening is greatest in the case of mass produced, hard varieties such as Cheddar, although both semi-soft and soft blue-vein varieties have received some attention.

Accelerated ripening is most simply obtained by increasing the maturation temperature but this potentially leads to the growth of pathogens and the development of atypical flavours. Various other approaches have been taken, including procedures involving the addition of enzymes related to ripening. This approach is currently the most advanced and 'enzyme-modified' cheese is in commercial production. In some countries, however, legal restrictions apply. Proteinases and/or peptidases from a number of sources have been used for accelerated ripening of hard cheese. These include mammalian tissues, the moulds *Aspergillus* and *Penicillium* and the bacteria *Bacillus subtilis*, *Lb. casei*, *Micrococcus* species and *Pseudomonas fluorescens*. Earlier problems of flavour and textural defects have been at least partly overcome by combining bacterial or fungal endopeptidases with exopeptidases which degrade the small bitter peptides. Further difficulties arise with the means of introducing enzymes into the ripening cheese. Microencapsulation, to separate enzymes and substrates until the cheese is made, offers the most satisfactory means of introducing enzymes. Microencapsulation may be obtained in a number of ways including liposomes, artificial vesicles with an aqueous, enzyme-containing core surrounded by concentric layers of lipid lamellae. Up to 90% of liposomes added to

---

BOX 7.3 **Ripeness is all**

Ripening continues, to a greater or lesser extent, to the point of consumption and the continued growth of large-scale retailing and the consequent need for lengthy distribution chains and extended shelf lives means that cheese may leave the ripening room in a relatively immature state. Quality problems may result from the ripening pattern being disturbed by poor temperature control during distribution and display.

milk are retained intact in the curd matrix at the end of cheddaring, but are disrupted in the maturing cheese.

The use of elevated levels of starter micro-organisms is a potentially attractive means of accelerating ripening. There are, however, inherent problems caused by excess acid production and while these may be overcome by in-vat neutralization with NaOH, this is seen as being technically crude. A more sophisticated approach is heat-shocking of the starter micro-organisms, which inactivates acid-producing capacity while retaining proteinase and peptidase activity. Heat-shocking is, however, an expensive procedure and the organoleptic quality of cheese produced using heat-shocked starters highly variable. A more satisfactory approach may be to use lac$^-$ strains of starters such as *Lactococcus* as a means of obtaining the necessary cell numbers without excess acid production.

7.2.5 Manufacture of soft cheese

Soft cheeses comprise a large group characterized by a high moisture content. Curds receive little, or no, cutting or scalding and whey is drained by gravity without pressing. Soft cheese may be either acid-set, in which the coagulum results entirely from acidification, or renneted.

Acid-set cheeses are technologically simple and are typified by lactic cheese. Lactic cheese is usually made with skim milk, although milk with a fat content as high as 25% can be used. Mesophilic starter cultures, usually containing *L. lactis* ssp. *lactis* or ssp. *cremoris* are used to produce an acid coagulum which may be lightly cut to assist drainage. Whey is drained by suspending the set curd in a cloth, when sufficiently dry the resulting cheese is salted and packaged. Herbs, etc., may be added at salting and the curd may be homogenized to impart a smooth texture.

White brined cheese is a special type of acid-set cheese common in eastern Europe and the Middle East. Traditional varieties are made using sheep or goats' milk but cows' milk is increasingly used. Mesophilic starter cultures comprising *L. lactis* ssp. *lactis* and ssp. *cremoris*, or thermophilic yoghurt starters comprising *Str. salivarius* ssp. *thermo-*

---

* An alternative approach to accelerated ripening involves the use of butterfat microcapsules containing extracts of *Br. linens*. Two capsules are used, the first containing cell extract and methionine to produce methanethiol, the second cell extract and cysteine to produce H_2S, which stabilizes the methanethiol.

CONTROL POINT: RIPENING CCP 2

Control

Temperature and humidity must permit normal ripening patterns.

Ripening should be for the correct time period.

Additional procedures such as piercing and surface smearing must be carried out correctly.

Monitoring

Monitor temperature and humidity in ripening store (where artificially controlled).

Apply formal stock management system.

Inspection of cheeses during ripening.

Piercing, smearing, etc., to be carried out by experienced personnel.

Verification

Quality of end-product.

Inspection of plant records.

---

philus and *Lb. delbrueckii* ssp. *bulgaricus* produce the coagulum which is drained in special moulds or bags. The curd is roughly cut into *ca.* 10cm cubes and salted by immersion in brine for *ca.* 12 h at 12 to 14°C. The salted curd is packed in a brine made from whey acidified to 0.36% lactic acid with 10–12% NaCl and matured for *ca.* 30 days.

Feta is a Greek brined cheese which is becoming increasingly popular in western Europe. Feta differs from other brined cheeses in being dry-salted, a very coarse salt being used for 24–48 h depending on ambient temperature. The young cheese is ripened for a few days before packing in brine.

Direct acidification is used in the manufacture of Queso blanco cheese and similar varieties. Traditionally lemon juice is used to set the coagulum, but citric and other acids may also be used. Methods of using direct acidification in place of starter cultures in production of some other varieties including Mozzarella and Cottage cheese have been devised, while white vinegar is used as a secondary acidulant in manufacture of the Italian Impastata. The curd of directly acidified cheese tends to have an open structure, often requiring the use of stabilizers and, in the case of cottage cheese, imparting a soft texture disliked by consumers. Yield, however, is higher than when starter cultures are used.

Whey cheese is made from the acid whey by-product of hard cheese manufacture. Whole or skim milk is added at levels of up to 10% to improve yield and the whey may be partially neutralized. The curd differs from that of other cheese in consisting primarily of whey proteins precipitated by heating for several hours at up to 90°C. The precipitated protein becomes enmeshed with fat globules and rises to the surface where it is removed by skimmers, packed into moulds and drained. Some varieties are lightly pressed.

In the manufacture of most varieties of renneted soft cheese, addition of rennet is made when the acidity is higher than in the harder varieties and thus the acid conditions play a more important part in formation of the coagulum. The simplest type of renneted soft cheese is unripened and resembles acid-set cheese such as lactic cheese with respect to organoleptic properties and short shelf life.

Quarg is a skim milk cheese production of which is now characterized by a very high level of automation. The starter culture (*L. lactis* ssp. *lactis* and ssp. *cremoris*) imparts an initial acidity of 0.7–0.8% lactic acid, rennet is added and the milk incubated at 30°C for 3–6 h in the short-set method and 16 h in the long-set. The coagulum is cut once, stirred and the whey separated either by drainage or by specially designed centrifugal separators. The curds are then blended, salt and, where appropriate, flavouring added and pumped through a continuous cooler before packaging.

Cottage cheese is highly popular as part of a 'healthful diet' and also lends itself to combination with a remarkably wide range of other foodstuffs (see page 315). The cheese differs from other soft cheese, however, in that the curd particles are separate and retain their own identity.

Most cottage cheese is made from skim milk which may be fortified with skim milk powder. Two basic manufacturing methods are used, the commonly used short-set in which 4–5% starter (*L. lactis* ssp. *lactis*) is added to milk at *ca.* 32°C to form a coagulum in 4–5 h, or the long-set in which milk is inoculated with 1% starter and incubated at *ca.* 21°C for 12–13 h. The coagulum is thus effectively acid-set although a small quantity of rennet is added to long-set and, in some factories, to short-set to firm the coagulum and assist formation of correctly textured curd.

The coagulum is cooked at temperatures between 53 and 65°C depending on the required curd firmness before draining and washing. Contact with cold water must be avoided immediately after cooking and most processes involve three separate washes at temperatures of 24–25, 10 and 2–4°C.

A number of types of cottage cheese exist, that produced in the UK being creamed with a separately prepared dressing. The dressing is prepared from cream fermented with *L. lactis* ssp. *lactis* biovar *diacetylactis* and *Leuc. mesenteroides* ssp. *cremoris*. The dressing is salted and may include added skim milk powder, to increase viscosity and the total solids content of the whole product, and stabilizers such as tragacanth and carrageenan. Sorbic acid, or other preservative, if permitted, may also be added.

Fromage frais is the generic name for a family of cheeses which include Boudon, Demi-Sel, Gervais and Neufchatel. Varietal differences lie in fat content, NaCl content, size, shape and packaging. Neufchatel undergoes a slight mould-ripening. The role of acid production in setting the coagulum is very limited and either no starter culture, or only a few drops, is added. Setting times may be as long as 48 h. In several varieties the curd is homogenized before packing by passing between granite rollers.

Ripened varieties of renneted soft cheese are made using non-starter micro-organisms as ripening agents. The best known are the surface mould-ripened types such as Camembert and Brie. Cows' milk is most common, but ewes' milk and goats' milk is also used in manufacture of varieties such as Chabris.

A great many varieties of surface mould-ripened cheese are produced, but most are of similar nature and share a similar basic technology typified by Camembert, the most important French variety.

Camembert is made from whole milk with species of *Lactococcus* as starters. Milk is inoculated with 1.5–2.5% starter culture and incubated for 1 h at 25–30°C before addition of rennet. After a further 70–75 min incubation the curd is ladled into perforated, cylindrical moulds and allowed to drain for *ca.* 6 h. The curd shrinks during drainage to about 50% of the original volume and the acidity rises to 0.6–0.7% lactic acid. The cheeses are then turned and drainage continues overnight at a minimum temperature of 22°C, the moisture content being *ca.* 60% when moulds are removed.

Control of temperature and humidity in the cheese room is an essential part of successful manufacture of Camembert and related cheeses, a balance being necessary to ensure adequate drainage without excessive acid production. Varietal differences may arise at this stage due to differing degrees of drainage. Brie, for example, is of higher moisture content than Camembert and ripens more rapidly.

Camembert is dry-salted at 24 h and held for a further 24 h at *ca.* 20°C. This is an important stage with respect to ripening, the dry surface of relatively high NaCl concentration inhibiting the growth of undesirable bacteria. The cheese is then sprayed with a suspension of spores of *Penicillium camembertii*, held for a further 24 h and transferred to the ripening room.

In addition to surface mould-ripened soft cheese, a bacterial smear-ripened variety Romadour is produced in Austria and south Germany. Romadour is effectively a high moisture variant of Limburger, made by a similar process and ripened by *Br. linens*.

7.2.6 Ripening of soft cheese

The ripening of soft cheese involves the same basic processes as that of harder varieties (Figure 7.3, page 303). However the use of non-starter micro-organisms as ripening agents is of greater relative importance.

---

* Scanning electron and light microscopy has been employed to illustrate in detail the microbial succession in St Nectare cheese. This cheese, which is similar to Brie, originated in the Auvergne region of France, but the samples studied were made at the Benedictine abbey of Regina Laudis, Bethlehem, Connecticut. The cheese is made from unpasteurized milk, no starter culture is added and the ripening microflora is entirely adventitious. The first two days of a 60 day ripening period were dominated by *Str. cremoris* and budding yeasts, primarily *Torulopsis* and *Debaryomyces*. *Geotrichum candidum* appeared shortly after the initial phase followed by zygomycetes of *Mucor* at 4 days. At day 20 the deuteromycete *Tricothecium roseen* was detected and from that point onwards *Brevibacterium* and *Arthrobacter* were seen amongst the fungal hyphae and yeast cells (Morcellino, N. and Benson, D.R. 1992. *Applied and Environmental Microbiology*, **58**, 3448–54).

Growth of *P. camembertii* during ripening is usually preceded by film yeasts and *Geotrichum*. There is controversy over the role of these organisms in creating conditions favourable for *P. camembertii*, but mould growth does not usually develop significantly until the pH value rises as a result of lactate metabolism by yeasts. Once growth of *P. camembertii* has been initiated, spread is rapid and development is maximum after 10–12 days. Shortly after softening of the cheese body becomes apparent. At this stage the cheese is wrapped and boxed.

The pH value rises during the later stages of ripening and is an important factor in the softening of the body of the cheese. The rise in pH value also permits the growth of non-acid tolerant micro-organisms derived adventitiosly during cheese making or, in the case of raw milk cheese, from the process milk. Such micro-organisms can make either a positive, or negative, contribution to flavour. In recent years, increasing use of pasteurized milk, together with improved general standards of hygiene, have meant that the numbers of adventitious micro-organisms are lower. It has been alleged that this leads to a bland cheese, lacking in some flavour notes. Recently, ripening cultures, representative of the adventitious microflora have become commercially available. These contain various mixtures of micro-organisms including species of *Micrococcus* and yeast. *Brevibacterium linens*, derived from the ripening environment, or added ripening culture, is a normal part of the microflora of mould ripened soft cheese, development being most extensive on the higher moisture types such as Brie. Excessive growth of *Br. linens* is, however, undesirable. Control of temperature and humidity is of particular importance during the first 10–12 days of ripening when *P. camembertii* is actively growing. The temperature is controlled at 11–14°C and the relative humidity at 85–90%. Ripening is delayed by low temperature and humidity and under extreme conditions the cheese may merely dehydrate without ripening being completed. At high temperatures and humidity the surface is likely to be wet and growth of both desirable and undesirable micro-organisms excessive. As with harder varieties the demands of supermarket retailing can cause problems, especially if the cheese enters a low-temperature distribution chain (temperatures below 4°C) before maximum growth of *P. camembertii* has occurred.

The rise in pH value is also important with respect to the growth of potential pathogens. Extensive growth of both diarrhoeagenic strains of *Escherichia coli* and *Listeria monocytogenes* can occur, especially in the crust of ripened soft cheese of high pH value.

7.2.7 Non-conventional cheese making techniques

Traditional cheese making involves a series of labour-intensive operations and mechanization is of considerable economic benefit. Mechanization has been applied to many aspects of the process from starter inoculation to handling of the mature cheese in the warehouse but essentially mimics manual procedures. Curd formation, however, largely remains a batch process and is seen as an obstacle to continuous cheese making and full automation. The initial approach to this problem involved the use, in soft cheese making, of a number of small vats which discharged curd onto moving belts (Alpma Fromat™ system). In a subsequent system, the Alpma Coagulator™, the curd was formed in a trough-shaped moving belt, divided into sections to act as a series of vats. Neither of these two systems represented true continuous curd making and in each case the mechanism of coagulum formation was the same as that of conventional manufacture.

A number of methods have been developed to exploit the Berridge cold renneting principle. In the Hutin–Stenne system concentrated milk is ripened with starter, cooled and renneted. The renneted milk is then mixed with hot water under turbulent flow conditions and forms a curd immediately. The curd particles formed in this way are fine and are passed through tubes to agglutinate into larger particles before drying-off the curd and moulding. This forms the basis of the APV Paracurd™, which is used commercially in the manufacture of some soft cheese varieties.

An alternative system, the Dutch NIZO system was originally designed for continuous production of Edam and Gouda cheese. Cold milk is inoculated with starter and rennet and held for up to 5 h at 2°C. The milk is then heated to 30°C in a continuous heat exchanger and the curd formed in a vertical separator. A rotating knife presses the curds onto a rotary drying screen and a curd washing stage may be introduced at this point. The curds are compacted in vertical tubes with perforated walls and filled into moulds.

Although the development of the Hutin–Stenne and NIZO systems aroused much interest, their further development, especially with regard to soft and semi-soft varieties, has been over-shadowed by the use of ultrafiltered milk.

Ultrafiltered milk offers a number of advantages (see page 288) which may be exploited in either conventional or non-conventional cheese making. The most common procedure involves preparing a precheese of a composition approximating to that of the final cheese. This comprises ultrafiltered milk together with starter, added either before or after ultrafiltration, and rennet. The degree of concentration of milk varies according to the variety of cheese, but four to fivefold is common. The lactose and mineral content may be reduced by diafiltration. Ultrafiltered milk is highly buffered and a higher level of addition of starter culture is required, but significantly less rennet is needed and clotting time decreases with increasing milk concentration.

No cutting or draining of the curd is required for soft cheese such as Quarg and cottage cheese, but some degree of syneresis occurs with semi-hard varieties such as Edam. The limited extent of syneresis may, however, result in texture differences between conventional cheese and that made with ultrafiltered milk.

Curd for Cheddar and other hard cheese may be made using ultrafiltered milk although this application is limited by the poor ripening properties of the cheese. A system for the continuous manufacture of cheddar cheese, the APV Sirocurd™, has been developed and is illustrative of the use of ultrafiltered milk in continuous cheese making. In the Sirocurd™ system milk is pasteurized and concentrated fivefold, it being important to minimize shear damage at this stage. Approximately 10% of the retentate is set aside to prepare a starter culture for subsequent production, while the remainder of the retentate is blended with starter and rennet in a recirculating loop device that feeds the precheese sequentially into static coagulating cylinders. The precheese remains at rest until coagulation is complete and is then expelled by incoming, uncoagulated precheese. The coagulum is then passed through a cutting grid into rotating drums where scalding and syneresis occurs. Movement is designed to simulate cheddaring. Approximately 8% of volume is expelled as whey during syneresis, the process being complete in 1 h. This permits only one generation of starter growth and a large inoculum is required.

Heat treatment and acidification of milk has also been investigated for cheese making and improves the yield. Potential advantages are less than for ultrafiltration but capital investment in filtration plant is avoided. Optimal conditions are a heat treatment at 90°C for 1 min and adjust-

---

* A commercial process, the Centriwhey™ system has been developed. In this system proteins are precipitated out of heated, acidified milk, concentrated by centrifugation and then added back to the process milk.

ment of the pH value to 6.2. Problems associated with bitterness may be overcome by reducing the rennet addition to 90% of conventional, but it is also necessary to extend the length of post-scald stirring from 60 to 120 min to reduce the moisture content sufficiently.

The manufacture of cheese from recombined milk, consisting of reconstituted skim milk powder and anhydrous milk fat, is of interest in countries where insufficient fresh milk is available. Cheese of satisfactory quality may be made, although textural characteristics may change. Milk powder should be subjected to only minimum heat treatment before drying, or powder prepared from ultrafiltered milk should be used.

7.2.8 Value-added cheese

The addition of herbs, chopped nuts or seeds such as caraway to the curds of lightly flavoured soft cheeses is well established practice, while the sage-containing variant of Derby is a traditional hard variety. In recent years the range of additions made to cheese has increased enormously. This may be illustrated by cottage cheese which is combined not only with traditional additives such as chives but such materials as prawns, ham and pineapple and salmon and cucumber. Additional stabilizers as well as colouring are usually necessary to enhance acceptability. A wide range of fruit-flavoured soft cheese based on Quarg and Fromage frais is also available.

The more strongly flavoured semi-hard and hard cheese varieties are, in general, less suitable for development of value-added variants. Cheddar and related cheeses may, however, be supplemented with various materials including chopped walnuts and pickles. Savoury or sour flavours have generally been thought most suitable as a complement to hard cheese, but recently a range of Stilton cheese with sweet-flavoured additives such as ginger has been introduced.

Cheese is a traditional accompaniment of alcoholic beverages and a natural extension of this association has been to combine the two in a single product. Both hard and soft types are involved and may be blended with a range of beverages including beer, whisky, port wine (or port wine colouring) and liqueurs.

Value-added cheeses have also been developed by combining two, or more varieties into a single cheese. Manufacture involves grinding each component and moulding in layers to produce the 'new' cheese. A similar principle is used in manufacture of cheese 'gateau' which has alternate layers of cheese and nuts or fruit.

Smoking has been used as a means of preserving some varieties of cheese, including some types of Provolone, for many years. More recently smoking, usually employing liquid smoke extracts, has been adopted as a means of 'adding value' both to traditional varieties, such as Cheddar, and to new 'designer' varieties. The search for novelty continues and, during 1992, a Scottish cheese was introduced which was smoked using the wood of old sherry barrels, previously used to store maturing whisky.

7.2.9 Nutritionally modified cheese

Although low-fat soft cheese varieties such as cottage cheese are an accepted part of a 'healthful' diet, there is a significant demand for hard cheeses of lower fat content. In the UK 'reduced fat' cheeses are legally defined as containing not more than 75% of the equivalent weight when no claim is made as to fat content, while 'low fat' contain not more than 50% of the equivalent weight when no claim is made as to fat content. Similar legislation exists in many other countries.

Attempts to make lower fat content hard cheeses using unmodified technology results in a final product of over-firm body which lacks the characteristic ripened flavour of the full fat equivalent and has bitter taste characteristics. Problems become greater as the fat content is reduced. The over-firm body results from changes in curd structure and texture may be improved by increasing the water content of the cheese within the limits imposed by legislation. This can be achieved by coarse cutting of the curd and lowering the temperature during scalding and stirring to 37°C in the case of reduced-fat 'Cheddar-type' cheese and to 35°C in the case of low-fat. Stirring should be reduced to a minimum and

---

BOX 7.4 To dream of cheese – toasted, mostly

In recent years, hot-eating cheese snacks have become a major growth area, especially in continental Europe. Such products are technically simple, basically consisting of a Mozarella-type cheese enrobed with breadcrumbs. In some cases a more healthful image is imparted by using a 'cheese' made from skim milk and vegetable oil.

the whey drained off immediately. The texture of lower fat cheese is also improved by the incorporation of undenatured whey proteins and thus use of ultrafiltered milk is beneficial. Further improvement may be obtained by use of acidification and diafiltration to remove calcium and reduce lactose levels with resultant decrease in body firmness and production of a desirable pH value.

Bitterness results primarily from a concentration of bitter peptides into the aqueous phase where sensory impact is greater. Peptide formation may be reduced by using starter cultures of low endopeptidase activity and restricting the quantity of rennet added and by ensuring a relatively high pH value in the curds at the draining and salting stages. Alternatively bitter peptides may be degraded by adding, at salting, *L. lactis* strains of high aminopeptidase activity or a debittering aminopeptidase preparation such as Accelase™.

Failure of lower fat hard cheese to develop typical flavour results directly from the low level of fat-derived flavour compounds and indirectly from the change in the pattern of flavour release in the mouth. The problem may be partially overcome by addition of a selected secondary microflora, but this may result in atypical characteristics such as a 'Swiss cheese-flavour' in Cheddar cheese. Alternatively enzymes may be added to enhance natural ripening.

At least some of the dietary objectives of low and reduced fat cheese may be met by substituting dairy fat with vegetable oils to produce a cheese of full fat content, but a high polyunsaturated : saturated fatty acid ratio. Several oils including soy and sunflower are suitable, but in most countries such 'filled' products cannot legally be called cheese.

Concern over dietary cholesterol has led to the development of a number of 'cholesterol-free' dairy products. Cheese made with 'cholesterol-free' milk suffers problems of soft curd and slow whey drainage, resulting in a low quality product of high moisture content and poor flavour and an alternative approach is to substitute buttermilk for part of the cheese milk. Buttermilk has been shown to lower cholesterol

---

* Fat-replacers are finding widespread application in the dairy industry. Cheese spreads have been successfully made in which dairy fat has been replaced with Stellar™, a fat-replacer made from modified corn starch, and a 'Cheddar-type cheese' has been successfully made using Simplesse™. Simplesse, however, is made from whey proteins, addition of which is expressly forbidden in the UK *Cheese Regulations* and products containing this fat-replacer cannot be described as 'cheese'.

level in rats and to nullify the hypercholesteraemic effect associated with cheese consumption. A similar effect in humans is yet to be proven.

Cottage cheese is an accepted part of a 'healthful' diet and it is not usually appreciated that the calcium content is low as a consequence of the solubility of Ca^{2+} ions in the low pH value whey. Calcium is of major importance in the prevention of osteoporosis and enrichment of low-calcium foods is considered to be a safer means of increasing calcium uptake than dietary supplements. Addition of calcium salts to cottage cheese leads to an unacceptable degree of bitterness, but this difficulty can be overcome by the co-addition of hydrocolloids such as guar gum.

Attempts to produce a low-sodium cheese by simply reducing the quantity of NaCl added during manufacture have been unsuccessful and result in a cheese of poor body which is prone to microbiological spoilage during ripening. Equally attempts to replace NaCl with 'salt substitutes' such as KCl have met with, at best, limited success. Suitable low Na^+ cheese has been made by blending a cheese base made from directly acidified milk (*cf.* medium-chain-triacylglycerol cheese) with a ripened cheese of normal Na^+ content.

7.2.10 Processed cheese and cheese spread

Processed cheese is usually made from a mixture of varieties of natural cheese, although a single variety may be used. The mix constituents are chosen to give the desired flavour and colour, and are then blended with emulsifying salts, usually di- or trisodium citrate, or tetrapotassium diphosphate and water. The mix is heated to 80–85°C for 5–8 min in a batch cooker, or extrusion cooked in an equivalent process. The cheese is then formed into portions or slices before packaging.

Cheese spreads are manufactured using similar principles but the water content is higher and additional dairy ingredients such as whey or skim milk powder are present. Stabilizers such as tragacanth may be added to

---

* In recent years medium-chain-triacylglycerol fat has become of considerable importance in the treatment of malabsorption syndromes including chyluria, steatorrhea, hyperprotolipidaemia and Whipple's disease. Medium-chain-triacylglycerols may be incorporated into the diet but conventional means often result in feelings of dissatisfaction by the patient. Cheese is seen as a more satisfactory means of introducing medium-chain-triacylglycerols and a suitable product may be made by blending a flavoured cheese base, prepared by direct acidification with HCl and heating to 70°C, with conventional cheese and medium-chain-triacylglycerol oil.

prevent separation during storage. A higher level of heat treatment is necessary for cheese spread, a batch process of 88–98°C for 8–15 min, or an equivalent extrusion process, being normal. The spread is then filled directly into tubes or containers.

Endospores survive the cooking process used for both processed cheese and cheese spread. Nisin or, less effectively, sorbic acid is added to control species of *Clostridium* including *Cl. botulinum*.

Both processed cheese and cheese spread have traditionally been combined with other foods such as chopped ham, pickles, etc. Some types are smoked after cooking.

7.2.11 Cheese analogues

Cheese analogues may be made from non-dairy, part-dairy or all-dairy ingredients (Table 7.8). With the exception of some non-dairy 'cheese', produced for vegans, cheese analogues are not usually considered to be consumer products, but are used in manufacturing. Cheese analogues are well established in the US, the major markets being for pizza toppings and school lunches. In most countries the labelling of cheese analogues requires a clear distinction to be made from cheese, the designation 'cheese product', being acceptable in the UK, for example.

Most cheese analogues are made by a blending procedure, similar to that used for conventional processed cheese. Varying the ingredients and the process conditions permits the characteristics to be tailored for a

Table 7.8 Typical ingredients of cheese analogues

Non-dairy
  soya oil
  soya protein
  artificial flavour
Part-dairy
  casein and/or caseinates
  soya oil
  enzyme-modified cheese and/or flavour
All-dairy
  casein and/or caseinates
  butter oil
  enzyme-modified cheese
  cheese

Note: All types contain NaCl, colour and stabilizer.

specific purpose. Pizza manufacture, for example, requires a cheese that shreds well, permits mechanical handling after shredding and has good melt properties, but flavour is unimportant. A typical process for a part-dairy cheese involves heating the soya oil to *ca.* 70°C, adding the stabilizer and blending in water to create an emulsion. The temperature is maintained at *ca.* 70°C and cheese, enzyme-modified cheese, salt and colour blended in. The pH value is then lowered by addition of acid, this stage being important in development of correct texture. The product is moulded while hot and cooled to 5°C.

An alternative process has been developed in the US for the manufacture of 'cheesebase', a generic name for a range of cheese products designed for ingredient use. Manufacture involves membrane filtration of milk and fermentation, using special starter strains, of the retentate. Acid coagulation is prevented by adjusting the ionic strength with NaCl. The fermented retentate is heated, the extent of the heat treatment determining the melting properties of the final cheesebase. The moisture content is then reduced by evaporation, under vacuum, in a scraped-surface heat exchanger. The final moisture content varies according to end-use, but is usually 35–40%.

Cheesebase is a very high viscosity paste and totally lacks the body and texture of conventional cheese. It is also very mild in flavour, although the addition of a ripening microflora after heating has been investigated. Theoretically cheesebase can be tailored to any specification and the ability to manufacture products of different melting point is a major advantage in the fast-foods market. Manufacture of cheesebase is also an efficient means of utilizing milk, yield being 16–18% higher than that of conventional cheese.

7.2.12 End-product testing

The importance of end-product testing varies according to the type of cheese. Chemical analysis for verification of composition is required, but in the case of ripened cheese this tends to be secondary to organoleptic assessment of quality. Formal grading schemes exist for varieties such as Cheddar, trained graders normally being used. Grading schemes based on compositional parameters have been devised, but success has been limited.

It is not usual practice to carry out microbiological examinations of ripened cheese on a routine basis, although examinations are required where doubts concerning safety exist (see page 341). Routine microbiological analysis of the end-product is, however, employed with unripened soft cheese to verify hygiene standards during manufacture and as a means of predicting a satisfactory storage life.

Cheese varieties such as Mozzarella are primarily used in cooking and in-use tests must be applied to determine melting, stretching and browning properties.

7.3 CHEMISTRY

7.3.1 Nutritional status of cheese

Cheese is recognized as being of dietary importance as a concentrated source of protein and, in many cases, of fat. Whey proteins are lost during cheese manufacture and the protein present in cheese is virtually entirely derived from casein. Despite this essential amino acids are usually present.

The fat content of cheese varies according to the fat content of the cheese milk and the variety of cheese. Whole milk cheeses of high fat content, such as Cheddar and Stilton contain 45–50% fat, while Edam and similar cheeses contain *ca.* 40%. Such cheeses are valuable fat sources for persons requiring a high energy diet. In contrast skim-milk cheeses have a very low fat content and often form part of a calorie-controlled diet. In common with all full-fat dairy products, the cholesterol content of many types of cheese is high, cheddar cheese containing *ca.* 70 mg/100 g.

Cheese is also a good source of vitamins and minerals, although vitamin C is lost during manufacture. Under some dietary circumstances, cheese is of particular importance as a source of minerals, especially calcium, iron and phosphorous.

With the exception of some very soft, unripened types, the lactose content of cheese is low and the product is suitable for consumption by lactase-deficient persons.

7.3.2 Chemical changes during curd formation

Conversion of milk from a fluid to a gel (coagulation) is a basic step common to all types of cheese. Gel formation is a consequence of protein destabilization and may be brought about either by acid

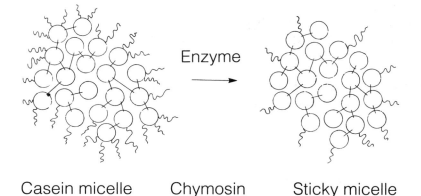

Figure 7.4 Destabilization of casein micelles by 'cutting the hairs'.

proteinases such as chymosin, the active component of rennet, quiescent acidification to a pH value close to the isoelectric point of the proteins, or by a combination of acidification and heating.

(a) Action of rennet and other milk clotting enzymes

Rennet coagulation involves two distinct stages, a proteolytic stage in which the casein micelle is destabilized by hydrolysis of κ-casein to yield para-κ-casein micelles, and a secondary, calcium-mediated, stage in which paracasein micelles undergo limited aggregation. The secondary stage requires quiescent conditions and a temperature in excess of 20°C.

Hydrolysis of κ-casein primarily involves cleavage of the peptide bond, Phe_{105}–Met_{106}, which is uniquely sensitive to hydrolysis by acid proteinases. This cleavage yields a para-κ-casein, common to all caseins and a macropeptide unique to each κ-casein component.

Although it is accepted that most, if not all, coagulating enzymes hydrolyse κ-casein at, or near, the Phe–Met peptide bond, it seems likely that there is some variation with respect to specificity and the extent of proteolysis which occurs during coagulation. Microbial proteases, especially those produced by *M. miehei*, are much less specific than chymosin and effect extensive non-specific hydrolysis of both κ-casein and para-κ-casein. This can adversely affect the quality

---

* The term 'macropeptide' is preferred to the earlier term 'glycomacropeptide', because some macropeptides contain no carbohydrate. Usage of the term 'glycomacropeptide' persists in some publications and can lead to confusion.

Table 7.9 Physico-chemical changes resulting from hydrolysis of κ-casein

1. Liberation of the highly charged, macropeptide C-terminal fraction of κ-casein from the surface of the micelle
2. Reduction in the charge at the micelle surface
3. Decrease in micellar solvation, resulting from a decrease in intermicellar repulsion and casein–water interaction, and an increase in intermicellar attraction
4. Increased sensitivity of the para-casein micelles to aggregation

---

of the finished cheese and restrict the application of microbial proteases.

A number of physico-chemical changes result directly from hydrolysis of κ-casein (Table 7.9). The most significant of these in the short-term is the liberation of the C-terminal part of κ-casein. This process, 'cutting the hairs' decreases the volume fraction and lowers stability, primarily through a reduction in steric repulsion (Figure 7.4). The paracasein micelles consequently show a concomitant mutual attraction. Viscosity of the milk falls to *ca.* 95% that of the starting milk, but then increases significantly even before any aggregation is visible, or present.

Aggregation results from intermicellar cross-linking *via* calcium binding to serine-phosphate groups and commences when *ca.* 86% of total κ-casein has been hydrolysed. Individual micelles, however, are only able to participate in aggregation when *ca.* 97% of their κ-casein has been hydrolysed. This is probably due to residual intact κ-casein 'hairs' providing sufficient repulsion to prevent permanent contact between adjacent micelles and/or overcoming the attractive van der Waals forces. Aggregation is, however, dependent on pH value and temperature and occurs at a lesser degree of κ-casein hydrolysis at lower pH values and higher temperatures.

A number of sophisticated methods have been used to study gel formation and indicate an ordered sequence of events which may be related to the visual rennet coagulation time (RCT, Table 7.10).

---

* Steric stabilization is often considered in terms of the 'average' situation involving all of the micellar surface. Steric stabilization is essentially, however, a localized function. For this reason, it is preferable to envisage the micelle geometrically and to estimate the probability, at any given level of proteolysis, that sufficient holes exist in the 'hairs' to allow the inner surfaces of the micelles to touch (Dalgleish, D.G. 1988. *Netherlands Milk and Dairy Journal*, 42, 341–3).

Table 7.10 Sequence of events during gel formation in renneted cheese

% RCT	
60	Micelles fully dispersed and non-aggregated
100	Gradual aggregation of paracasein micelles in aggregates of three or four
200	Aggregates fuse to form strands
300	Continuous three-dimensional network formed by overlapping and crossing of strands

Note: % RCT = percentage of visual rennet coagulation time.

(b) Formation of acid gels and combined acid/rennet gels

The mechanism of formation of acid gels is similar to that of gel formation in fermented milks (Chapter 8, pages 375–7). The heat treatment applied to cheese milk is, however, often less severe than that applied to yoghurt and this has consequences with respect to gel structure, acid cheese gels being generally less rigid and more prone to syneresis than those of yoghurt. In comparison with rennet gels, acid gels are much less likely to exhibit syneresis since the viscous reaction to stress (part of visco-elastic behaviour) is less significant. This probably results from the relatively permanent structure formed in acid gels, in comparison with that of rennet gels which are more easily rearranged.

Manufacture of some cheeses involves formation of a combined acid/rennet gel. pH value is critical in determining the behaviour of such gels, at pH values below 5.15 the characteristics are largely those of an acid gel, while at pH values above 5.15 the characteristics are those of a rennet gel. The acid gel characteristics are, however, modified by the presence of rennet and continuing proteolytic activity (at temperatures above 15°C) means that the final gel stiffness is less.

Combined acid/rennet gels are notable for the very strong microsyneresis, which occurs at pH values above 5.15 but not below. This probably stems from differences in the casein particle structure and relaxation behaviour of the interparticle bonds. These differences are particularly marked at higher temperatures, probably due to a change in the relative contribution of of the different interaction products, rather than to a change in the nature of the bonds.

7.3.3 Syneresis

Considerable contraction of rennet gels occurs on cutting, leading to expulsion of water from the curd. It is well known that the rate of syneresis is controlled by a number of factors, including pH value and cooking temperature, but the process itself remains poorly understood. Three possible mechanisms have, however, been proposed, changes in solubility, re-arrangement of the paracasein network and shrinkage.

Changes in solubility have formed the basis of mechanisms explaining shrinkage in polymer gels. Rennet curd is primarily a particle gel and while small regions around flocculated micelles may exist as a polymer gel, this is unlikely to be of significance. Changes in solubility may, however, be of importance in acid-set curd.

Re-arrangement of the network of paracasein particles provides a partial explanation of changes during syneresis of rennet curds. Re-arrangement involves the formation of a more compact network with an increased number of bonds. A more compact configuration is, however, difficult to obtain because of immobilization of paracasein into the network. Despite this an increase in the number of bonds could occur due to van der Waals and electrostatic attractive forces between flocculated micelles. There may also be limited cross-linking resulting from thermal motion of the gel strands.

Each of the above mechanisms would increase tensile stresses on the gel strands, potentially leading to breakage and formation of new bonds. Breakage of strands could also result from external pressures, and it is likely that re-arrangement becomes of increasing importance as syneresis proceeds.

Shrinkage of paracasein particles and thus of the gel is not thought to be an important mechanism under normal circumstances. Significant shrinkage does, however, occur if the pH value is lowered, or the temperature raised.

7.3.4 Changes during subsequent cheese making procedures

Continuing acidification has a number of important functions in addition to increasing the rate of syneresis. The calcium content of cheese falls with reducing pH value as a consequence of the solubilization of colloidal calcium phosphate and its loss in the whey. The calcium content of the curd has a major effect on texture. Low-calcium curds such as those of Cheddar-type cheeses have a crumbly texture in contrast to the rubbery, elastic texture of high-calcium Emmental or Edam types. pH value also determines the quantity of chymosin, but not of fungal pepsins or rennins, retained by the curd. This, in turn, affects the rate of proteolysis during ripening and the final quality of the cheese.

Fat is considered to play no direct role in curd formation. Fat globules, however, become enmeshed in the curd network and modify properties throughout cheese making and in the final cheese. A number of mechanistic explanations have been proposed, it appearing that the fat globules act as 'buffers' to prevent the curd network becoming too rigid. Fat also plays an important role in modifying the salt flux through the curd following salting and consequently preventing excessive shrinkage of the cheese matrix. The flux occurs by diffusion and it appears that fat acts by physically blocking some of the pores through which the flux occurs.

7.3.5 Chemical changes during cheese ripening

(a) Role of methanethiol as key flavour compound in cheddar and related cheeses

Ripened cheese contains a wide range of compounds which contribute, either positively or negatively, to flavour and aroma. Flavour compounds include peptides and amino acids, free fatty acids, methyl ketones and esters of fatty acids, sulphur-containing compounds such as methanethiol, hydrogen sulphide and dimethyl sulphide, acetaldehyde, diacetyl and alcohols including ethanol and methanol. In the past it has been common to define organoleptic characteristics in terms of a 'component balance' theory and it is true that many of the compounds have limited, or no, flavour impact, but contribute background flavour notes. More recently, however, it has been appreciated that volatile sulphur compounds, especially methanethiol, appear to play a pivotal role in determining the flavour of Cheddar and Cheddar-type cheese and it has been possible to correlate the Cheddar character with the presence of this compound. The precise role of methanethiol remains undefined but it now seems likely that methanethiol does not contribute directly to flavour, but that reactions and interactions between methanethiol and other compounds results in formation of flavour compounds.

The origin of methanethiol is obscure but it seems unlikely that, in Cheddar cheese, micro-organisms are directly involved. It has been postulated that methanethiol is produced by non-enzymic chemical reactions involving addition or substitution reactions between H_2S and casein or methionine. Starter micro-organisms are, however, involved to the extent of producing the reducing conditions which both favour the production of methanethiol and are necessary for its stability.

In contrast to Cheddar, micro-organisms are involved in methanethiol production in smear-ripened cheeses. *Brevibacterium linens* is an established producer of methanethiol which results from the breakdown of the side chains of amino acids such as cysteine. It is probable that at least part of the methanethiol is converted to thioesters, by a process involving interaction between *Br. linens* and non-starter micro-organisms such as micrococci.

(b) Proteolysis

Proteolysis is a major factor in the ripening of cheese and affects both the flavour and texture of the end product. The relative importance of proteolysis to lipolysis and glycolysis varies with the type of cheese. Proteolysis is essential to the development of flavour and texture in Cheddar and related cheeses, is of equal importance to glycolysis in Swiss-type cheeses, but secondary to lipolysis in hard, Italian-type cheeses such as Romano.

Proteolytic enzymes are derived from three sources, rennet or other coagulant, plasmin and both starter and non-starter micro-organisms. The main proteolytic pathway appears to involve primary degradation of paracasein by residual coagulants to yield polypeptides which are further degraded by bacterial proteinases and peptidases to peptides and amino acids. Chymosin has only limited activity against paracasein, but this is significant over extended ripening periods. Other coagulating enzymes have much greater proteolytic activity, which can result in differences in the ripening pattern.

Beta-casein is also degraded, hydrolysis involving the combined action of coagulating enzymes, bacterial enzymes and plasmin. This is thought to be the major role of plasmin, although this enzyme may also be involved in degradation of paracasein and continuing hydrolysis of polypeptides. The role of plasmin has, however, been the subject of some controversy. The activity of plasmin is highly pH-dependent and some workers consider that plasmin plays a significant part in the ripening of high pH value cheeses such as Emmental and Gouda, but not

---

* The use of elevated levels of plasmin has been considered as a means of accelerating the ripening of cheese. Plasmin has a number of advantages over other enzymes (*cf.* page 306) in that it is fully incorporated into the cheese by binding onto casein micelles. This reduces losses in whey to a very low level. Plasmin is, however, very expensive and its use at present would be uneconomical. It has been suggested that the cost could be lowered significantly by use of genetically modified micro-organisms (Farkye, N. and Fox, P.F. 1992. *Journal of Dairy Research*, 59, 209–16).

in the low pH value Cheddar and similar types. Other studies, however, have suggested that plasmin has significant activity in Cheddar-types.

The immediate effect of proteolysis on the texture of hard and semi-hard cheese is a softening due to weakening of the casein network. The effect is, however, strongly modified by pH value. In Edam cheese, for example, the high pH value permits proteins to exist in a matrix which softens with increasing proteolysis. Proteolysis also proceeds more rapidly in the centre of the cheese where the NaCl content is lowest. In contrast the low pH value Cheddar-type cheese becomes crumbly with increasing proteolysis as interstitial water is bound by ionic groups.

The total extent of proteolysis is greater in soft, surface-ripened cheeses than in hard and semi-hard types and the situation with respect to textural changes is more complex. In mould-ripened types, significant proteolysis does not occur until residual lactose is utilized by glycolysis, which also raises the pH value and stimulates plasmin activity. Proteolysis further raises the pH thus tending to soften the structure. Further softening at high pH values is caused by calcium phosphate precipitating on the surface of the cheese, a calcium gradient being established from the surface to the cheese centre. A similar sequence of events is probable in bacterial smear-ripened cheese. In either case, the direct contribution of surface micro-organisms to proteolysis in the body of the cheese is usually limited since the enzymes diffuse for only a short distance. Surface proteolysis is, however, involved in raising the pH value in the body of the cheese since hydrolysis products diffuse greater distances. The intracellular proteinases of *P. roquefortii* are of greater significance than the extracellular, the contribution to overall proteolysis increasing markedly between weeks 10 and 16 of ripening as a consequence of release of enzymes due to mycelial lysis and/or leakage.

Peptides have a number of possible taste qualities and have been placed in five categories; sour, sweet, salty, umami, bitter. The contribution made by peptides may be either pleasant, or unpleasant depending on the predominance of one or more types. In many cases peptides contribute to the overall taste spectrum, rather than acting as flavour impact compounds. Water-soluble peptide fractions, for example, have been associated with 'brothy' notes in the flavour of both Cheddar and Swiss cheese.

Peptides are also associated with defects due to bitter flavours in cheese. Bitterness is generally considered to be caused by polypeptides with a high content of hydrophobic residues. An important route of bitter peptide production involves the cell wall proteinases of *Lactococcus*. The situation is complicated by the fact that there is a very wide variation in the quantities of bitter peptides produced by different strains of *Lactococcus*, and that both chymosin and plasmin have been implicated in bitter peptide production. Variability between studies of the problem suggests that other factors may be involved.

Bitter peptides are present in many cheeses, but become of significance only at a total concentration above threshold levels. In many circumstances, bitter peptide formation is readily recognized as a fault, but bitter, and the related astringent, flavours may also be considered to make a positive contribution to the organoleptic quality of cheese when present at low levels.

Amino acids are also important in determining cheese flavour and proline is a flavour impact compound in Swiss cheese. *Propionibacterium* is the major source of proline-releasing peptidases, other flavour compounds characteristic of Swiss cheese being produced by *Lb. helveticus* and probably by other starter lactobacilli. It is probable that patterns of amino acids have a particular association with specific cheese varieties, the dominant amino acids in the Greek Kopanisti cheese, for example, being alanine, γ-aminobutyric acid, leucine and valine.

Amino acids are themselves further metabolized by micro-organisms during ripening of cheese. The amino acid catabolic activity of *Br. linens*, for example, is responsible for the production of important flavour compounds in Limburger cheese. These include 3-methyl-1-butanol, phenylethanol and 3-methylthiopropanol derived from leucine, phenylalanine and methionine respectively. Extensive amino acid breakdown also occurs in mould ripened cheese, products being volatile compounds such as ammonia, aldehydes, acids and amines.

(c) Lipolysis

In most types of cheese lipolysis plays a secondary, although not necessarily insignificant, role. For obvious reasons the relative importance of lipolysis is determined primarily by fat content and the extent

---

* The dominant free fatty acids of Romano cheese reflect the species of milk used in its manufacture. Sweet, fruity notes are provided in cows' milk Romano by 2-methylbutanoic acid and 2-ethylbutanoic acid, while 4-ethyloctanoic acid provides a 'goaty' character in Romano cheese made from a mixture of cows' and goats' milk. The 'sheepy' character of ewes' milk Romano results from a combination of 4-methyloctanoic acid and 4-ethyloctanoic acid with cresols and 3,4 dimethylphenol (Ha, J.K. and Lindsay, R.C. 1991. *Journal of Food Science*, **56**, 1241–8).

to which lipolysis is encouraged during manufacture and ripening. The use of rennet preparations with a high level of lipase activity, microbial lipases, or pregastric enzymes during the making of Italian cheese, such as Romano, results in high levels of free fatty acids, which dominate the flavour spectrum.

In the soft Greek cheese Kopanisti, the total free fatty acid content can be as high as 50 g/kg, most of which comprises short chain acids (C_4 to C_8). Oxidation of free fatty acids leads to formation of further flavour compounds, methyl ketones. Although lipolysis is not a dominant reaction during maturation of surface mould-ripened cheese, *P. roquefortii* has strong lipolytic activity and produces fatty acids, methyl ketones, the most important of which is 2-nonanone, and other flavour compounds.

Free fatty acids have been associated with flavour defects in some types of cheese, including 'cowy' taints in low quality Cheddar. The fungal note in the aroma of Camembert and Brie is provided by oct-1-en-3-ol, while phenylethanol is important in all surface mould-ripened cheeses.

(d) Glycolysis and related reactions

Metabolism of residual lactose results in a number of changes in ripening cheese including the racemization of L-lactate to D-lactate (see page 339). Metabolism of lactate to CO_2 and H_2O by surface moulds of cheeses such as Brie and Camembert is significant in raising the pH value and stimulating both proteolysis (see page 328) and growth of *Br. linens*.

In Swiss cheese metabolism of L-lactate to propionate, acetate and CO_2 by *Propionibacterium* is important in eye formation and in production of the characteristic flavour. Metabolism involves an initial oxidation of lactate to pyruvate, part of which is further oxidized to acetyl-CoA and CO_2. Acetyl-CoA is converted to acetate, the reaction yielding adenosine triphosphate. Formation of propionate involves a reductive randomizing pathway, which balances the oxidative formation of acetate and CO_2.

---

* The primary natural habitat of *Propionibacterium* is the rumen of cattle and other ruminants, where lactate is available as a result of the metabolism of other microorganisms, and the skin of humans and other animals. Depending on species, *Propionibacterium* is anaerobic, or micro-aerophilic, but contains both cytochromes and catalase, which are features normally associated with aerobic life.

In Cheddar and Dutch cheeses, lactate may be metabolized to acetate by the non-starter organism *Pediococcus*. High levels of acetate are considered beneficial in Dutch cheese, but to be a fault in Cheddar.

7.3.6 Chemical analysis of cheese

Chemical analysis is required during cheese manufacture. Composition of the cheese milk is standardized on the basis of crude protein:fat ratio or casein:fat ratio. In the past, the protein content was often determined by the Kjeldahl technique, but this has largely been replaced for routine analysis by dye binding methods or near-infrared spectroscopy. Casein determination requires a modified acidity titration, the Walker formol casein test being used on a routine basis in Australia and New Zealand.

Older techniques such as the Gerber and Babcock methods are still in use for determination of the fat content of cheese milk, but in large operations have been replaced by instrumental methods such as the Milko-tester and near-infrared spectroscopy.

Process control in cheese making is heavily dependent on the development of acidity. This is often expressed as percentage lactic acid, although other acids are involved. For many years acidity was determined by titration and there is still debate over the relative merits of titratable acidity and pH value. The use of pH value as control parameter has the advantage that it is possible to follow the process from the cheese milk to the pressed cheese. pH value is also more convenient to determine, but adoption was delayed by the unreliability of early instrumentation. These problems have been overcome for several years, although it is necessary to be aware of sources of error such as poisoning of the surface of the electrode and variations in temperature. The high buffering capacity of milk can also cause problems in interpretation of results.

Fat and protein content of the finished cheese can be determined by NIR spectroscopy, although definitive methods such as the Rose-Gottlieb method for fat and the Kjeldahl method for protein must be used on some occasions. Lactose content may also be determined by near-infrared spectroscopy and moisture by near-infrared spectroscopy or drying. Total solids content is determined gravimetrically and NaCl by silver nitrate titration, or by use of an ion-selective electrode.

7.4 MICROBIOLOGY

7.4.1 Cheese as an environment for growth of micro-organisms

(a) Hard and semi-hard varieties

Hard and semi-hard cheese varieties place a selective pressure on micro-organisms as a consequence of low pH value and high lactic acid concentration, a high, but variable, NaCl content in the aqueous phase and a low redox potential. The a_w in the body of the cheese varies from *ca.* 0.97 to 0.94 in high moisture cheese such as Gouda, to 0.9, or below, in very fully aged cheddar cheese. Aged cheese of some very hard varieties may have an a_w below 0.85 and can be considered to be an intermediate moisture food.

Despite intrinsic selective factors, cheese does permit the growth of a relatively wide range of micro-organisms. In general, Gram-negative bacteria are most affected, conditions favouring the growth of Gram-positive species of bacteria, yeasts and moulds.

(b) Soft varieties

Cheese such as cottage cheese, which are of relatively high pH value and high moisture content support growth of a wide range of micro-organisms. Conditions are, however, sufficiently inhibitory to delay initiation of growth during the early stages of storage.

Soft, unripened, high acid varieties are inhibitory to most bacteria and resemble fermented milks in selecting for yeasts and moulds. The situation is similar in young surface-ripened soft cheese, but the elevation of the pH value during ripening much reduces environmental stresses and permits growth of a wide range of bacteria. In some varieties, however, the NaCl content is sufficiently high to inhibit growth of Gram-negative bacteria.

7.4.2 Cheese and foodborne disease

Although cheese is generally considered to be a low-risk food, both hard and soft types have been associated with significant outbreaks of foodborne disease in recent years. The situation is complicated by the ability of some pathogenic micro-organisms, including *Listeria monocytogenes* and some types of *Escherichia coli* to grow in some soft, but not semi-hard or hard, varieties. In all cases, however, the use of unpasteurized milk, insufficient growth of starter micro-organisms and post-pasteurization contamination are major risk factors.

(a) Hard and semi-hard varieties

Hard and semi-hard varieties of cheese have been implicated in a number of outbreaks of *Salmonella* food poisoning. In most cases the source of the organism has been unpasteurized milk, *Salmonella* being able to survive manufacturing operations and even being capable of limited multiplication in the curd of low-acid (pH > 4.95) cheese. The organism dies during maturation and a period of 60 days at not less than 4.4°C was previously thought to ensure the safety of raw milk cheese by eliminating *Salmonella*. Many strains of *Salmonella*, however, acquire enhanced resistance to low pH value conditions by the process of acid adaptation and small numbers persist for periods significantly longer than 60 days.

Although raw milk cheese is most commonly associated with *Salmonella* food poisoning, a very large outbreak in Canada, which affected an estimated 10 000 persons was caused by pasteurized milk cheese. In this case pasteurization was inadequately controlled and permitted the survival of the causative serovar *S. typhimurium* phage type 10.

Listeria monocytogenes infection has not been associated with consumption of hard or semi-hard cheese but, in some cases, the organism is able to survive cheese making and persist in the final product for a considerable length of time. It seems likely that relatively minor differences in cheesemaking practices can have significant effects on the survival of *L. monocytogenes* in a given variety.

Hard and semi-hard cheese has been responsible for a number of outbreaks of staphylococcal enterointoxication. The organism may be derived either from human sources or from milk including that drawn from the apparently uninfected quarters of a mastitic udder. The underlying cause is starter failure leading to slow acidification, under

---

* Acid adaptation promotes the persistence of *Salmonella* in foods and is probably an important survival mechanism in the environment. Adaptation is triggered by external pH values of 5.5–6.0 and maintains the intracellular pH value above 5.0–5.5. Acid adapted cells have increased resistance to organic acids in fermented dairy products and survive better during fermentation and ripening. Increased survival in cheese may also be partially mediated by other adaptive responses including those induced by heat, starvation and osmotic stress (Leger, G.J. and Johnson, E.A. 1992. *Applied and Environmental Microbiology*, **58**, 2075–80).

which conditions *Staph. aureus* can grow and elaborate enterotoxins. Growth and enterotoxin production may continue in the curd and is favoured by high temperatures and high levels of NaCl. The organism is usually killed during ripening but the enterotoxin persists.

Reported outbreaks of staphylococcal enterointoxication have involved only Cheddar-type and Swiss-type cheeses, but a wider range of cheese including Stilton and Lancashire varieties have been implicated in outbreaks that were not officially investigated. The importance of ensuring that fermentation proceeds correctly is recognized by experienced cheese makers and problems due to *Staph. aureus* are now less common. Problems do, however, occur in small-scale production especially where starter cultures are not used. Such processes are considered inherently unsatisfactory.

(b) Soft cheese

In common with harder varieties, soft cheese has been implicated in outbreaks of *Salmonella* food poisoning. Contamination of the curds, or of the finished cheese, appears to be a significant risk where cheese is made on a farmhouse scale under relatively primitive conditions. An outbreak of salmonellosis involving the Swiss Vacherin Mont d'Or variety was attributed to contamination of partly or fully ripened cheese by piglets housed adjacent to the cheese factory. Soft cheese has also been implicated as a cause of staphylococcal enterointoxication, although accounts of outbreaks are largely anecdotal and lack a full epidemiological investigation.

Diarrhoeagenic *Escherichia coli* has been responsible for food poisoning in Brie and similar varieties of soft cheese. Enteropathogenic *E. coli* was responsible for a large number of outbreaks in the early 1970s associated with French cheese imported into the US. The organism is able to grow in the cheese during ripening and poor temperature control together with poor hygiene were underlying factors. Brie was also the vehicle of infection in an outbreak of food poisoning caused by enterotoxigenic *E. coli* in Holland. Enteroinvasive *E. coli* has also been responsible for cheeseborne food poisoning, a single outbreak affecting more than 380 persons. The source of the organism was inadequately treated river water used for cleaning plant.

Escherichia coli is not uncommon in soft cheeses and may be present in large numbers. Strains present are generally considered non-pathogenic, although enterotoxigenic *E. coli* has been isolated during surveys. The significance of *E. coli* in soft cheese is uncertain, although recognized diarrhoeagenic strains should be absent. The ability to grow in the ripening cheese does, however, mean that *E. coli* has no function as an index organism in cheese of this type.

Listeria monocytogenes is also able to grow in Brie and similar types of cheese during ripening, although the organism appears to be restricted to the outer crust. The first known outbreak, which affected 86 known persons after consumption of Jalisco brand Mexican-style cheese in the western US, led to soft cheese being placed under strict surveillance. *Listeria monocytogenes* was isolated from a number of types including Brie, where the resulting product withdrawal, during 1986, affected almost 60% of imports into the US. A number of cases of listeriosis have been associated with soft cheese and while most of these have been isolated, an outbreak involving Vacherin Mont d'Or affected more than 50 persons.

The use of unpasteurized milk in cheese making is of greatest general concern with respect to listeriosis. Problems have not, however, been restricted to raw milk cheese but, despite concern over the heat resistance of *L. monocytogenes*, recontamination appears to be of greatest concern. Investigation of the Jalisco cheese outbreak suggested strongly that the finished cheese was contaminated by raw milk.

Dairy products have only rarely been implicated in botulism, cheese being the only dairy product in which growth of *Cl. botulinum* has led to illness (*cf.* yoghurt, Chapter 8, page 381). A single outbreak of type B botulism, affecting more than 80 people in France and Switzerland, followed consumption of Brie which had been stored at a high temperature for an excessive period of time. Processed cheese spread has also been a vehicle of botulism, the most recent outbreak involving *Cl. botulinum* type A in Argentina during 1974. At least six persons were affected with three deaths. The spread was of relatively high pH

BOX 7.5 **Folsom prison blues**

The US judiciary takes a rigorous approach to persons whose actions, or inactions, have been a significant underlying cause in an outbreak of food poisoning. The vice-president of the company producing Jalisco brand cheese was imprisoned for 60 days with a further two years' probation and fined $9300.

value and a_w level and had been subject to temperature abuse during storage.

7.4.3 Production of biogenic amines in cheese

The biogenic amines histamine and tyramine can be present in mature cheese as a result of decarboxylation of the amino acids histidine and tyrosine respectively. In sensitive persons biogenic amines can cause a critical increase in blood pressure together with headaches, flushing and sometimes rashes. Gastrointestinal disturbances may also occur and in a large outbreak involving Stilton cheese, predominant symptoms were sudden onset vomiting, abdominal pain and, in some cases, diarrhoea.

The non-starter organism *Lb. buchneri* appears to play a major role in decarboxylation of histidine. Histamine is also produced, albeit in relatively small quantities, by strains of *Lb. fermentum*, *Lb. helveticus*, *L. lactis* and *Enterococcus faecium* isolated from Swiss cheese. *Lactobacillus helveticus* and *L. lactis* are of particular significance due to their role as starter organisms. The incidence of histidine decarboxylating strains is, however, relatively low and even where present, significant quantities of histamine are produced only under certain conditions. These include high temperature and low NaCl concentration. Cheese usually contains only low levels of free histidine and proteolysis appears to be the initial step in histamine formation. For this reason problems of histamine toxicity occur most commonly with varieties of cheese such as Stilton and Swiss-types which undergo extensive proteolysis during ripening.

In practice it is not possible to exclude histidine decarboxylating

---

* *Streptococcus zooepidemicus*, a β-haemolytic species of Lancefield group C, has previously been rare in man. In recent years, however, the organism has been the cause of a small number of outbreaks of severe disease associated with consumption of raw milk (see Chapter 2, page 44) and raw milk cheese. Persons with an underlying predisposing condition are usually affected and symptoms include septicaemia, endocarditis and meningitis. The death rate may be high.

Queso blanco cheese made from raw milk was responsible for 16 known cases of *Str. zooepidemicus* infection and two deaths in an outbreak in New Mexico, USA and has also been a vehicle for brucellosis. The cheese was made on a small scale with no effective process control and under very poor hygienic conditions for sale amongst Mexican immigrants. The outbreak illustrates the need for registration and inspection of all food producers, irrespective of the size of the operation. In practice, however, this can be very difficult especially in such situations where local entrepreneurs establish 'back yard' operations to meet the demand within immigrant communities for particular foodstuffs which are not readily available at a low price.

bacteria from cheese, but ripening at temperatures below 7°C appears to be the most effective means of control. Relatively simple methods for assay of histamine are available and methods have also been developed for the detection of histidine decarboxylating bacteria. These may be used in the investigation of possible cases of illness associated with biogenic amines, or for verification of the status of suspect batches of cheese.

7.4.4 Spoilage of cheese

(a) Hard and semi-hard varieties

Spoilage of hard and semi-hard varieties of cheese is of two main types, surface growth of micro-organisms, usually moulds, and gas production due to the growth of micro-organisms in the body of the cheese.

Mould growth produces highly visible spoilage which, in severe cases, is accompanied by extensive proteolysis and lipolysis. *Penicillium* is responsible for spoilage in 60–80% of cases, *Aspergillus* also being a common contaminant. Mould spoilage can be a major particular problem with prepackaged cheese and rigorous precautions are required at packing plants. These include sterile filtration of air, ultraviolet disinfection of handling surfaces and, where permitted, anti-mycotic coating of packaging material. The incidence of mould spoilage has been much reduced by widespread use of vacuum and modified atmosphere packaging. Incipient growth present at the time of packaging may still,

---

BOX 7.6 **Dead man's finger**

Retro-thinkers, who campaign vociferously for a return to traditional cheese making and 'real' cheese, would do well to consider the serious quality and spoilage problems which affected the cheese industry during the 'golden years' of farmhouse production. A particular defect of Cheddar cheese in Somerset during the 1930s was colloquially (and aptly) known as dead man's finger due to the appalling stench of affected cheese, which even cattle refused to eat. There can be no doubt that dead man's finger and other, less spectacular, spoilage resulted from poor technological control of cheesemaking, poor hygiene and storage under inadequate conditions. (Davis, J.G. 1983. *Journal of Applied Bacteriology*, **55**, 1–12).

however, develop sufficiently to cause visible spoilage and leaking packs is a continuing problem. Gamma-ray irradiation has been proposed as a means of control.

Thread mould is a sporadic problem caused by growth of mould, possibly in association with yeasts, in folds and wrinkles of plastic film packaging. The fault results in black, dark brown or green spots, or threads and is associated with formation of free whey released from the cheese when vacuum packed. The incidence is related to manufacturing technology and is most common in cheese made in a tower block-making system. Relatively few moulds are capable of thread spoilage, but these are widespread in factory environments. Species of *Cladosporium* and *Penicillium* are most common, but *Phoma* may also be involved. Species of *Candida* are the most common yeast.

The high incidence of mould spoilage of cheese has led to concern over the possibility of mycotoxin production. Surveys have produced differing results, but it seems probable that *ca.* 20% of common spoilage moulds (*Penicillium* and *Aspergillus*) produce potentially toxic metabolites.

Yeasts and bacteria can also develop on the surface of hard cheese especially where the surface is moist. Spoilage involves production of slimes, discoloration and 'rots' due to proteolysis. There may also be accompanying off-flavours. In some varieties of cheese, such as white Stilton, the role of yeasts as spoilage organisms is equivocal since yeasts are almost invariably present and may be considered to contribute desirable flavours. Surface discoloration is usually due to pigmented micro-organisms such as *Aureobacterium liquefaciens, Br. linens* or, less commonly, carotenoid producing strains of *Lb. brevis*. A particular type of pink discoloration affecting the nitrate-containing Gouda cheese, however, has been attributed to reduction of nitrate to nitrite by bacteria, predominantly *Micrococcus*, derived from storage shelving. At pH values between 5.2 and 6.7 nitrite reacts with the annatto dye in the coating of the cheese to produce a pink compound.

Spoilage of hard cheese by internal gas production occurs either in the curds or young cheese ('early blowing'), or during maturation ('late blowing'). Members of the *Enterobacteriaceae* are common causes of early blowing and in extreme cases gas production takes place during fermentation, but other micro-organisms have been implicated including yeasts and species of *Bacillus*. Improvements to hygiene and production control have much reduced the incidence of early blowing, although members of the *Enterobacteriaceae* may be present in sufficient numbers, in the absence of visible gas formation, to impart a faecal taint.

Late blowing is due to gas production by species of clostridia capable of fermenting lactate to butryic and acetic acids, hydrogen and carbon dioxide. Late blowing is a major problem with varieties such as Emmental, Gouda and Edam, but can be a fault in other varieties such as Cheddar. *Clostridium butyricum* is most commonly implicated, but other species including *Cl. tyrobutyricum* and *Cl. sporogenes* are also involved, especially at higher pH values. Endospores are present in the cheese milk and only very low levels are necessary to cause major spoilage problems. The ultimate source is the feed, especially poor quality silage and brewers' grains, and the problem is consequently greater during winter feeding.

In addition to well defined spoilage patterns a number of micro-organisms have been involved in producing taints in cheese. It should be appreciated that in some cases incidents were isolated, but have been widely discussed due to their unusual nature, or technological significance. *Candida*, for example, has been implicated in spoilage of cheddar cheese due to production of high levels of ethanol, ethyl acetate and ethyl butyrate, which imparted a 'fermented yeasty' flavour. Spoilage appeared within six months, the cheese being of high moisture and low NaCl content. *Kluyveromyces marxianus* has also been implicated in the 'gassy' spoilage of Parmesan cheese.

Non-starter lactic acid bacteria have been associated with a number of taints and also produce visual and textural defects due to precipitation of calcium lactate crystals. Strains of *Lactobacillus*, *Leuconostoc* and *Pediococcus* capable of racemization of L-(+) lactic acid to D-(−) lactic acid are responsible for the defect. Dead cells of starter bacteria may serve as nuclei for crystal growth, which is also markedly faster at high temperatures. Formation of calcium lactate crystals is thus particularly associated with the use of high temperatures to accelerate ripening. Crystal formation is, however, much reduced by vacuum packing.

The role of starter micro-organisms in producing bitter peptides is well

---

* Although *E. coli* is associated with faecal taints at numbers in excess of 10^6 cfu/g, its presence at lower numbers (*ca.* 10^5 cfu/g) results in a distinctive 'sharp' flavour. This flavour is considered highly desirable by some consumers who, mistakenly, identify the sharpness with maturity.

known, but continuing metabolism of starters persisting into ripening can result in other flavour defects. Some strains of *L. lactis* ssp. *lactis* are capable of forming 'fruity' flavoured esters such as ethyl butyrate and ethyl hexanoate *via* reactions between ethanol and butyric or hexanoic acid. Conversion of amino acids such as leucine to the corresponding aldehyde 3-methylbutanol by transaminase and decarboxylase enzymes of *L. lactis* ssp. *lactis* can lead to formation of 'malty' flavours.

(b) Soft varieties

High acid cheeses are normally spoilt by yeasts and moulds, but bacterial spoilage is important in higher pH varieties such as cottage cheese. Gram-negative bacteria including *Pseudomonas fluorescens*, *Ps. putida* and *Enterobacter agglomerans*, derived from wash water or added ingredients, are most common, although species of *Enterococcus* may also be involved in spoilage. *Pseudomonas* may be controlled by acidification to pH 4.5, but *Ent. agglomerans* is able to grow at pH values as low as 3.8. A wide range of yeasts have been implicated in spoilage of cottage cheese and other unripened types such as quarg. Yeast populations as high as 10^6–10^7 per gram are not uncommon in soft cheese during refrigerated storage. Species of *Candida*, *Cryptococcus*, *Kluyvera*, *Pichia*, *Sporobolomyces* and *Torulopsis* have all been implicated in spoilage, usual spoilage patterns involving flavour and aroma defects, gassiness and growth of visible colonies.

Perceived spoilage in surface ripened soft cheese is often a consequence of ripening having proceeded to a greater extent than acceptable to the individual. On some occasions, however, the normal ripening flora may be overgrown by undesirable micro-organisms. These are usually moulds, but bacteria are occasionally involved. Yeasts are normally thought to contribute to ripening, but may also be involved in occasional spoilage.

7.4.5 Microbiological examination of starter cultures

Unless direct-in-vat cultures are used, some degree of microbiological control is required. Where starters are propagated from commercially supplied mother cultures, this may be restricted to simple activity tests,

---

* Mould ripened soft cheese sometimes has a very slight, but distinctive flavour of celluloid. This fault results from the synthesis of vinyl benzene (styrene) by *P. camembertii*. Synthesis occurs only when substrate exhaustion leads to the mould entering the starvation state during ripening. Inadequate temperature control and removal of lactose and lactate by excessive washing of the curd are underlying technological factors (Spinnler, H.E. *et al.* 1992. *Journal of Dairy Research*, **59**, 533–41).

Microbiology

but where starter production is undertaken entirely within the factory, more extensive testing is required including determination of the presence of phage.

Starter activity is usually determined by a simple acidification test. The most common method is to monitor the pH value of milk inoculated with the starter during incubation at an appropriate temperature. This does not correlate directly with either cell numbers or lactic acid production but gives a good indication of activity during cheesemaking.

Determination of the presence of phage is an essential part of some culture management systems (see page 290). Several methods exist and it is advisable that at least two methods should be used in parallel to increase the likelihood of phage detection. Two basic methods exist, the plaque assay and the inhibition of acid production, but the plaque assay is not suitable for use with mixed strain starters.

In the continuing battle with bacteriophage, it is easy to forget that starter cultures are subject to contamination with other microorganisms. It is usually possible to detect contaminants by simple tests including direct microscopic examination and by streaking onto non-selective media. Both carbohydrate-containing and non-carbohydrate-containing media should be used and examination of colonies should be accompanied by microscopic examination and determination of simple biochemical properties such as catalase-activity. Indications of the presence of contaminants may also be obtained by examination of broth cultures. Such indications include excessive gas production, pellicle or sediment formation and unusual odours.

7.4.6 Microbiological examination of cheese

In hard cheese *Staph. aureus* is considered to be the organism of primary concern and, while the product has been associated with salmonellosis, routine examination for this organism is not considered worthwhile. Standard cultural methods using Baird-Parker selective medium are satisfactory, but it must be recognized that enterotoxin may be present in the absence of recoverable cells. Although the development of kit systems has simplified enterotoxin detection, extraction from foods remains difficult, especially in high fat and high protein foods such as cheese, and enterotoxin assays cannot be undertaken on a routine basis. As an alternative the use of the thermonuclease (TNase) test, which indicates the *likely* presence of enterotoxin is widely recommended. Different methods for the TNase test vary considerably in sensitivity. The commercial 'Staphynuclease[C]' kit, which is based on an antibody inhibition assay is, however, generally suitable. Cheeses which test positive for TNase should be assayed for enterotoxin.

Staphylococcus aureus is considered to be a lesser problem in soft cheese and it is not usual practice to test for the organism. It is advised, however, that each type should be reviewed individually and testing applied if evidence suggests a significant risk.

Until the recognition of *L. monocytogenes* as a foodborne pathogen, diarrhoeagenic strains of *E. coli* were considered to present the greatest risk in soft cheese. Under normal circumstances, however, routine examination for the organism is not justified, but methodology should be available for use if necessary. Isolation of diarrhoeagenic *E. coli* is complicated by the atypical phenotype expressed by many strains, including failure to ferment lactose at 37°C. Methods have been developed, the most satisfactory for use in safety assurance being a membrane technique involving resuscitation on minerals modified glutamate agar for 4 h at 37°C followed by transfer to selective tryptone bile agar for 20 h at 44°C. This method will recover both diarrhoeagenic and non-diarrhoeagenic strains, although it is necessary to be aware of the possibility that cheese, like other dairy products, may contain strains of *E. coli* unable to grow at 44°C or to produce indole. Methods for detection of enterotoxigenic *E. coli* using DNA hybridization techniques and the polymerase chain reaction have been developed, but are not currently suitable for use in quality control laboratories. If required, toxins produced by enterotoxigenic strains may be detected using commercial kits.

The association of soft cheese with *L. monocytogenes* requires that methods for detecting this organism should be available. Some manufacturers, especially those supplying multiple retailers, examine for *L. monocytogenes* on a routine basis. The value of this is dubious and it is regrettable that, in some cases, end-product testing for the *L. monocytogenes* has diverted attention from control of the organism at manufacturing level. A very large number of media and methods have been developed for *L. monocytogenes* in recent years, but Oxford agar incubated at 30°C for up to 48 h is recommended as selective medium. Enrichment in FDA broth at 30°C for 24 h is most suitable for dairy products. Alternatively use may be made of commercially available rapid detection kits based either on ELISA or DNA hybridization.

Routine examination of hard cheese, or surface-ripened soft cheese for spoilage micro-organisms is not usually considered necessary. Microbiological analyses are, however, required on unripened types of soft cheese. Similar criteria and methods to fermented milks are employed (see page 384), although counts for *Pseudomonas* using CFC medium are sometimes included when examining cottage cheese. In many cases spoilage micro-organisms are present only in very small numbers immediately after manufacture and the predictive value of microbiological analysis at this time is often low.

EXERCISE 7.1.

You have been engaged as a cheesemaker by a small company manufacturing Cheddar cheese by traditional methods. For many years the product has been of consistently high quality commanding a significant premium price. Since the death of your 88 year old predecessor, however, the quality of the cheese has been highly variable, the texture varying from 'rubbery' to 'exceptionally crumbly'. On joining the company you find that your predecessor kept few records and relied entirely on empirical methods of process control. Development of acidity, for example, was monitored by placing a hand in the cheese vat and judging the 'feel'. Consider the possible underlying causes of quality inconsistency and draw up an outline plan of investigation of your specific problem. Design a quality assurance programme stipulating control and monitoring procedures for each critical stage in the Cheddar manufacturing process.

EXERCISE 7.2.

Define the ideal characteristics of an accelerated cheese ripening system, including technical, commercial and consumer acceptability/safety in your considerations. To what extent is it likely to be possible to incorporate all of the ideal characteristics in a single system? Discuss the potential future role of genetic modification in the development of accelerated cheese ripening systems.

EXERCISE 7.3.

Heat treatment of milk for acid-set cheese varies from no treatment, through thermization to full pasteurization. What is the effect of different heat treatments on the structure and properties of the gel? To what extent do differences affect the final character, including organoleptic quality, of the cheese?

EXERCISE 7.4.

Pregnancy is a recognized risk factor for listeriosis and in a number of countries, pregnant women are warned against consumption of foods, including some types of soft cheese, which are known to have been associated with listeriosis.

1. To what extent do you consider that 'health warnings' are effective as part of a national strategy of reducing morbidity due to foodborne disease?
2. What are the practical difficulties of ensuring that the warnings reach the persons at risk, without causing unneccessary alarm amongst the population as a whole?
3. Consider the development of possible alternative strategies to reduce the incidence of foodborne listeriosis during pregnancy. Could similar strategies be used for other high-risk groups, such as the immunocompromised or must a separate strategy be applied to each group?

8
FERMENTED MILKS

OBJECTIVES

After reading this chapter you should understand
- The different types of fermented milk
- The role of micro-organisms in their manufacture
- The processing technology
- The major control points
- The nutritional and therapeutic properties of fermented milks
- The chemistry of flavour development
- The physico-chemical changes in milk during the manufacture of yoghurt
- The structure of yoghurt and other fermented milks
- Microbiological hazards and patterns of spoilage

8.1 INTRODUCTION

A wide range of fermented milks exists, although many are similar with respect to technology. Fermented milks may be classified in a number of ways but a system based on the type of starter micro-organism used is generally satisfactory (Table 8.1).

Until relatively recently production was usually concentrated in particular regions. The wider-scale popularity of fermented milks, especially yoghurt, first arose from interest in their proposed life-prolonging characteristics but this market was not sustained. The development in the 1950s of fruit and flavoured yoghurt, however, resulted in this product becoming of major importance in the dairy industries of western Europe, the US and other non-traditional markets. Since then the number of types of yoghurt and yoghurt-based foods has increased further and there has been a revival of interest in yoghurts and other fermented milks as a means of promoting health.

At the same time the appeal of other fermented milks such as kefir has

Table 8.1 Classification of fermented milks

Mesophilic lactic fermentation
  Cultured buttermilk
  Cultured cream
  Filmjolk
  Scandinavian ropy milks
Thermophilic lactic fermentation
  Yoghurt
  Acid buttermilk
'Therapeutic lactic fermentation'
  Acidophilus milk
  Yakult products
  Acidophilus–Bifidus (AB) yoghurts
  Proprietary therapeutic products
Lactic/yeast fermentation
  Kefir
  Koumiss
Lactic fermentation/mould ripening
  Viili

BOX 8.1 A niche in the market-place

Perceptions of yoghurt vary amongst consumers according to nationality. In the US, for example, much of the yoghurt is a mild, dessert-like product, which would not be recognized by most European consumers. Within the overall US market, however, four major niches may be recognized, which require yoghurts of very different properties: childrens' yoghurt (an extremely mild and sweet product), breakfast yoghurt, dessert yoghurt and health yoghurt (high acid with pH value as low as 3.5).

---

been increased by modifications resulting in less strongly flavoured products which are also compatible with modern packaging technology. The various varieties of yoghurt, however, continue to dominate the market for fermented milks.

8.2 TECHNOLOGY

The technology of fermented milks is relatively straightforward, small-scale manufacture requiring only simple equipment. The need for consistency and low production costs associated with large-scale operations means that a higher level of control and generally more sophisticated equipment are required, although the basic manufacturing principles remain unchanged. Despite the wide range of fermented milks, the technology is similar, differences, in most cases, being restricted to the type of starter culture and the total solids content of the milk.

8.2.1 General role of starter micro-organisms in the manufacture of fermented milks

(a) Types of micro-organism

The most commonly used starter micro-organisms are members of the group commonly referred to as lactic acid bacteria (LAB). These bacteria and their use as starter micro-organisms are discussed in greater detail in Chapter 7, pages 276–88. In contrast to cheese manufacture, however, micro-organisms other than LAB are also used.

Revived and increased interest in the therapeutic properties of fermented milks has led to the use of the intestinal bacterium *Bifidobacterium* in starter cultures. Yeasts also have a role as starter cultures in fermented milks. Starter cultures used in the production of two alcoholic fermented milks, koumiss and kefir, contain yeasts which act in conjunction with LAB. *Kluyveromyces marxianus* var. *marxianus* and *K. marxianus* var. *lactis* are used as starter cultures for koumiss, while the kefir grain (see page 367) contains *Candida kefyr* together with one or more other yeasts.

(b) Technologically important properties of starter cultures

The major technologically important properties of starter micro-organisms in fermented milks are the same as those in cheese. With the exception of viili and closely related products, however, there is no

---

* *Bifidobacterium* comprises irregular, gram-positive, asporogenous, rod-shaped bacteria which metabolize carbohydrates by means of the unique fructose-6-phosphate phosphoketolase pathway. Bifidobacteria require as specific growth factors ('bifidogenic factors') the carbohydrates N-acetylglucosamine, found in human milk, and lactulose, found in heated milk. There is considerable variation in growth response to the gaseous environment, some strains are obligately anaerobic while others, which appear to possess weak catalase activity, tolerate oxygen in the presence of carbon dioxide. There are currently more than 18 recognized species of *Bifidobacterium*, of which *B. breve* and *B. longum* are most commonly used as starter cultures (Scardovi, V. 1986. In *Bergey's Manual of Systematic Bacteriology*, vol. 2 (eds Sneath, P.H.A., Mair, N.S., Sharpe, M.E. and Holt, J.G.). Williams and Wilkins, New York, pp. 1418–34).

manufacturing stage corresponding to ripening and no role played by micro-organisms other than starters. For this reason production of flavour compounds during fermentation is of greater importance in fermented milks than in most types of cheese. Diacetyl (see Chapter 7, page 279) and acetaldehyde are major flavour compounds, acetaldehyde being produced by most starter LAB. Pathways of formation vary according to species, *Lactococcus* and *Leuconostoc* possessing threonine aldolase which mediates the formation of glycine and acetaldehyde from threonine, but sugars being the major important precursor in *Lactobacillus* and *Str. salivarius* ssp. *thermophilus*. Many possible routes exist, however, and are summarized in Figure 8.1.

(c) Probiotic and therapeutic properties associated with starter micro-organisms

Historically interest in the probiotic and therapeutic properties of starter micro-organisms stems from the observations of Metchnikoff on the longevity of Balkan peasants and his development of theories

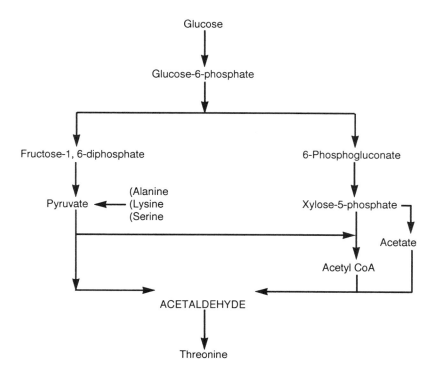

Figure 8.1 Pathways of acetaldehyde formation.

Table 8.2 Probiotic and therapeutic properties associated with starter micro-organisms in fermented milks

Property
maintenance of normal intestinal microflora
Micro-organism
Bifidobacterium spp.; *Lactobacillus acidophilus*
Proposed mechanism
(a) production of inhibitors
(b) stimulation of host immune system

Property
alleviation of lactose maldigestion
Micro-organism
general property of fermented milks
Proposed mechanism
(a) reduction in lactose content of product
(b) auto-digestion of lactose by starter derived β-galactosidase
(c) unknown cause(s)

Property
anti-carcinogenic activity
Micro-organism
Bifidobacterium spp.; various lactic acid bacteria
Proposed mechanism
(a) removal of dietary procarcinogens
(b) stimulation of host immune system

Property
reduction of serum cholesterol levels
Micro-organism
Bifidobacterium bifidum; *Lactobacillus acidophilus*
Proposed mechanism
not known

Property
nutritional enhancement
Micro-organism
Bifidobacterium bifidum; *Lactobacillus acidophilus*
Mechanism
(a) synthesis of B-complex vitamins (*B. bifidum* only)
(b) increased calcium absorption

Property
alleviation of effects of renal malfunction
Micro-organism
Bifidobacterium spp.; *Lactobacillus acidophilus*
Mechanism
reduce level of toxic amines

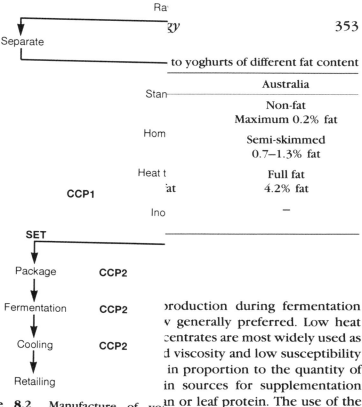

Figure 8.2 Manufacture of yoghurt. ²Stabilizers, etc.

controlled by ensuring that the
the temperature of storage is su
bacterial or milk-derived protease
Milk may be full cream or of red
usually standardized to ensure
preferences are met. The descrip
content and any legislative stand
8.4).

Standardization can involve skim
use of standardizing centrifuges.
common commercial practice to
This improves the body of the fi

to yoghurts of different fat content

	Australia
	Non-fat Maximum 0.2% fat
	Semi-skimmed 0.7–1.3% fat
	Full fat 4.2% fat
	—

production during fermentation
generally preferred. Low heat
centrates are most widely used as
d viscosity and low susceptibility
in proportion to the quantity of
in sources for supplementation
in or leaf protein. The use of the
means of utilizing local materials
ed by either vacuum evaporation,
d, where fresh milk is scarce,
used.

whey protein powder is usually
hanger fouling. The milk may also
e subsequent starter growth and
resence of dispersed gases in the
iently, de-aeration may be com-

re texture, decrease susceptibility
dule formation. Increasing homo-
y but also increases susceptibility
o define an optimum pressure for

onventional vacuum concentrated milk
usceptibility to syneresis. These problems
ively a high quality yoghurt is obtained by
ed by reverse osmosis.

each type of yoghurt produced, but pressures used are typically 15–20 MPa at *ca.* 65°C. Single-stage homogenization is most common, but a second stage treatment at 4 MPa may be applied.

Milk for yoghurt manufacture is almost invariably heat treated. For many years a treatment of 80–85°C for 30 min has been considered optimal, but in practice the treatment applied varies considerably from an approximation of HTST pasteurization to a full UHT process. Heat treatment is important in increasing the viscosity of the yoghurt and improving texture. Milkborne pathogens such as *Salmonella* and *Campylobacter* are eliminated and the number of adventitious bacteria, some of which may interfere with growth initiation by the starter culture, reduced. A full UHT process is required where milk contains large numbers of endospores especially if the starter consists of slower growing bacteria such as *Bifidobacterium*. Heat treatment also stimulates growth initiation by starter bacteria by reducing the oxygen content of the milk and, according to the heat treatment applied, may either stimulate or inhibit subsequent growth.

Milk may be heat treated in either a batch or continuous process. A temperature of 85°C for 30 min is used in the batch process and, where stirred yoghurt is being made, the milk may be heat treated, cooled and fermented in the same vessel (multipurpose processing tank). This system produces a high quality yoghurt, but involves a lengthy production cycle, is of low productivity and expensive in terms of building space required and energy costs. Continuous heat treatment at 90–95°C for 5–10 min is often preferred, either tubular or plate heat exchangers being used.

(b) Other ingredients

Although many consumers prefer yoghurts to be 'additive-free', stabilizers may be added to stirred yoghurts to improve viscosity, and body and to reduce susceptibility to syneresis. Stabilizers also improve mouth-feel and permit reduction in calories while maintaining the organoleptic quality. Stabilizers are hydrocolloids including gelatin and carbohy-

---

* Stimulation and inhibition occurs in cycles and appears to be related to changes in levels of amino-nitrogen and, possibly, other compounds such as formic acid.
Stimulation: heat treatment in range 60°C/30 min to 70°C/40 min.
Inhibition: heat treatment in range 72°C/45 min to 82°C/>10 min, or 90°C/1–45 min.
Stimulation: heat treatment 90°C/>60 min, or 120°C/15–30 min.
Inhibition: heat treatment 120°C/>30 min.

CONTROL POINT: MILK AND MILK TREATMENT CCP 1

Control

Milk must be of good initial quality and free from antibiotics.

Milk must be stored at correct temperature and for not longer than the stipulated period.

Ensure heat treatment correctly applied and at least equivalent to pasteurization.

Homogenization pressure must be correct.

Ensure milk correctly standardized.

Additional milk protein sources to be of good quality and suitable for the purpose.

Monitoring

Determine microbiological quality of incoming milk and ensure freedom from antibiotics.

Temperature of milk should be monitored by thermograph and a formal procedure instituted to control milk storage.

Heat treatment equipment should be fitted with a thermograph and the process monitored by trained and experienced personnel.

Monitor homogenization pressures throughout process.

Standardization should be supervised by trained and experienced personnel. Chemical analysis may be appropriate at this stage.

Obtain additional protein sources from reputable supplier. All must be certified 'antibiotic-free' and conform to any other predetermined specification.

Verification

Product quality.

Chemical analysis.

Examination of thermograph and other plant records.

drates such as pregelatinized starch, agar, guar gum, pectin and carrageenan. Gelatin and starch may be used at concentrations of up to 1%, but concentrations of other stabilizers should not exceed 0.3 to 0.5% otherwise flavour may be adversely affected. The use of hydrocolloids has increased in recent years in response to reduced usage of added milk solids due to high cost. Careful selection of stabilizers also permits a range of products to be prepared from a single base formula.

Sweeteners, colouring and flavouring are usually added after pasteurization to avoid thermal degradation. Addition may be made pre- or post-fermentation. Yoghurts were originally sweetened with sucrose but this is at variance with the 'healthy' image imparted even to artificially flavoured and coloured types which bear little resemblance to the yoghurts of tradition. A number of alternative sweeteners have been proposed including high fructose corn syrup, saccharin, glucitol (sorbitol) and aspartame. Raw cane sugar and honey are used where a 'healthy' image is desired. It is also possible to use decolourized, deflavoured, deacidified and deodourized fruit concentrates. Apple and pear concentrates are widely used in Europe and tropical fruit concentrates in the US. Consumer trials have shown a preference for glucitol and aspartame, saccharin being considered unsuitable due to bitterness. High fructose corn syrup, however, lowers the viscosity less than glucitol or aspartame and, especially in the United States, is readily and cheaply available.

Fruit and, less commonly, nuts are usually supplied as heat-treated purees in large cans, or in bulk containers for direct connection to the yoghurt handling line. Fruit purees originally used resembled a jam, but a lighter puree of 30–50° brix is now preferred, the necessary body being obtained through the presence of stabilizers. Care must be taken to ensure that the stabilizers used do not interfere with flavour-release. Addition to stirred yoghurt is made after fermentation, but in the case of

---

* Stabilizers are usually added before homogenization and heat treatment and this has the obvious advantage of destroying vegetative microbial pathogens and potential spoilage organisms. Increasingly, however, stabilizers are added after pasteurization, with a carrier such as flavouring or colouring. This has advantages from a technical viewpoint, but it is necessary to introduce stringent criteria for the microbiological status of the stabilizer.

* The reduction of water activity resulting from pre-fermentation addition of sweeteners has implications for growth of starter cultures. In general terms low water activity favours *Str. salivarius* ssp. *thermophilus* over *Lb. delbrueckii* ssp. *bulgaricus* and can lead to quality problems resulting from unbalanced growth. The response to different sweeteners varies, both starters are inhibited by glucitol at concentrations in excess of 7%, for example, while fructose favours the growth of *Lb. delbrueckii* ssp. *bulgaricus* over *Str. salivarius* ssp. *thermophilus*.

set yoghurt a layer of fruit in a viscous gel is placed into the container before addition of the inoculated milk. Sulphur dioxide is still used as a preservative in fruit and occasionally causes off-flavours in fruit yoghurts.

(c) Fermentation

Fermentation usually involves the combined growth of *Lb. delbrueckii* ssp. *bulgaricus* and *Str. salivarius* ssp. *thermophilus*, although *Lb.*

CONTROL POINT: ADDITIONAL INGREDIENTS CCP 1 OR 2

Control

Ensure ingredients of good quality and suitable for intended purpose.

Ingredients to be added after heat treatment should conform to microbiological criteria with respect to absence of pathogens and general quality.

Ingredients must be stored under good conditions and used within a stipulated time period.

Monitoring

Obtain ingredients from a reputable supplier and ensure conformity to predetermined specifications.

Microbiological examination of 'high-risk' ingredients added after heat treatment.

A formal inspection plan should be instituted for canned goods.

Inspection of storage facilities.

Verification

Quality of end-product.

Physical properties of end-product (stabilizers).

Inspection of plant records.

helveticus ssp. *jugurti* is sometimes used in place of *Lb. delbrueckii* ssp. *bulgaricus*. The relationship between the starter components is synergistic, *Str. salivarius* ssp. *thermophilus* being stimulated by amino acids and peptides released from casein by the *Lactobacillus* which, in turn, is stimulated by formic acid produced by the streptococci.

Traditional mixed strain cultures have now been replaced in large-scale manufacture by single or multiple strain defined cultures used in rotation. Problems due to bacteriophage activity are relatively rare, although bacteriophage-insensitive mutants have been used in Australia. Most modern large-scale production uses the 'short-set' method, in which starter culture is added at 2% (v/v) permitting the fermentation to be completed within 4 h at an incubation temperature of 40–42°C. At this point the acidity will be 0.90–0.95%. A small amount of yoghurt is still made using the long-set process, in which starter culture is added at 0.5% (v/v) and the fermentation continued for 14–16 h at 30°C.

Direct-in-vat inoculation is widely used in large-scale production, especially in the US. Cultures may be supplied either as superconcentrated frozen suspensions or in freeze-dried form. Freeze dried cultures are more convenient in use and are technically superior due to the ability to dry blend the two components to give very fine control of the final properties of the yoghurt.

Set yoghurt is fermented in the final retail container, incubation taking place either on a batch basis in water baths or a temperature controlled room, or continuously, during progression through a heated tunnel. Stirred yoghurt is fermented in bulk either in multipurpose tanks or in dedicated fermentation vessels. Slow to medium speed stirring is applied for no more than 5–10 min. This produces the required body and texture and also slows fermentation, reducing the risk of overacidification.

---

* Formate production may require metabolic adaptation and takes place during the late exponential and early stationary phase of growth. Formate production is mediated by pyruvate:formate lyase:

Pyruvate + Coenzyme A   Formate + Acetyl Coenzyme A

and the reaction may be used for ATP generation *via* acetyl coenzyme A synthesis of compounds requiring acetyl groups, or generation of C_1 fragments for synthesis. This can result in considerable variation in the quantities of formate available to the *Lactobacillus* even when growth conditions are normal (Perez, P.F. *et al.* 1991. *Journal of Dairy Science*, 74, 2850–4).

To some extent the final consistency of yoghurt is dependent on process factors, but the role of starter micro-organisms in producing the desirable smooth, viscous consistency through the production of slime should not be overlooked. Slime is also important in reducing the need for stabilizers, in flavour retention and in producing a glossy appearance. Careful control is, however, needed to avoid a 'ropy' texture. Slime is usually composed of amino-sugar containing extra-cellular polysaccharide (see page 378), the quantity produced and the structure varying from strain to strain.

Therapeutic yoghurts differ from the conventional type in the starter micro-organisms used, other aspects of production technology being similar. *Lactobacillus acidophilus, B. bifidum* or *B. longum* and, less commonly, *Lb. casei* are used and are available as commercial cultures in various combinations. Cultures are usually propagated in monoculture and direct-in-vat inoculation may be used. Therapeutic yoghurts can be prepared using therapeutic starter cultures alone or in the presence of normal starter bacteria. Acid production by therapeutic starters is slow in the absence of normal starter bacteria and strict precautions must be taken against contamination and overgrowth by undesirable bacteria. Therapeutic activity may also be enhanced by the presence of normal starter cultures.

It is essential that starter cultures used in therapeutic yoghurts are able to survive transit through the stomach, remain active in the presence of bile and have the ability to colonize the intestine. A minimum of 10^6 viable cells/ml (the therapeutic minimum) has been considered necessary for therapeutic activity, but a more realistic number is probably *ca.* 8×10^6 cells/ml. Both *Bifidobacterium* and *Lb. acidophilus* are acid-sensitive and, to ensure the therapeutic minimum is present in the final product, it is necessary either to use an inoculum level of 10–20%, or to maintain the final pH value above 4.6 either by buffering the yoghurt, or by terminating incubation at a pH value of 4.9–5.0.

Proprietary therapeutic cultures such as Biogarde$^{(R)}$ (*Str. salivarius* ssp. *thermophilus, Lb. acidophilus* and *B. bifidum*), Bioghurt$^{(R)}$ (*Str. salivarius* ssp. *thermophilus* and *Lb. acidophilus*) and Bifighurt$^{(R)}$ (*B. bifidum*) may also be used as starter cultures for the therapeutic

---

* The starter culture may also be a factor in nodule production and this fault is less common where *Lb. helveticus* ssp. *jugurti* is used as starter in place of *Lb. delbrueckii* ssp. *bulgaricus.* Culture tendency towards nodulation is, however, exacerbated by using an excessively high inoculum, an inactive culture, or poor growth conditions.

> **BOX 8.2 Innocence and health**
>
> It is considered misleading to describe yoghurts as having health promoting properties unless the therapeutic minimum number of viable cells is present at point of sale. Strains of *Bifidobacterium* used in some commercial products neither survive gastric transit, nor product acidity during storage. Problems of survival are particularly acute in yoghurts produced for the north American market, where acidity is associated with 'healthiness' and the pH value is as low as 3.5. In contrast the mild nature of therapeutic yoghurts produced for the European market is a major attraction and can be the basis for preference over conventional yoghurts.

versions of other fermented milks. The starters may also be cultured separately and incorporated into non-fermented products. Examples of such 'bifid-amended' products include milk, milk powder, butter and frozen desserts.

(d) Post-fermentation processing

In most cases post-fermentation processing of yoghurts is restricted to cooling, addition of fruit puree, etc., and packaging. Excessive agitation and shear stress reduces the viscosity of yoghurt and should be avoided during handling, although application of shear stress is beneficial in reducing nodulation. For this reason, stirred yoghurt may be 'smoothed' before cooling by passage through a fine mesh screen (texturizer). This is effective in reducing of nodulation but, unless very carefully controlled, the process leads to a reduction in viscosity and leakage of free whey.

Cooling should be carefully controlled since too rapid a rate leads to syneresis. Common industrial practice involves a first stage cooling to 15–20°C, addition of fruit, etc., if applicable and second stage cooling to below 5°C in a cold-store. Tunnel incubators for set yoghurt incorporate a cooling section, while stirred yoghurt may be cooled either in the fermentation vessel or, more efficiently, using a continuous heat-exchanger.

Ideally filling of yoghurt into final containers should follow immediately after cooling. If this is not possible, storage should be as short as possible

CONTROL POINT: FERMENTATION CCP 2

Control

Fermentation to proceed to predetermined acidity (pH value).

Monitoring

Monitor starter activity before addition to milk (see Chapter 7, page 341). Addition of starter to be made by trained and experienced personnel.

Monitor temperature of milk throughout fermentation.

Follow course of fermentation by determination of acidity (pH value).

Verification

Quality of end-product.

Microbiological examination to determine number of starter organisms and ratio between types.

Inspection of plant records.

---

and should never exceed 24 h. A storage temperature of 10–20°C is optimal to avoid loss of viscosity and syneresis. Yoghurt may be packaged in a number of containers including foil-capped glass bottles or, most commonly, plastic pots capped with metal foil or plastic, 'snap-on' lids. The latter type may be fitted with a safety collar to prevent tampering with the contents. Low viscosity yoghurt may be gravity-filled, but piston fillers are required for other types. The filling nozzle should have a large orifice and the speed of operation be as low as operating requirements permit to minimize shear-stress.

(e) Special types of yoghurt

Drinking yoghurt is essentially stirred yoghurt which has a total solids content not exceeding 11% and which has undergone homogenization to further reduce the viscosity. Flavouring and colouring are invariably

CONTROL POINT: POST-FERMENTATION TREATMENT CCP 2

Control

Ensure efficiency of cooling and that any predetermined patterns of cooling are followed.

Handling procedures must avoid high shear stresses.

Ensure even distribution of added ingredients and consistent filling.

Monitoring

Cooling equipment to be fitted with thermographs and cooling monitored by experienced personnel.

Monitor operation of equipment involved in handling of yoghurt.

Monitor operation of equipment used for making additions to and for filling of yoghurt.

Verification

Quality of the end-product.

Measurement of level of addition and consistency of fill.

Inspection of thermographs and other plant records.

Maintenance of equipment on a regular basis.

---

added and one type is slightly carbonated. Heat treatment may be applied to extend the storage life. Two processes are used, a HTST

---

* The low pH value (4.0–4.5) of yoghurt means that a relatively mild UHT process is adequate to ensure microbiological stability. The pH value is also close to the isoelectric point of the caseins, leading to problems of chemical instability. The most satisfactory solution is to use highly methoxylated pectin as stabilizer. In the yoghurt, the pectin associates with calcium, acquiring a positive charge which, in turn, leads to an association between the pectin and caseins. As a result, the caseins also acquire a positive charge, repelling the pectin and establishing a system of ionic stabilization.

pasteurization coupled with aseptic packaging to give a life of several weeks at 2–4°C and a UHT process coupled with aseptic packaging to give a life of many weeks at room temperature.

Concentrated (strained) yoghurt is produced in a number of countries and may be known as labneh (Middle East), skyr (Iceland) and shrikhand (India). Products such as ymer (Denmark) are similar but are fermented by mesophilic starter cultures in conjunction, in some cases, with yeast. Concentrated yoghurt may be considered as intermediate between conventional fermented milks and high moisture, unripened soft cheese such as quarg and it is notable that suggested culinary uses as a spread or a salad dip are more usually associated with soft cheese.

The traditional process for preparation of concentrated yoghurt, straining in cloth bags, is still used on a small scale, but is laborious and subject to microbiological contamination. The most widely used commercial process involves centrifugal separation of skim milk yoghurt to produce a concentrated base, which is then recombined with butter oil or cream to the desired fat content. This process is used to produce 'thick and creamy' Greek-style yoghurts with a total solids content of *ca.* 24% and a fat content of *ca.* 10%. More recently there has been much interest in the use of ultrafiltration either to concentrate yoghurt into the final product or to concentrate the milk prior to fermentation. The latter system, however, leads to a product of deficient body and flavour and the most satisfactory procedure is to homogenize and concentrate yoghurt direct from the fermentation tank at a temperature of 40–45°C.

Heat-treated (pasteurized) yoghurt is intended to be stable at room temperature for periods of *ca.* 3 months. The yoghurt may be heated either in a plate heat exchanger at 75–80°C for 15 s or 'heat-shocked' in-carton at 58°C for 5 min. Higher than usual levels of stabilizer may be required with a possible adverse effect on organoleptic quality.

Frozen yoghurts vary widely in nature. As originally conceived, the product can be prepared from conventional set or stirred yoghurt, although an elevated level of sugar and stabilizer is required to maintain the coagulum during freezing and storage and a small quantity of cream may be added to improve 'mouth-feel'. It is also possible to replace the

---

* Heat-treatment of yoghurt is precluded by law in some countries, it being required that the product contains 'abundant and viable' organisms. Some strains of *Lb. delbrueckii* ssp. *bulgaricus* may survive pasteurization and thus permit legislation to be circumvented, but over-acidification is likely to lead to quality problems.

milk solids with whey protein concentrate. The yoghurt is then either frozen in a blast-freezer to at least −20°C, or frozen with aeration in an ice-cream freezer. Other types of frozen yoghurt, however are effectively a frozen, low-fat dessert (see Chapter 9, page 389), in which yoghurt acts as a source (with skim milk and buttermilk) of milk protein. In products of this type very little, or none, of the yoghurt character is retained.

Traditional dried yoghurt is produced by simply concentrating conventional yoghurt by boiling and sun-drying, the dried product being reconstituted before consumption. This process is not suitable for large-scale manufacture and for this purpose either freeze drying or spray drying may be used. Freeze drying, however, is an expensive process and most interest lies in spray drying. Low-cost spray dried yoghurt is seen as a means of supplying protein to nutritionally deprived areas, but in industrialized nations it is largely used as an ingredient in infant foods, baked goods and confections. The properties required for successful ingredient use vary and it is not possible to combine these in a single yoghurt product. In most cases, therefore, it is necessary to tailor the processing parameters according to the proposed end use.

8.2.3 Other fermented milks

(a) Acidophilus milk

Acidophilus milk is a traditional therapeutic milk fermented with *Lb. acidophilus*. Skim or whole milk may be used, a heavy heat treatment of *ca.* 95°C for 1 h, a tyndallization process, or UHT treatment being applied to reduce the microbial load and favour the slow-growing *Lb.*

BOX 8.3 **Milk comes frozen home**

Although technologically distinct from ice cream, frozen aerated yoghurt is superficially similar and is retailed in a very similar market. In some countries frozen yoghurt is the fastest growing sector of the frozen dessert market and is gaining market share at the expense of ice cream and low-fat frozen desserts. Specialist outlets have been established within restaurants, etc., at which flavouring or fruit is added at point-of-sale, enabling a very wide choice to be offered. This assists overall sales by establishing a high quality image.

> **BOX 8.4 Under pressure**
>
> The use of very high pressures, usually in the range 120–140 MPa, to inactivate micro-organisms has attracted much interest in recent years. The major advantage lies in the fact that, while the process is highly effective in inactivating micro-organisms, the 'quality carrying' molecules, vitamins, flavour/fragrance compounds, pigments, amino acids, etc., are unaffected. High pressure treatment is most advanced in Japan, where the process is used on a commercial scale in production of jam, fruit juice and yoghurt. The relative economic success of high pressure-treated yoghurt is not, however, known. (Mertens, B. 1992. *Food Manufacture*, **November**, 23–4).

acidophilus. For the same reason precautions, including daily sub-culturing, are necessary to ensure a high initial level of starter activity. Milk is inoculated at a level of 2–5% and incubated at 37°C until coagulated. Some acidophilus milk has an acidity as high as 1% lactic acid, but for therapeutic purposes 0.6–0.7% is more common.

(b) Cultured buttermilk

Small quantities of cultured buttermilk, the 'acid' or 'Bulgarian' type are fermented by *Lb. delbrueckii* ssp. *bulgaricus* and, possibly *Str. salivarius* ssp. *thermophilus*, the product resembling yoghurt. The more common type, however, is fermented by a mesophilic starter comprising *Str. lactis* ssp. *lactis* together with some, or all of, *Str. lactis* ssp. *lactis* biovar *diacetylactis*, *Str. lactis* ssp. *cremoris* and *L. mesenteroides* ssp. *cremoris*. Buttermilk may be used, but skim or whole milk is now more common. A widely used process involves heating the milk to *ca.* 80°C, de-aerating, homogenizing, heating to *ca.* 95°C and cooling to 20–25°C before addition of the starter culture. Starter is added at a level of 1–2%, fermentation proceeding for 16–20 h at *ca.* 25°C, to an acidity of 0.9% lactic acid. Separation may occur in the finished product and gelatin is sometimes added as stabilizer, although attention to process

---

* *Lactobacillus acidophilus* possesses an alcohol dehydrogenase and metabolizes acetaldehyde to ethanol. For this reason acidophilus milk may be considered tasteless in comparison with other fermented milks and attempts have been made to widen the appeal by co-fermentation with yoghurt starters or *Bifidobacterium*. The distinction between these products and yoghurt is, however, very blurred indeed.

control is considered a superior means of preventing this fault. Cultured buttermilk is largely a consumer product in Europe, but in the US very large quantities are produced for ingredient use, especially in bakery goods.

(c) Cultured cream

The fat content of cultured cream is standardized between 12 and 30% depending on the required properties. The starter culture is similar to that used for cultured buttermilk, but *Leuconostoc* is usually omitted. Standardized cream is heated to 75–80°C before homogenization at high pressure and temperature (above 13 MPa at 60°C) to improve texture. The inoculation and fermentation conditions are similar to those for cultured buttermilk, but the fermentation is stopped at an acidity of 0.6% lactic acid. The consistency of cultured cream is adversely affected by post-fermentation handling and inoculation and fermentation may take place in the final retail pack.

(d) Kefir

Kefir is a foamy effervescent drink which is usually made from whole milk heat treated to *ca.* 95°C for 5 min. This process denatures whey proteins and improves product consistency, homogenization also in general use in modern practice. A portion of the process milk is used to prepare the inoculum, being mixed with kefir grains and incubated at 20–25°C for *ca.* 20 h. At the end of this period the kefir grains are removed by sieving and rinsed in cold water for re-use. The cultured milk is then used as starter at a level of 3–5%, the main fermentation also being at 20–25°C for *ca.* 20 h. The kefir is then held for several hours during which time the coagulum stabilizes ('ripens'), the final product containing 0.9–1.1% lactic acid and 0.5–1% ethanol.

In recent years the nature of kefir grains has been investigated in some detail. The grains are gelatinous granules some 2–15 mm in diameter which consist of a mixture of micro-organisms grouped in a highly organized manner (Figure 8.3). The micro-organisms present vary but may include *L. lactis* ssp. *lactis* and ssp. *cremoris, Lb. acidophilus, Lb. kefir, Lb. kefiranofaciens, Lb. casei, Candida kefyr, Kluyveromyces marxianus* var. *marxianus* and species of *Saccharomyces* including *Sacch. cerevisiae*. Under conditions of regular sub-culturing, the grains proliferate in milk over many generations, the character and properties remaining unchanged. Kefir grains appear to consist of a matrix of which *ca.* 50% is the glucose and galactose containing carbohydrate, kefiran.

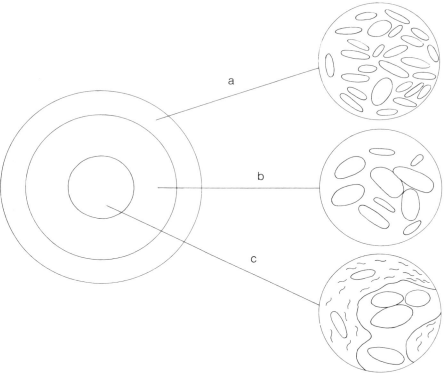

Figure 8.3 Structure of the kefir grain: schematic diagram. Rod-shaped bacteria are dominant in the peripheral layer (a). Below this an intermediate zone exists, with a higher number of yeasts (b). The centre of the grain largely consists of yeasts embedded in a kefiran matrix (c).

The matrix develops as convoluted sheet-like structures in which kefiran-producing lactobacilli are embedded. Non-kefiran producing lactobacilli and yeasts each predominate on separate sides of the sheets. Kefiran is produced by *Lb. kefiranofaciens* in the centre of the grain where growth is favoured by anaerobic conditions and the presence of ethanol. *Lb. kefiranofaciens* is thus responsible for propagation of the grains which does not occur in the absence of this organism, although non-propagable grains retain kefir-producing capacity. *Lb. kefir*, the

---

* Although kefir is popular in eastern Europe, consumption elsewhere is limited by the harsh taste of the traditional product. Variants have been developed in some western European countries and in both Germany and Sweden a modified kefir is produced using a starter culture consisting of micro-organisms derived from the kefir grain rather than the kefir grain itself. Such starters are suitable for direct-in-vat inoculation.

most common lactobacillus in kefir itself is present only in small numbers at the surface of the grain.

(e) Koumiss

Koumiss is traditionally made from mares' milk, but similar products are made from whole or skimmed cows' milk containing added sucrose. *Lactobacillus acidophilus*, *Lb. delbrueckii* ssp. *bulgaricus* and the yeast *Kluyveromyces marxianus* var. *marxianus* or var. *lactis* are used as starters and added to cooled, heat-treated (90–93°C for *ca.* 5 min) milk at a level of 10 to 30%. This gives an initial acidity of *ca.* 0.45% lactic acid, the inoculated milk being incubated at 26–28°C until the desired levels of lactic acid and ethanol are present. These are determined by market requirements, and vary from 0.6% lactic acid:0.7% ethanol to 1.0% lactic acid:2.5% ethanol. After incubation the product is cooled to *ca.* 15°C and agitated to ensure a smooth consistency and to slightly aerate. In many cases, after bottling, the koumiss is held for some time at ambient temperature to allow CO_2 to accumulate.

(f) Scandinavian ropy milks

Scandinavian ropy milks have a characteristic texture which is both sticky and yet easily cut by a spoon. A number of types are produced but the technology and microbiology are similar and typified by filmjolk. Milk is heated to 78–80°C, de-aerated, homogenized and re-heated to 90–95°C for 3 min. After cooling to *ca.* 20°C, 1–2% of a starter culture containing *Leuc. mesenteroides* and 'ropy' strains of *L. lactis* ssp. *lactis* and *L. lactis* ssp. *lactis* biovar *diacetylactis* is added. Incubation continues for 20 h when the acidity is 0.8–0.9% lactic acid, the coagulum is broken by stirring and the product cooled and packed. In some variants incubation takes place in the final pack, while traditionally made taetmjolk is thickened by addition of herbs.

One type of ropy milk, the Finnish viili, is the only mass produced fermented milk which is ripened by a mould, *Geotrichum candidum*. Milk of fat content 2.5–4.0% is used, the milk being heated but not homogenized. During fermentation, in retail containers, the fat rises to the top of the milk and forms a layer on which a 'felt' of mould mycelium develops. Spores of *G. candidum* are added with the starter culture, but the mould plays no role in the fermentation.

(g) Yakult$^{(R)}$ and Yakult Miru-Miru$^{(R)}$

Yakult products are commercially developed therapeutic fermented milks manufactured, and largely consumed, in Japan. Relatively little information is available concerning the technology of the products but

Yakult is fermented by a strain of *Lb. casei* ssp. *casei* with properties similar to *Lb. acidophilus*. The process milk is of low solids content containing only 1.1% fat, 1.2% protein and 1.1% lactose. However other carbohydrates are present at a level of 14%. Yakult Miru-Miru is made from process milk of similar composition to whole cow's milk but containing additional sugars at a level of 6.1%. The milk is fermented by a mixture of *B. bifidum*, *B. breve*, *Lb. acidophilus* and *Lb. casei* ssp. *casei*.

8.2.4 Dietary modification of fermented milks

Many fermented milks are of low fat content and thus perceived as 'healthy' foods without specific modification and low-cholesterol products have also been proposed. Further, many yoghurts fermented with *Lb. acidophilus* or *Bifidobacterium* are of low lactose content and suitable for consumption by lactose-intolerant persons. Lactose-reduced versions of non-therapeutic fermented milks have, however, been produced either by reducing the lactose content of the process milk by chemical or enzymatic hydrolysis, or by fermentation with thermophilic starter organisms. Fermented milks fortified with iron, calcium and dietary fibre have also been produced.

8.2.5 Fermented milk analogues

Production of fermented milk analogues based on soya milk have been seen as a means of increasing soya protein consumption and thus meeting the dietary needs of both the underfed of developing nations and the overfed of western Europe and the US. As with soya milk itself, however, problems have been encountered due to an unacceptable 'beany' flavour and flatulence associated with the presence of oligosaccharides such as stachyose. The coagulum forming properties of soya milk also differ from those of mammalian milk due to the predominance of albumins and globulins.

---

* A number of attempts have been made to improve the flavour of fermented milk analogues by fortification with various materials including whey protein concentrate, non-fat dried milk, lactose, sucrose and fructose. It is difficult to make comparisons on the basis of different trials, but it appears that fortification with sugars is most effective both in improving flavour and, by increasing fermentation of stachyose, reducing flatulence. Viscosity may, however, be lower. A number of starter bacteria have been found suitable for fermenting soya milk, the ability to ferment sucrose being of prime importance. Commercial freeze-dried cultures of *Lb. acidophilus*, *Lb. delbrueckii* ssp. *bulgaricus* and *Str. salivarius* ssp. *thermophilus* were suitable, although strain variation may be expected.

8.2.6 End-product testing

Chemical analysis of fermented milks is required to ensure that compositional standards are met and that the product is of the correct level of acidity. Organoleptic properties, including consistency, are also of major importance. Instrumental methods can be used, in some cases, to determine physical parameters such as coagulum strength, but better results may be obtained using either a taste panel or skilled individuals.

Microbiological analysis is used as a means of verifying the fermentation has proceeded correctly and that general hygiene standards have been satisfactory.

8.3 CHEMISTRY

8.3.1 Nutrients in fermented milks

Most work on the nutritional status of fermented milks concerns yoghurt, and while other products are similar some variations occur. The fermentation process itself affects the levels of nutrients present, but to a considerable extent the nutrients present in the finished product reflect those of the starting milk. Skim milk yoghurt, for example, has a lower content of fat and fat-soluble vitamins than whole milk yoghurt, while the practice of fortifying starting milk with dried skim milk powder, or concentrating the starting milk, results in an increase in the content of proteins and other water-soluble constituents. Loss of heat-labile vitamins occurs during heating of the starting milk and differences in manufacturing practice may account for variations in the content of B-group vitamins. The species of milk used has an obvious effect and some contribution is made by added ingredients. There is thus considerable variation from one type of yoghurt to another.

(a) Changes in nutrients resulting from fermentation

Fermentation reduces the lactose content during yoghurt manufacture, although where starting milk is fortified or concentrated, the lactose content of the final product can be higher than in liquid milk.

In unfortified yoghurt the protein and total amino acid content and composition is similar to that of starting milk. Some casein degradation occurs, although the extent is small compared with cheese, accounting for no more than 1% of the total protein. Some of the products of casein degradation are utilized by the starter micro-organisms while others accumulate in the yoghurt as free amino acids. The quantities of free amino acids accumulating varies according to species being 33 mg/100 g in goats' milk yoghurt, 23 mg/100 g in cows' milk and 18 mg/100 g in ewes' milk. These quantities represent increases over the starting milk of 160, 400 and 500%, respectively. In cows' milk the total content of essential amino acids rises from 1.2 mg/100 g to 4.77 mg/100 g with a particularly large increase in proline, followed by serine, alanine, valine, leucine and histidine. Levels of free amino acids continue to rise during storage as a consequence of continuing *Lb. delbrueckii* ssp. *bulgaricus* metabolism. A limited degree of lipolysis occurs during manufacture of yoghurt and other fermented milks. The extent varies according to the milk species, but both concentration and pattern of free fatty acids is affected. Small quantities of volatile fatty acids are also formed during fermentation and amino acid degradation.

The activities of micro-organisms during yoghurt fermentation have important consequences for vitamin content, which are superimposed on differences due to composition of the starting milk and thermal processing. Vitamins are initially metabolized and then synthesized by starter micro-organisms. There is little net effect on levels of thiamine, riboflavin, nicotinic acid, pantothenic acid or biotin, although the two latter vitamins are usually slightly reduced in concentration, while others may be increased. Levels of folic acid are increased by *ca.* 100% and there is usually an increase in choline content. In each case the increase is largely due to synthesis by *Str. salivarius* ssp. *thermophilus*. Vitamin B_{12} is required by the *Lb. delbrueckii* ssp. *bulgaricus* and levels are reduced.

(b) Effect of fermented milks on digestibility and bioavailability of nutrients

Discussion of the nutritional properties of fermented milks can be complicated by the possibility of the product possessing specific therapeutic properties and both alleviation of lactose maladsorption and nutritional enhancement have previously been mentioned in this context (page 350). In addition fermented milks are recognized as being more digestible in the general sense and, in the case of yoghurt, the specific rate of protein digestion is twice that in raw milk. This stems partly from the partial breakdown of protein in the yoghurt and partly from the physical structure of its 'soft' coagulum. This is very different in nature to the clot which normally forms when milk enters the stomach and permits ready interaction between stomach enzymes and their substrates.

In the past it has been suggested that fermented milks increase the bioavailability of calcium and other minerals and trace elements. A systematic study with rats, however, showed that, in yoghurts with low lactose levels, the bioavailability of calcium, magnesium and zinc was reduced, although the reduction was of no dietary significance. Iron absorption was also less from yoghurt than from milk due to the shortened intestinal transit time, but bioavailability of phosphorus was greater in yoghurt.

8.3.2 Flavour of fermented milks

In many modern fermented milks, especially yoghurt, the flavour is derived from fruit and other additives and intrinsic flavour is of little significance. In plain varieties, however, the characteristic flavour is largely derived from starter culture metabolites. A large number of compounds are produced which may contribute to flavour (Table 8.5), but lactic acid and the carbonyl compounds acetaldehyde and diacetyl are of overwhelming importance. Lactic acid is a major flavour component in all fermented milks and the quantity present may dictate acceptability, excess amounts resulting in flavour defects. The relative importance of acetaldehyde and diacetyl varies according to the starter culture used. In the case of yoghurt and similar products acetaldehyde is dominant, being detectable when the pH value of the yoghurt falls to 5.0. Maximum levels are present at pH 4.2 and the compound stabilizes at *ca.* pH 4.0. Levels of acetaldehyde are increased by addition of milk solids and by milk heat treatments stimulatory to starter cultures (see page 354). Milk species also affects acetaldehyde production, highest levels being present in cows' milk yoghurt and lowest in ewes' milk. Some loss of acetaldehyde occurs during storage, especially when fat is present at only low levels.

---

* The role of hydrocolloids, added to flavoured and some other fermented milks as stabilizers, in human nutrition is often overlooked, but Tamime and Robinson (1985) consider that this is '... an aspect of yoghurt consumption in industrialized countries that should not go unnoticed'. Hydrocolloids are often referred to as 'unavailable carbohydrate'. They have no direct nutrient role, although some stabilizers are considered to have a role in depressing blood cholesterol levels. The major role in human nutrition stems from their properties as bulking agents. Stabilizers have been shown to stimulate intestinal peristalsis and thus avoid some risks of colonic malfunction; to absorb some potentially toxic microbial metabolites formed in the large intestine and to delay the diffusion of nutrients to the cell wall. This latter factor is of particular significance to the lactose-intolerant and those prone to post-prandial hyperglycaemia since the rate of entry of lactose and glucose into the blood is reduced (Tamime, A.Y. and Robinson, R.K. 1985. *Yoghurt, Science and Technology*. Pergamon Press, Oxford).

Chemistry

Table 8.5 Flavour compounds of fermented milks

Non-volatile acids
  lactic, pyruvic, oxalic, succinic
Volatile acids
  formic, acetic, propionic, butyric
Carbonyl compounds
  acetaldehyde, acetone, acetoin or diacetyl
Miscellaneous
  amino acids and/or constituents formed by thermal degradation of protein, fat and lactose

From Tamine, A. Y. and Robinson, R. K. 1985. *Yoghurt: Science and Technology*. Pergamon Press, Oxford.

Diacetyl is present in organoleptically significant quantities only when *L. lactis* ssp. *lactis* biovar *diacetylactis* or *Leuconostoc* spp. is present in the starter culture. Diacetyl is thus an important character impact compound in cultured buttermilk, some cultured cream and many Scandinavian ropy milks. The carbonyl acetoin, which is also present in significant quantities, is essentially tasteless, but can be oxidized to diacetyl. Although ethanol is formed by lactic acid bacteria, the quantity is very small and has little, or no, role in the flavour of most fermented milks. The obvious exceptions are the alcoholic fermented milks such as kefir and koumiss in which ethanol produced by yeasts contributes to the overall organoleptic characteristics. Yeasts such as *Sacch. cerevisiae* can also convert amino acids to volatile compounds through transamination and decarboxylation reactions and it is probable that these volatiles, and esters and acetals arising from their reaction with ethanol, contribute to the flavour of alcoholic fermented milks. In the case of kefir, variations in the type of yeast present in the granule may account for starter-related differences in flavour.

The mould *Geotrichum candidum* develops as a non-starter organism in viili and plays a major role in the development of the characteristic flavour of that product. Growth of the mould is largely on the fat and free fatty acids and methyl ketones are important flavour compounds.

Although the flavour of yoghurt is largely derived during fermentation, there can be a contribution from the milk and other ingredients of the starting mix. The milk species, for example, modifies the overall flavour perception, while Maillard products in poor quality milk powder may introduce flavour defects. Lactones are produced from fat during heating of whole milk yoghurt and may contribute to flavour as may the formic acid-like compound responsible for stimulation of *Lb. delbrueckii* ssp. *bulgaricus*.

8.3.3 Physico-chemical changes to milk constituents during yoghurt manufacture

(a) Homogenization

Homogenization affects the fat phase of milk which plays no direct role in formation of the yoghurt coagulum. The reduction of size, and increase in number, of fat globules resulting from homogenization does, however, modify the subsequent gel. In the first place the adsorption of 'small' fat globules onto casein micelles increases viscosity and the effective total volume of suspended matter. Secondly syneresis is decreased due to the increase in hydrophilicity of casein micelles as a result of protein : protein and casein : fat globule membrane interactions. There is also an additional effect resulting from denaturation of some proteins.

(b) Heating

Heating of yoghurt mix has an important technological role in modifying the properties of proteins and contributing to the formation of a stable coagulum. There is also a redistribution of calcium, magnesium and phosphorus ions between the soluble and colloidal form which tends to reduce coagulation time.

The major direct consequence of heating on the proteins is the denaturation of the whey proteins and the interaction with κ-casein. For a number of years it was thought that the interaction involved only β-lactoglobulin, but it is now recognized that α-lactalbumin is also involved (Figure 8.4). A number of ultrastructural studies have been made and a number of workers have described the formation of a chain matrix of micelles which results in a fine network of protein and an effective immobilization of the aqueous phase. In milk heated at 95°C for 10 min filamentous appendages, apparently consisting of denatured β-lactoglobulin, were observed attached to the casein micelles. The appendages increased the firmness of the coagulum by inhibiting micellar contact and fusion but tended to 'diffuse' after fermentation. It is important to appreciate, however, that binding between whey proteins and casein is highly dependent on both pH value and the level of Ca^{2+} ions and that natural and normal variations can result in completely different casein micelles.

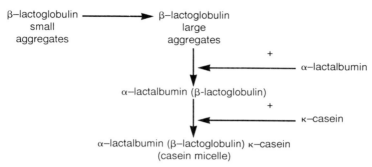

Figure 8.4 Interactions between casein and whey proteins during heating of milk. Based on Tamine, A.Y. and Robinson, R.K. 1985. *Yoghurt: Science and Technology.* Pergamon Press, Oxford.

A further important effect of heating is the increase in hydrophilicity of the proteins which reduces syneresis and increases gel firmness. This is a consequence of the covalent attachment between κ-casein and β-lactoglobulin which results in a new surface structure with fewer exposed hydrophobic groups. Maximal hydration is obtained by heating milk to 85°C for 30 min, the yoghurt produced after this treatment exhibiting true thixotropic behaviour. Increasing the severity of heat treatment further increases hydrophobicity and leads to syneresis and a poor quality yoghurt.

(c) Coagulum formation

The yoghurt coagulum is primarily of the acid type similar to that of acid-set cheese (page 324). There are important differences in that while syneresis is desirable in cheese manufacture for curd formation, it is undesirable in yoghurt and other fermented milks. In either case earlier theories of gel formation based on the concept of aggregation of the micelles following neutralization are considered over-simplistic and an alternative model based on simulated yoghurt manufacture has been developed. This involves the concept of the casein micelle as a 'skeleton' of α_{s1}-casein with bound β-casein and amorphous calcium phosphate surrounded by a stabilizing layer of κ-casein. As the pH value falls during fermentation amorphous calcium is released by acidification and complexation, but the 'skeleton' is retained. Disaggregation is followed by aggregation which is initiated by β-casein as soon as the pH value is sufficiently low and the two major caseins carry opposite charges. Further lowering of the pH value is followed by contraction, when the β-casein becomes positive in charge and the α_{s1}-casein remains negative, and ultimately, network formation. Rate of acidification is important in network formation, fast acid production leading to precipitation (*cf* acid casein).

Later work has further increased understanding of acid gel formation, although many aspects remain to be resolved. An optical method involving light reflection was used to study acid milk coagulation and demonstrated a progressive structural rearrangement of proteins occurring as the pH value fell during fermentation. Beta- and κ-casein were re-incorporated into the micellar framework, forces of hydration and electrical repulsion overcome and further attractions created that initiated formation of a three-dimensional protein network. This study was developed further, using milk acidified with glucono-δ-lactone, it being shown that the collapse of the hairy κ-casein layer of the casein micelle at low pH value is the major factor leading to aggregation. At temperatures above 30°C this leads directly to aggregation, while at temperatures of 15–20°C micellar protein solubilization occurs followed, possibly, by re-incorporation of solubilized casein and aggregation (Figure 8.5).

The microstructure and physical properties of the yoghurt coagulum are affected by fortification of the starting milk and changes in the level of protein and the proportion of casein and non-casein nitrogen protein, the latter parameter being of particular importance in determining physical properties. A casein to non-casein protein ratio of 3.2–3.4:1.0 is considered to give the highest quality coagulum, lower ratios resulting in a soft coagulum and syneresis and higher ratios in a rough, coarse coagulum.

Damage by proteolytic enzymes has a deleterious effect on the yoghurt coagulum. Treatment of milk with bacterial proteases, which hydrolyse κ-casein, resulted in a coagulum of increased firmness, syneresis and apparent viscosity, while plasmin, which hydrolyses β-casein, resulted in reduced firmness, apparent viscosity and syneresis. In each case the coagulum had a lower water holding capacity and degree of hydration.

Hydrocolloids are added to some fermented milks as stabilizers and act by forming a network of linkages between milk constituents and themselves. This property arises either from the presence of a salt able to sequester Ca^{2+} ions or from the presence of negatively-charged groups such as hydrogen or carbonyl radicals. Water binding capacity of the coagulum is increased by stabilization of the protein network to retard free movement of water, increasing the level of hydration of constituents (primarily proteins) and binding of the water of hydration.

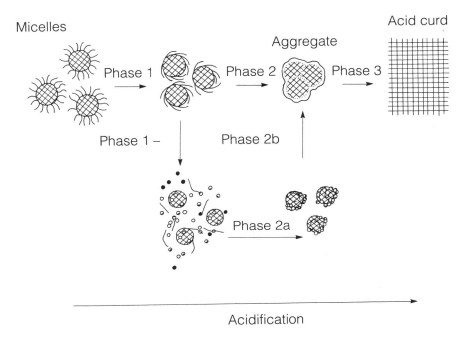

- o Micellar casein monomer
- ⌒ Casein monomer of the micellar hairy layer

Figure 8.5 Sequence of events during formation of the yoghurt coagulum. Redrawn with permission from Banon, S. and Hardy, J. 1991. *Journal of Dairy Science*, **75**, 935–41.

Although fat has no direct role in coagulum formation, work with filled dairy gels showed fat globules to be interdispersed in the protein network and to increase firmness by hydrostatic restriction of the network adjacent to globules. The situation in fermented whole milk products may be similar.

Nodulation is a serious fault in yoghurt, which appears to be associated with proteins. Nodules contain *ca.* 82% protein and *ca.* 4% phosphorus, but compared with the surrounding coagulum have only traces of fat and lactose. Electron microscopy has shown the nodules to be of a definite, smooth-surfaced compact structure and to have no contact with the casein of the coagulum network. Formation is associated with *Lb. delbrueckii* ssp. *bulgaricus* in the starter culture and also with a high whey protein content, it being postulated that associated changes in micelle structure are a predisposing factor to nodule formation.

8.3.4 Role of microbial slime formation in the structure of fermented milks

In yoghurt slime producing strains of *Str. salivarius* ssp. *thermophilus* play an important role in stabilizing the protein network. Extracellular polysaccharides are probably of greatest importance, although lipotechoic acid may also be involved. The nature of the extracellular polysaccharide varies from strain to strain and while several workers have reported that glucose and galactose are major constituents, the ratios of the two sugars vary. The amino sugar *N*-acetylgalactosamine has been reported and small amounts of arabinose, mannose, rhamnose and xylose may also be present. A distinct type of extracellular polysaccharide composed of rhamnose and galactose in the ratio 1:1.47 has also been isolated from a commercial yoghurt culture.

Electron microscopy has shown starter micro-organisms to be present in 'pockets' in the yoghurt coagulum, with filaments of polysaccharide attaching the organisms to the coagulum network (Figure 8.6).

The role of slime in determining the viscosity and texture is more obvious in the case of Scandinavian ropy milks. There does, however,

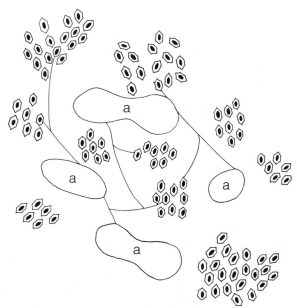

Figure 8.6 Relationship between starter micro-organisms and the yoghurt coagulum. Cells of starter micro-organisms (a) are attached to each other and to the yoghurt structure by gum-like polysaccharide filaments.

appear to be a considerable variation in the nature of the slime. The major component of slime produced in the Swedish langfil, for example, was a deacylated lipotechoic acid and surface-exposed lipotechoic acid was also considered to be involved in slime formation in viili. Slime forming strains of *L. lactis* ssp. *lactis* have been shown to have fewer membrane antigens and to possibly have defects in cell wall biosynthesis.

Slime production by other starter strains examined involves products such as phosphopolysaccharides. Analyses of slimes have tended to produce differing results, probably as a consequence of different analytical methods but a slime comprising 47% protein and 29% carbohydrate has been considered fairly typical. A novel phosphopolysaccharide from *L. lactis* ssp. *cremoris* has been described which comprised 21% protein and 42% carbohydrate and had a carbohydrate to phosphorus ratio of 5.2:1. The major sugars were rhamnose, glucose and galactose in the ratio 1:1.45:1.75, the slime carrying a negative charge and being able to form complexes with protein.

The characteristic texture of ropy milks results from a network of slime which enmeshes the cells of the starter micro-organisms and links protein micelles to form casein conglomerates. The rheological properties of the bacterial slime are superimposed on those of the milk protein coagulum.

8.3.5 Chemical analysis of fermented milks

(a) Ingredients

Testing is required to ensure that the milk is free of antibiotics (see Chapter 2, page 86) and, where the milk is to be concentrated or fortified, determination of the total solids content permits accurate control of the solids content of the mix. The fat content of milk should also be determined during manufacture of whole milk products, or for standardization. The increasing popularity of goat and ewes' milk products and the relative scarcity and high price of the starting milk means that partial substitution with cows' milk is a possibility. In cases where the provenance of the milk is not known the species may be determined using ELISA based techniques.

The quality of skim milk powder can directly affect the quality of the finished product and chemical analysis is required (Table 8.6). Where milk is concentrated before fermentation, the concentration process

Table 8.6 Chemical analyses required for skim milk powder used in fortification of starting mix for fermented milk production

Acidity
Fat content
Moisture
Scorched particles
Solubility

and the finished product should be controlled as outlined in Chapter 3, page 111.

(b) End-product testing

The total solids and fat content should be determined to verify that compositional standards have been met and it is usual to verify the pH value. It is also common practice to ensure that the product is of the required physical characteristics. Viscosity is of greatest importance and may be determined by a number of methods with stirred and fluid products. Viscometers provide the most accurate measurement, but non-instrumental techniques are acceptable providing that conditions are standardized. Such methods include flow rate through a tube or down an inclined plane, or time for a sphere or plummet to pass down a cylinder containing the product. In very small-scale operations subjective techniques such as spooning are adequate providing personnel are experienced.

Set products are more difficult to assess since many methods applicable to stirred destroy the set coagulum. Falling sphere methods are suitable but the most satisfactory results are obtained using a penetrometer.

8.4 MICROBIOLOGY

8.4.1 Fermented milks as an environment for micro-organisms

Fermented milks, in general, are of low pH value and high lactic acid concentration and are thus a highly selective environment favouring the growth of yeasts and moulds as spoilage micro-organisms. For this reason sorbate and benzoate are permitted preservatives in some countries, but their effectiveness may be limited by the emergence of resistant strains. Secondary selective pressure is exerted by oxygen availability which restricts the development of moulds and nonfermentative yeasts, while the addition of added sugars favours the growth of fermentative yeasts. The solute level in some cases is sufficiently high to produce a small, but significant, lowering of the a_w level. This is not sufficient to restrict the growth of yeasts or moulds, although the behaviour of starter micro-organisms may be affected (see page 356). The addition of humectants to lower the a_w level further has, however, been investigated as a means of extending the storage life of yoghurts. Low-fat yoghurts appear to be less suitable as a growth medium than full-fat, but the reason is not known.

8.4.2 Fermented milks and foodborne disease

Fermented milks are only rarely associated with foodborne disease, although consumption of products such as yoghurt containing a large number of yeasts can lead to digestive disturbances. The pH value of fermented milks is too low, and the lactic acid concentration too high, to permit the growth of vegetative pathogens and death of non-growing cells is likely to be rapid. *Campylobacter*, for example, is rapidly killed in the presence of lactic acid, while *Salmonella* is usually considered to be destroyed, or inactivated, in fermentations where the lactic acid concentration exceeds *ca.* 1% and the pH value is lower than 4.55. Caution is required, however, since the results of survival experiments may, for a number of reasons, be highly variable. Investigations into the survival of *Listeria monocytogenes*, for example, an organism usually considered to be sensitive to lactic acid, in yoghurt have variously shown that death is very rapid, or that the organism persists for 3–9 days at pH 4.1 depending on inoculum level. It is also necessary to consider that habituation to acidic conditions may permit enhanced survival of organisms such as *Salmonella*, while the increasing trend to 'mild' products buffered at a relatively high pH value inevitably increases the possibility of survival of vegetative pathogens. It is not possible, therefore, to depend on pH value and lactic acid content to ensure the safety of fermented milk products and heat treatment of starting milk and protection from recontamination are considered essential for safety.

An outbreak of botulism due to consumption of hazel-nut yoghurt during 1989 provided an exception to the otherwise very good safety record of fermented milks.

The underlying cause, however, lay not with the yoghurt making, but underprocessing of the canned hazel-nut puree used to flavour the yoghurt and subsequent growth and toxigenesis by *Clostridium botulinum* in the puree. Process parameters used were based on requirements

> **BOX 8.5 Every physician hath his favourite disease**
>
> The outbreak of botulism associated with hazel-nut yoghurt was the largest in the UK. One of the characteristic features of the outbreak was the large number of cases where diagnosis was seriously delayed. This may be partially attributed to the atypical symptoms, which included sore throat, drowsiness and fever. At the time, however, there was both considerable publicity given to the outbreak and a climate of major public concern over food-borne disease in general. A number of patients raised the possibility of botulism, but despite this hospitalization was delayed by up to 2 weeks by the refusal of medical practitioners to diagnose the disease.

for fruit of significantly lower pH value than hazel-nut and no account was taken of the effect of substituting aspartame for sucrose. Control procedures at the yoghurt plant were, however, inadequate and a lack of visible spoilage was taken to indicate the safety of individual cans, despite the fact that on an overall basis the number of 'blown' cans was high.

8.4.3 Spoilage of fermented milks

Most work concerning the spoilage microbiology of fermented milks has been concerned with yoghurt. Most fermented milks, however, have a similar spoilage pattern and yoghurt may thus be generally considered representative. For obvious reasons most attention is given to recognized spoilage organisms such as yeasts, but it is also important to appreciate the role of the starter microflora in deterioration of organoleptic quality in the absence of overt spoilage. Numbers of starter micro-organisms remain in excess of 10^8 cfu/g for long periods at temperatures 10°C or less, but numbers can decline rapidly at higher

---

* *Lactobacillus delbrueckii* ssp. *bulgaricus* is responsible for undesirable changes during storage and a number of means of prevention have been suggested. These include pre-growth of *Lb. delbrueckii* ssp. *bulgaricus* followed by thermal inactivation and inoculation with *Str. salivarius* ssp. *thermophilus* and the use of a membrane dialysis fermenter. In this system the two starter components are cultured separately, the cultures being dialysed against each other in hollow cellulose acetate fibres to permit interchange of stimulatory compounds. This technique, which is currently developed to a laboratory scale only, has also been applied to production of cultured buttermilk (Klaver, F.A.M. *et al.* 1992. *Netherlands Milk and Dairy Journal*, 46, 31–44).

temperatures. Continuing post-fermentation metabolism by starters can limit life due to excess acidity and proteolytic activity resulting in bitterness and syneresis. Bitterness may also result from the death of starter micro-organisms possibly due to the release of proteolytic enzymes.

Yeasts are the most important spoilage organisms and are most commonly associated with fermentation leading to gas production. Such spoilage is readily recognized by 'doming' of foil lids and even burst containers. Genera such as *Kluyveromyces marxianus* and *Saccharomyces spp.* are most commonly involved involved in fermentative spoilage and yoghurts containing added sugars are usually affected. Fruit purees contaminated by unsatisfactory handling procedures are often considered to be the major source of contamination, although chocolate-flavoured yoghurt is also highly susceptible to fermentative spoilage.

Oxidative yeasts are also important spoilage organisms in yoghurt. Growth is limited by availability of oxygen and may be restricted to the air/yoghurt interface. In such cases moist, flat colonies, or a film of growth may develop. Thin walled polystyrene containers, however, permit sufficient air to enter the pack to support the growth of oxidative yeasts in the bulk of the yoghurt. In such circumstances oxidative yeasts predominate near the pack walls and fermentative yeasts in the centre of the pack. A wide range of oxidative yeasts have been isolated from yoghurt including species of *Candida, Debaryomyces, Metschnikowia, Pichia, Rhodotorula, Torulaspora, Trichosporon* and *Yarrowia*.

Mould growth at the yoghurt/air interface leads to development of visible mycelial 'mats' or 'buttons'. A wide range of species have been isolated including *Absidia, Alternaria, Aspergillus, Micelia, Monilia, Mucor, Penicillium, Pullaria* and *Rhizopus*. *Aspergillus* is of particular significance as isolates have included aflatoxin-producing strains of *A. flavus*. Although moulds are usually of secondary importance to yeasts, mould growth is sometimes a more significant problem in yoghurts held for extended periods at temperatures of *ca.* 0°C before transfer to retail display.

8.4.4 Microbiological analysis

Microbiological analysis is employed in manufacture of fermented milks to ensure that starter culture components are present in the correct proportions. This is of particular importance in yoghurt where there is a symbiotic relationship between the two components. A ratio of *Lactobacillus* to *Streptococcus* of 1:1 is theoretically required in the yoghurt starter, although, in practice, the greater tendency of chains of *Streptococcus* to break up is compensated for by use of a ratio of 1:2.7. The relative numbers may be determined by colony count, the commercially available L-S differential medium, which permits differentiation between *Lactobacillus* and *Streptococcus* on the basis of colonial morphology, reduction of tetrazolium trichloride and hydrolysis of casein being most suitable. Colony counts are, however, time consuming and direct microscopic examination, which provides information directly before use is more common.

Enumeration of starter organisms is also used in end-product testing as a means of assessing freshness at point of sale. Yoghurt should contain both *Str. salivarius* ssp. *thermophilus* and *Lb. delbrueckii* ssp. *bulgaricus* at numbers in excess of 10^8 cfu/g to be considered satisfactory, colony count techniques being considered most appropriate at this stage.

A selective medium for *Bifidobacterium* is required to monitor numbers in 'therapeutic' fermented milks. The reference medium contains neomycin, paromomycin, nalidixic acid and lithium chloride as selective agents, but preparation of this medium is difficult and for routine use an alternative based on lithium chloride and sodium propionate is preferred.

End product testing for non-starter micro-organisms is usually restricted to determination of colony count on a non-carbohydrate containing nutrient agar and enumeration of yeasts. Yeasts may be enumerated by colony count, preferably using a neutral pH selective medium such as Rose Bengal–chloramphenicol (RBC) agar, or by a modified direct epifluorescent technique. The number of yeast present immediately after production, however, is very low and the predictive value of enumeration at this stage is often poor. *Ad hoc* methods involving pre-incubation at 25 or 30°C are used by some producers.

Colony counting methods using media such as RBC can also be used for enumeration of moulds, but are notoriously inaccurate. An enzyme-linked immunosorbent assay method has been described, which has much improved predictive value. Application of such methods, however, is limited by availability of suitable antisera.

Although microbiological standards for fermented milks sometimes include *Staphylococcus aureus*, examination for specific pathogens is not considered necessary under normal circumstances. Determination of 'coliforms' may be used as a means of assessing post-process contamination. Death of the organisms due to the acidic conditions means that determinations must be made immediately after packaging and results interpreted with caution.

EXERCISE 8.1.

In many countries the legislative situation concerning therapeutic yoghurts is poorly defined and even where legislation exists it is usually based on the number of starter organisms present at the end of the recommended storage life. Do you consider this to be sufficient to protect consumers, or should more stringent criteria be applied? Do you consider the evidence supporting therapeutic activity to be sufficient to allow claims concerning this activity to be made on packaging or in advertising?

EXERCISE 8.2.

It is possible to gel milk (sweet curdling; *cf.* page 74) by use of starches and other hydrocolloids. Consider the structure of gels produced by starches and pectins, both widely used in yoghurts. What interactions with the acid gel are likely? What are the consequences of interactions for the stability and texture of yoghurt?

Information concerning the structure of hydrocolloid gels may be obtained from Oakenfell, D. 1987. *CRC Critical Reviews in Food Science and Nutrition*, **26**, 1–25.

EXERCISE 8.3.

You are employed as a microbiologist for a company manufacturing yoghurt, primarily for sale under supermarket own-labels. In response to publicity concerning food poisoning, the Chief Food Technologist of a major customer has *demanded* that end-product tests on yoghurt should include examinations for enterohaemorrhagic *Escherichia coli* (O157: H7), *Listeria monocytogenes*, *Salmonella* and *Shigella*. You consider this demand to be absurd and are asked by your management to produce a paper supporting your position. Do you consider that monitoring of any of these organisms in the environment of the plant would be justified as part of your HACCP-based safety management system?

9

ICE CREAM AND RELATED PRODUCTS

OBJECTIVES

After reading this chapter you should understand
- The nature of ice cream and related products
- The role of the various ingredients
- The calculation of recipes
- The processing technology
- The major control points
- Physico-chemical changes during processing
- The structure of the final ice cream
- Microbiological hazards

9.1 INTRODUCTION

The creation of iced desserts, by mixing snow with fruit and fruit juices, is an ancient culinary practice which probably originated in China. Iced desserts appear to have been introduced to Europe in the late 13th century, but for over 500 years were made only as novelties for the overfed aristocracy. Larger-scale manufacture, and sale in restaurants, began in the 19th century with the invention of a hand-operated ice cream maker and the growth of an affluent middle class. Wholesale manufacture began in 1851 after which the industry expanded rapidly, especially after the introduction of mechanical refrigeration and the development of power operated ice cream making equipment.

Ice cream comprises a number of related products, which primarily differ in the relative quantities of ingredients rather than in manufacturing technology (Table 9.1). Terminology varies somewhat in different countries, but a working classification is possible.

Ice cream is commercially the most important and contains fat and milk

Table 9.1 Typical composition of ice cream and related products

	Fat (%)	MSNF[1] (%)	Sugar (%)	Emulsifier/stabilizer (%)
Ice cream				
standard	10	11	14	0.5
premium	15	10	17	0.3
super premium	17	9.25	18.5	–
Milk ice	4	12	13	0.7
Sherbet	2	4	25	0.6
Water ice	–	–	30	0.5

[1] Milk solids-non-fat.

solids-non-fat in accordance with legislation. Legislation varies considerably, but minimum requirements in the UK are listed in Table 9.2. Minimum quantities are commonly considerably lower than those used in commercial practice. In many countries non-dairy fat is not permitted in ice cream and products made with non-dairy fats must be described by an alternative name such as, in the US, mellorine. The only European Community countries which permit the use of non-dairy fats in ice cream are Iceland, Portugal and the UK. In these countries the term 'dairy' ice cream is applied where all the fat is milk fat.

Milk ice is made from milk, usually without added fat. Non-dairy fat is not permitted and in most countries a minimum fat content of 2.5–3.0% is stipulated.

Parev ice (Kosher ice cream) usually has a stipulated fat content of 10%, but must contain no dairy fat or any other milk derived ingredient.

Custards are made with eggs or egg yolk solids and in the US must

Table 9.2 United Kingdom product standards for ice cream

	Minimum fat (%)	Minimum MSNF[1] (%)
Ice cream	5.0	7.5
Ice cream and fruit	5.0	7.5
or	7.5	2.0[2]

[1] Milk solids-non-fat.
[2] In this case the minimum content of milk fat and milk solids-non-fat must be 12.5%, expressed as a percentage of whole product including fruit, fruit juice, etc.

Introduction

> **BOX 9.1 Taken from the French**
>
> Custards are sometimes described as 'French ice cream' or 'French custard ice cream', but differ from the genuine French product, *glace aux oeufs*, which must contain at least 7% egg yolk solids. Traditional milk ice, containing cornflour, may also be described, incorrectly, as 'custard'.

contain at least 1.4% (w/w) egg yolk solids.

A number of other frozen products exist which contain little, or no fat or dairy ingredient. These, however, are marketed alongside ice creams and generally considered to be part of the same product group.

Water ice is made from dilute fruit juice with sugar and added acid. Stabilizer and added colours and flavours may also be present. Water ices may be frozen with, or without, incorporation of air and may be hard frozen or retailed as a semi-frozen slurry (e.g. Slush puppy^{TM}). Water ice hard frozen onto a stick is referred to as an 'ice lolly' in the UK and may be combined with ice cream as a 'split'.

Sherbets and *Sorbets* are both similar to water ice, but contain small quantities of milk solids and, in some cases, whipping agents to give a higher overrun.

Low fat frozen desserts are a complex group of products. Most are formulated to have similar properties to conventional ice cream, but have a fat content below the legal minimum permitted in ice cream. A complex stabilizer system is required to provide body. Intense sweeteners, such as Aspartame^{TM}, and fat replacers, such as Simplesse^{TM}, may be used.

---

* Aerated desserts such as mousse are of similar composition to ice cream, but the structure is stabilized by high levels of stabilizer rather than by freezing. In the past mousses have been retailed as chilled products, but the use of UHT sterilization, in conjunction with aseptic packaging, has permitted the development of the 'ambient mousse' using nitrogen as aerating agent to minimize oxidative deterioration. Products of this nature are important in allowing ice cream makers to more fully utilize production plant outside the peak summer season and to reduce dependence on a hot summer to stimulate sales. Equally some dairies have installed aerating equipment to enable mousse production as a means of extending product range and adding value to milk.

In recent years, large quantities of ice cream have been sold in bulk packs, although sales of hand-held products such as wafer bars and cornets are still significant. Ice cream is also an ingredient of novelty products such as ice cream cakes and pies. The market has been dominated by 'standard' quality ice cream, 85% of that sold in the UK during 1991 being made with non-dairy fat. Three separate trends, however, are having a significant effect on the ice cream market; the development of 'premium' and 'super-premium' ice cream, the development of ice cream-based versions of chocolate confectionery bars and the production of low-fat, low-calorie, 'healthful' ice cream-like products (frozen desserts).

'Premium' and 'super-premium' ice cream were originally developed in the US, but is now available in the UK, where development has been stimulated by a Milk Marketing Board initiative encouraging the use of fresh cream. Such ice creams are seen as products for the adult market, the main target group being ABC1 23–24 year olds. The growth of this sector is illustrated by the UK ice cream sales during 1991, when total sales fell by 5%, but sales of premium category products rose 11.5%.

BOX 9.2 **By their fruits ye shall know them**

The terms 'premium' and 'super-premium' are not legally defined, however 'premium' may be considered to have 14–16% butterfat, an overrun of up to 70% and to contain stabilizers. 'Super-premium' is considered to have 16–18% butterfat, an overrun no greater than 30% and to contain no stabilizer or emulsifier (in contrast the overrun of 'standard' quality ice cream may be as great as 120% and require both emulsifiers and stabilizers. Both 'premium' and 'super-premium' ice cream are also characterized by 'extravagant' flavour combinations and the use of added ingredients such as whole nuts, shortcake, etc., to provide textural contrasts. Vanilla flavours such as vanilla pear or vanilla fudge remain most popular, other flavours tending to fall into two categories; 'brown combinations' such as butterscotch brittle, coffee fudge and peanut butter bites; and 'fruited varieties' such as strawberry cloud, blueberries and cream and banana crunch. Ice creams of this type are considered to extend the total ice cream market rather than to compete with existing 'standard' quality ice cream.

Ice cream-based chocolate confectionery bars have taken a significant share of the market for hand-held products. Ice cream of this type was introduced by a multinational confectionery manufacturer, who successfully exploited brand loyalty by use of existing product names. Established ice-cream makers have now entered the market, but generally lack existing brand names and have to establish identity for each product. A similar use of brand extension is now being made in the iced lollipop market with a product derived from a popular brand of fruit flavoured sweets.

'Healthful' frozen desserts follow a well established trend, although product development has been given an impetus by the availability of fat replacers and, where permitted, new artificial sweeteners. 'Healthful' products are, to some extent, complementary to 'premium' and 'super-premium' ice cream, which tend to be eaten as a treat. Strong competition, however, comes from related dairy products such as frozen yoghurt.

9.2 TECHNOLOGY

9.2.1 Basic ice cream products

The basic technology of ice cream production is summarized in Figure 9.1.

(a) Ingredients

Ingredients of ice cream and their functions are summarized in Table 9.3.

Table 9.3 Ingredients of ice cream and their principal functions

Ingredient	Principal functions
Fat	Provides flavour, body, texture and mouth-feel
Milk solids-non-fat	Provides body, texture and contributes to sweetness and air incorporation
Sugar	Provides sweetness and improves texture
Flavouring	Provides non-dairy flavours
Colouring	Improves appearance and reinforces flavours
Emulsifiers	Improve whipping quality and texture
Stabilizers	Improve mix viscosity, air incorporation, texture and melting qualities
Value-added ingredients	Provide additional flavour and enhance appearance

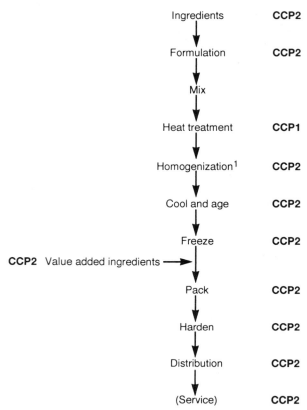

Figure 9.1 Basic process: manufacture of hard ice cream. [1] Emulsification may be used as an alternative in small-scale systems.

In each case a wide range of materials will fulfil the ingredient function.

Whole fresh milk is the most suitable source of both fat and solids-non-fat (SNF) and gives a better flavour than more highly processed sources. The fat and SNF content, although adequate for milk ice, requires supplementation for ice cream. Concentrated whole milk is increasingly used but results in a slight 'processed' flavour.

Fresh cream is the most suitable of concentrated fat sources and imparts a rich character to the end product. Fresh cream, however, is highly perishable and expensive. Plastic cream and frozen cream may also be used, but the end-product is of rather lower quality. A good quality ice cream may be made from butter, sweet cream, unsalted being most suitable, or anhydrous milk fat. Use is also made of milk fat fractions (see page 256), which have the additional advantage of being associated with phospholipids such as lecithin. Milk fat fractions can be 'tailored' to suit specific applications and wider use may be anticipated especially for speciality ice cream. The use of low-cholesterol milk fractions in dietary ice cream has also been proposed.

Milk fat promotes desirable textural properties, contributes a subtle flavour and is a good synergist for added flavours, although there is some lowering of the whipping rate. As such milk fat is used for higher quality ice creams, but acceptable quality ice cream may be made from vegetable fats. Coconut, palm, palm kernel or, less commonly, cottonseed and soya oil are used, either singly or as a blend. The oil is partly hydrogenated to produce a fairly sharp major melting peak at 28–30°C. It is also necessary to ensure that all of the fat is melted below 37°C to avoid a persistent 'fatty' mouth-feel.

Fat replacers are now available which replace all, or part of the fat while retaining the essential creaminess and body. Up to 50% of milk fat may be replaced by maltodextrin of less than 3 dextrose equivalents, derived from rice (low dextrose equivalent rice-solids), while complete fat replacement is possible with Simplesse™.

Solids-non-fat may be also obtained from a number of sources in addition to milk and that contributed by cream or other fat source. In all cases protein is the key component, functional activities being water binding and emulsification. Concentrated milk is very widely used and increasing use is being made of concentrated whole milk. Sweetened condensed milk has been popular in the past, but is difficult to handle and can give ice cream of a rough texture due to lactose crystallization.

Spray dried skim milk powder is used by many ice cream manufacturers and has the advantage of a relatively long storage life under good conditions. Medium heat powder (see page 132) is most suitable with respect to emulsification, foaming and water absorption. Whole milk powder may also be used as a source of both SNF and butter fat, but is highly prone to oxidative deterioration, which adversely affects the quality of the end-product. Ice cream may also be produced with dried, sweet cream buttermilk, which improves the whipping qualities and enhances flavour.

---

* Coconut, palm and palm kernel oil resemble milk fat in containing a relatively high proportion of saturated C_{12} to C_{16} fatty acids. These oils are preferred from a technical viewpoint since a high proportion crystallizes during ageing. This produces an ice cream of better eating quality and good storage stability.

Whey protein products are of increasing interest as sources of SNF. Whey powder can be used successfully as a partial replacement for skim milk powder, although the quantity used is limited by the high mineral content, which results in excessive saltiness, and the high lactose content, which can lead to lactose crystallization. These problems may be overcome by use of demineralized, delactosed whey powder, although this product is expensive. Lactose-hydrolysed whey can be used to replace up to 75% skim milk powder and contributes to product sweetness. Ice cream produced using lactose-hydrolysed whey has a soft texture popular with consumers. A further possibility is to use whey protein concentrates. The use of whey products in ice cream has been associated with atypical flavours and it is essential that material used must be of the highest quality.

A number of proprietary sources of SNF are now available comprising blends of skim milk powder, whey powder and, in some cases, casein. These products are of greater convenience during formulation, but offer no technological advantage over SNF blended in-house.

Extensive use of dry ingredients involves rehydration in water. The ingredient role of water in some ice cream is often overlooked, but it is necessary to ensure that water is of satisfactory chemical and microbiological quality.

Dairy SNF contributes towards sweetness, but is insufficient and additional sweetener is required. The quantity added is largely dictated by market preferences but sugars also play an important role in producing body and texture. Sucrose is still the most widely used sweetener and may be used alone, or in combination. The sugar is relatively cheap and available either in granulated or crystalline form, or as a syrup.

Hydrolysed corn sweeteners are also available in either a powder, or as a

---

* The increasing cost of conventional sources of solids-non-fat has stimulated interest in alternatives. Considerable promise is shown by ultrafiltration retentates which, used at 25%, produce an ice cream of better body, flavour and texture than whey protein concentrate and which has very good storage quality and heat shock stability (Lee, F.Y. and White, C.H. 1991. *Journal of Dairy Science*, 74, 1170–80).

* In many ways syrups are more convenient to handle than granulated or crystalline sugar. However the high viscosity at low temperatures means that syrups must be handled at 35–45°C, careful control of the upper temperature being required to prevent deleterious changes to colour and flavour. Mould and yeast growth on the upper surface of syrups may be a problem and requires special precautions such as filtration and ultra-sterilization of the air in the headspace of the storage tank.

syrup, the composition of the powder reflecting that of the syrup from which it was prepared. Corn syrups contain a variable quantity of dextrose, maltose and dextrins and are often used at a level of 30% in conjunction with sucrose. Corn syrups improve the body of ice cream, but properties vary according to the degree of hydrolysis and thus the dextrose equivalent. In general terms, increasing the dextrose equivalent leads to a decrease in mix viscosity, freezing point, fat destabilization and firmness. Corn syrups refined by ion-exchange tend to produce a better quality ice cream than syrups refined by carbon filtration.

More specialized sweeteners derived from corn starch are finding increasing application in ice cream manufacture. Dextrose is of value in preventing build-up of ice crystals and in suppressing lactose crystallization and, like high fructose corn syrup markedly lowers the freezing point and is suitable for use in the manufacture of ice cream for service direct from the deep freeze. High fructose corn syrup is also of importance in the manufacture of dietetic ice cream, while some manufacturers consider high maltose corn syrup to be of value in producing an ice cream of good body and flavour.

Lactose-hydrolysed whey products are now widely and cheaply available and the successful substitution of up to 50% of sucrose with hydrolysed and hydrolysed-isomerized syrups from whey ultrafiltration permeate has been demonstrated. This use would appear to complement the use of whey ultrafiltration retentate as a source of SNF.

Although artificial sweeteners are forbidden as ice cream ingredients in many countries, their use is permitted in the related frozen dairy desserts. The use of artificial sweeteners presents difficulties in that, unlike sugars, there is no beneficial effect on body and texture, resulting in a product that is crumbly, coarse and flaky. Attempts have been made to overcome this problem by the use of bulking agents such as polydextrose and maltodextrin, but none are fully satisfactory and some may even lead to intestinal disturbances. An alternative approach is to raise the SNF content, but this is likely to lead to lactose crystallization. Enzymatic hydrolysis of lactose in the SNF using β-galactosidase has, however, been successfully used in the formulation of Aspartame^{TM}

---

* The manufacture of sophisticated ice cream products, such as confectionery bars, often requires highly specialized ingredients. A high maltose-dextrose syrup is used as sweetener in Mars^{TM} frozen confections and provides good body without excess sweetness, provides stability against crystallization and maintains the high melting point required in hand held ice cream (Jackson, B. 1991. *Food Processing*, April, 25–6).

Table 9.4 Types of stabilizer

Protein[1]
  gelatin
Seaweed derivatives
  carrageenan (Irish moss, *Chondrus crispus*)
  alginates (*Laminaria* ssp.)
  agar-agar (*Gelidium amansi*)
Cellulose derivatives
  carboxymethyl cellulose
  microcrystalline cellulose
Gums
  guar (*Cyamopsis* spp.)
  locust (*Carob, Ceratonia silqua*)
  xanthan (*Xanthomonas campestris*)

[1] Proteins derived from other ingredients have some stabilizing activity, especially if mix is UHT processed.

sweetened frozen dairy desserts. The reaction products, glucose and galactose, reduce the quantity of added sweetener required, but increase the number of calories in the product.

Stabilizers are used in virtually all ice creams except super-premium, and improve mix viscosity, air incorporation and the body, texture and melting properties of the finished ice cream. Stabilizers also increase perceived creaminess and minimize the effects of temperature variations during storage (heat-shock). The amount and type vary according to the composition of the mix, the nature of other ingredients, process parameters and projected storage life. In general terms, mixes of high solids content, processed by UHT require less stabilizer than mixes of low solids content processed by HTST pasteurization.

A wide range of materials have been used as stabilizers (Table 9.4). Seaweed derivatives, especially carrageenan and alginate, are widely used, carrageenan being effective in preventing precipitation of whey proteins ('wheying-off') and syneresis which occurs with some stabilizers. Propylene glycol alginate has a good stability to acids and is also highly effective in mixes processed by HTST pasteurization.

* Carrageenan has been reported to be a cause of ulcerative colitis and hyperplastic changes in the gastric mucosa. Carrageenan has been shown to enter the tissues of experimental animals and may have tumour-enhancing properties. During experimental work, however, carrageenan was supplied as a fluid and it is possible that these findings do not apply when carrageenan is ingested as part of a food (Marcus, R. and Watt, J. 1981. *Lancet*, I, 338).

Cellulose derivatives are also widely used in ice cream. Carboxymethyl cellulose is most common and produces a dry ice cream of good body and texture. There is, however, a tendency to wheying-off, although this may be overcome by use in combination with a small quantity of carrageenan. Microcrystalline cellulose is less commonly used but, in combination with gum-type stabilizers, is highly effective in mixes containing whey proteins.

Gum-type stabilizers are usually obtained from plants. Locust and guar gum are widely used in ice cream and promote good viscosity and texture, although some wheying-off may occur. Guar gum has the advantage of hydration in cold water, but locust gum requires a temperature of *ca.* 70°C for *ca.* 15 min and is not suited to UHT or HTST mix processing. Some use is also made of xanthan gum produced by the bacterium *Xanthomonas campestris*.

Stabilizers are often prepared as proprietary, preblended formulations. In some cases stabilizers and emulsifiers are combined as homogeneous 'integrated' products, but such preparations are disliked by many manufacturers. In the US use is sometimes made of ice-cream 'improvers', which consist of a blend of stabilizers together with a casein coagulating enzyme such as rennet or pepsin. Partial coagulation improves body and texture, but the reaction must be controlled by heating. The melted ice cream often has a curdled appearance and the benefits of 'improvers' are doubtful except, possibly, with ice cream of low solids content. Antioxidants may also be incorporated with stabilizer but, despite being made from natural materials such as oat flour, the use of antioxidants is forbidden in many countries.

Emulsifiers are used to improve the whipping quality of the mix and to produce an ice cream of smooth, dry texture. Emulsifiers also facilitate the manufacturing process. Egg yolk, of which the active ingredient is lecithin, was the original emulsifier. Egg yolk, either frozen or dried, remains in use in custards and in some high-quality 'natural' ice cream, where it contributes other desirable qualities. Egg yolk is only a moderately effective emulsifier, best performance being obtained when butter is used as the fat source.

---

* *Xanthomonas campestris*, is a gram-negative, obligately aerobic, rod-shaped bacterium, which is classified with the family *Pseudomonadaceae*. *Xanthomonas campestris* is economically important as a pathogen of brassicas and a wide range of other plants. A large number of pathovars exist which produce widely different symptoms, sometimes on the same host. Extracellular polysaccharides (xanthan gums) appear to play an important role in the pathology of non-systemic infections such as leaf spot.

> **BOX 9.3 Advice is seldom welcome**
>
> The use of raw eggs in small-scale ice cream making in the home and in catering premises led to a number of outbreaks of salmonellosis due to contamination of the eggs with *Salmonella enteritidis* phage type 4. Outbreaks continued long after the risk associated with raw eggs was well known. In one domestic outbreak, which affected guests at an elaborate dinner party, the hostess had been aware of the hazards of uncooked egg products but had been misled by the erroneous 'advice' of a totally unqualified cookery writer.

Glyceryl monostearate, of monoacylglycerol content 40–60%, subsequently became very widely used as an emulsifier. More recently highly effective emulsifiers such as polyoxyethylene glycol and glucitol esters have been introduced, although their use is not universally permitted. The desire to use 'natural' ingredients has also led to the suggestion that monoacylglycerol emulsifiers should be replaced by *Psyllium* husk fibre.

Colouring and flavouring is added to almost all ice cream. Artificial colours and flavours have been widely used in the past but there is now a continuing trend to natural, or nature-identical materials. Vegetable fat has little intrinsic flavour and it is necessary to add flavouring of sufficient quality to overcome the bland property without appearing in excess. In contrast milk fat has intrinsic flavour which can interfere with the effect of added flavouring. It is also necessary to balance flavour and sweetness, plain ice cream usually containing *ca.* 15% sugar and fruit-flavoured ice cream containing 17–18%. Acidulants such as citric acid may also be an ingredient of ice creams flavoured with acid fruits such as rhubarb. Chocolate ice is flavoured and coloured with 2–3% high quality cocoa powder and requires a separate mix.

The temperature of serving has an effect on flavour perception, flavours being less strong at lower temperatures and this must be considered when formulating ice cream for serving direct from the deep frozen state.

A wide range of added value ingredients are used in ice cream manufacture and include fruit chunks, nuts, chocolate chips, etc. Added value ingredients may be incorporated into the body of the ice cream or used as a coating. Liqueurs have achieved some popularity as added value ingredients in high quality ice cream.

(b) Mix formulation and calculation

A number of factors must be considered when determining mix formulation. Legal standards must obviously be met, although in a number of countries an ice cream of minimum legal composition would, in any case be of unsatisfactory quality.

The proposed market, and the preference of consumers within that market, have an obvious bearing both on the composition of the ice cream and the nature of the ingredients. Ice cream of relatively low solids content containing vegetable oils as fat source is perfectly acceptable for bulk packs, but not as a premium product. Equally ice cream novelties designed for children are likely to be sweeter and of a lurid colour when compared with ice cream intended for a sophisticated adult market. Economic considerations, however, interact with market considerations and it is common practice to use the cheapest possible ingredients compatible with quality standards.

The type of equipment available, especially the freezer, influences the formulation. A high total solids level is necessary to provide body in ice cream frozen by high overrun continuous freezers. Conversely a high solids content can lead to an excessively heavy body in ice cream frozen in vertical freezers which incorporate only a small quantity of air.

A fundamental requirement of formulation is that mixes must be balanced. In the first place this involves ensuring that the correct fat:sugar ratio is obtained to prevent a 'fatty' mouth-feel. This ratio varies

---

BOX 9.4 **Whiskey in the jar**

The marketing success of liqueur-flavoured ice cream depends heavily on the brand image of the liqueur. In many cases the packaging used for the ice cream resembles that of the liqueur. In some cases, including ice cream flavoured with the whisky-based Drambuie^{TM}, the brand association is further reinforced by use of an ice cream container constructed as a facsimile of the characteristic Drambuie^{TM} bottle.

CONTROL POINT: INGREDIENTS CCP 1 OR 2

Control

Ensure ingredients of good quality and suitable for intended purpose.

Ensure ingredients stored under correct conditions and used within a stipulated period.

Ensure ingredients added after heat treatment conform to microbiological criteria with respect to absence of pathogens and general quality.

Monitoring

Obtain ingredients from reputable supplier.

Monitor storage conditions and stock rotation. Thermographs should be fitted where refrigerated storage is required.

Verification

Quality of end-product.

Inspection of thermographs and other plant records.

Microbiological examination of ingredients added post-heat treatment.

---

according to the type of freezer (Table 9.5). The second balance required is the total solids to water ratio. If this is too high there is a risk of sandiness and rough texture due to lactose crystallization. If too low, large ice crystals may form resulting in a 'glassy' or 'icy' texture, while the product would also be of insipid taste and weak body. To obtain the correct ratio it is necessary to calculate the quantity of SNF which will 'absorb' water remaining in the mix when all other solid ingredients are present. The calculation is based on the assumption that milk SNF will absorb *ca.* six times its weight of water. Then the maximum quantity of SNF will be:

$$100 - \text{Percent of non-SNF solids}/7.$$

Thus the optimum SNF for a mix for a vertical freezer containing 7% fat,

Table 9.5 Approximate fat and sugar content of ice cream mixes

	Fat (%)	Sugar (%)
Vertical batch freezer	6	12
	7	12.5
	8	13
Horizontal batch freezer	9	13.5
	10	14
Continuous freezer	10	14
	12	15

Based on Rothwell, J. 1986. *Ice Cream Manufacture*. J. Rothwell, Reading.

12.5% sugar and 1% combined emulsifier/stabilizer would be:

$$100-(7 + 12.5 + 1)/7 = 11.35\ (11.4)\%.$$

Planning and balancing a basic formula therefore involves five stages:

1. Decide on the percentage of fat required.
2. Finalize the fat:sugar balance (Table 9.5).
3. Decide on rate of use of stabilizers and emulsifiers (usually as recommended by manufacturer).
4. Decide on rate of use of additional ingredients such as cornflour.
5. Calculate the SNF required to balance the mix.

The basic formula determines the functional composition of the mix, but does not provide the actual quantities of mix ingredients. This requires a knowledge of the composition of the ingredients with respect to their fat, SNF and sugar content. It should also be appreciated that the basic formula may be achieved by many different combinations of ingredients.

The simplest situation is that in which all of the fat and all of the SNF is derived from single ingredients (e.g. butter and skim milk powder). Butter has a fat content of 80% and skim milk powder a SNF content of 97% (the small amounts of SNF contributed by butter and fat contributed by skim milk powder may be ignored). Thus for the basic formula discussed above the quantities of each ingredient may be determined by simple arithmetic:

Fat 7% to be supplied by butter (80%)
 7 x 100/80 = 8.7 units
Sugar 12.5% to be supplied by granulated sugar (100%)
 = 12.5 units
Emulsifier/ stabilizer 1% to be supplied at 100%
 = 1.0 unit

SNF 11.4% to be supplied by skim milk powder (97%)
11.4 x 100/97 = 11.7 units

Water must obviously be used in this recipe and the quantity may be calculated by subtracting the total weight of dry ingredients from 100. The weight of water may then be converted to volume according to the units in use.

The situation is more complex where the SNF is obtained from two sources and the use of Pearson's square method is recommended (*cf.* Chapter 5, page 193). In this example SNF is obtained from liquid skim milk (9% SNF) as well as skim milk powder (97% SNF), while other ingredients remain unchanged. As a first step the weights of the known ingredients (butter, sugar and emulsifier/stabilizer) are subtracted from 100:

$$100-(7 + 12.5 + 1) = 79.5$$

The 'unknown' part of the mix thus totals 79.5 units consisting of a mixture of fresh skim milk and skim milk powder which must supply SNF to a final concentration of 11.4%.

The SNF content of the fresh skim milk and the skim milk powder are entered onto Pearson's square at positions a and d, respectively (Figure 9.2). The SNF required is then entered into the centre of the square (e). This value is a percentage of the proportion of the mix contributing SNF:

$$11.4/79.5 \times 100 = 14.3$$

The quantity of skim milk powder may then be calculated by subtracting a from e and entering the value at c:

$$14.3-9 = 5.3$$

and the quantity of fresh skim milk by subtracting e from d and entering the value at b:

$$97-14.3 = 82.7$$

These two values represent the parts by weight of skim milk powder and fresh skim milk respectively. Actual weights are then calculated:

Skim milk powder
79.5 x 5.3/88   = 4.8 units
Fresh skim milk
79.5 x 82.7/88   = 74.7 units

The use of two fat sources requires more complicated calculations. An

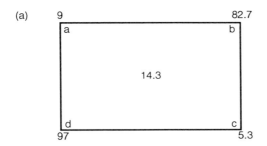

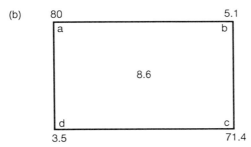

Figure 9.2 Use of Pearson's square in formulation of ice cream. (a) Single fat source, two sources of solids-non-fat; (b) Two sources of fat and solids-non-fat.

algebraic method is available, but requires solving three simultaneous equations with consequent risk of error. Algebraic methods are, however, better suited for use with computerized formulation. In other cases the 'serum point' method, which again involves the use of Pearson's square, is often preferred.

In this example liquid whole milk serves as a source both of fat (3.5%) and SNF (9%). Additional fat is obtained from butter and additional SNF from skim milk powder. The stabilizer/ emulsifier remains unchanged.

The first stage of the calculation involves determining the quantity of serum by subtracting from 100 the combined weight of all non-serum containing ingredients (in this case fat, sugar and emulsifier/ stabilizer):

$$100-(7 + 12.5 + 1) = 79.5$$

At an SNF content of 9% this quantity of serum contains:

$$79.5 \times 9/100 = 7.15$$

This is less than the required SNF of 11.4 by:

$$11.4 - 7.15 = 4.25$$

The difference is supplied by skim milk powder (97%). This contains 88 parts (97–9) SNF in addition to that provided by serum. The weight of skim milk powder to be added is therefore:

$$4.35/88 \times 100 = 4.83 \text{ units}$$

Total sum of non-fat ingredients is:

Skim milk powder	4.83	units
Sugar	13	units
Stabilizer/emulsifier	1	unit
	18.83	units

The remainder of the mix therefore consists of a mixture of milk and butter which will jointly contribute all of the butterfat. Quantities of each ingredient are calculated using Pearson's square. The fat content required:

$$7/81.17 \times 100 = 8.6(2)$$

is placed at the centre of the square (**e**). The fat content of the butter (80%) is placed at **a** and the fat content of the milk (3.5%) at **d**. The parts of milk required may then be calculated by subtracting e from a and entering the answer at c:

$$80 - 8.6 = 71.4$$

The parts of butter required may be calculated by subtracting d from e and entering the answer at b:

$$8.6 - 3.5 = 5.1$$

These two values represent the parts by weight of fresh milk and butter, respectively. Actual weights are then calculated:

Fresh milk
$79.5 \times 71.4/88$ = 64.5 units
Butter
$79.5 \times 5.1/88$ = 4.6 units

(c) Mix blending

In small-scale operations mix blending is an entirely manual operation. Low temperature–long time batch pasteurization is used and ingredients are usually blended in the pasteurization vat during heating. Pasteurization vats are usually equipped with an agitator or, in some cases, a blender-emulsifier to facilitate dispersion and mixing of ingredients.

In large-scale operations using HTST or UHT of the mix it is necessary to blend the mix before heat treatment. Separate blending vessels are used and, since blending is a batch process, several blending vessels are required to ensure a continuous product flow to the pasteurization or UHT plant. In modern plant, metering of liquid ingredients and weighing of solid ingredients is a highly automated process which is usually under microprocessor control.

Dispersion of dry ingredients may present problems and pre-liquefaction into slurries is sometimes used. In general, however, direct incorporation of dried ingredients is preferred and blending tanks are fitted with highly efficient turbine agitators. Dispersion can also be aided by introducing solid ingredients into the liquid in the pipe feeding the blending tank. Dried milk powder is difficult to wet, disperse and hydrate at temperatures below 35°C, although this problem may be overcome by the use of hydration tanks. Ingredients such as guar gum cannot, however, be used without heating the mix.

(d) Heat treatment

In the vast majority of countries heat treatment of ice cream mix, to a level sufficient to destroy vegetative pathogens, is mandatory. The minimum permitted heat treatment varies from country to country and, in the US, from state to state. In the UK, four minimum temperature/time treatments are permitted, 65.6°C/30 min, 71.1°C/10 min, 79.4°C/15 s and 148.8°C/2 s. More severe heat treatments have been suggested to ensure destruction of *Listeria monocytogenes* (*cf.* liquid milk, page 90), but may result in burnt or cooked flavours. The minimum UK heat treatments are, in fact, currently considered adequate to ensure destruction of *Listeria monocytogenes*, but less rigorous treatment used elsewhere may require re-evaluation.

---

* The choice between hot and cold blending before continuous heat treatment is important with respect to overall energy efficiency. Cold blending permits regenerative heating and cooling over a much wider temperature range than hot blending and, contrary to many impressions, leads to significantly reduced energy costs. This may be illustrated by an operation in which raw dairy ingredients at 6°C and liquid sugar at 45°C are used in a 74 to 26% preparation to blend at *ca.* 16°C. Energy savings over hot blending are 35.6% in steam and 51.7% in refrigeration (Mitten, H.L. and Neirinkx, J.M. 1986. In *Modern Dairy Technology, vol. 2. Advances in Milk Products* (ed. Robinson, R.K.). Elsevier Applied Sciences, London, pp. 215–59).

CONTROL POINT: MIX BLENDING CCP 2

Control

Correct quantities of ingredients to be used.

Dry ingredients to be fully dispersed.

All ingredients to be fully blended.

Monitoring

Check calibration of weighing and measuring equipment.

Monitor temperatures where hot blending used.

Monitor hydration of milk powder.

Supervision of blending and of mix by experienced personnel.

Verification

Quality of end-product.

Inspection of plant records.

Specialist examination and maintenance of weighing and measuring equipment and, where fitted, microprocessor-based systems.

Low temperature–long time processes involve batch heating in a steam- or water-jacketed vat and are suitable only for small-scale operations. Plate heat exchangers (see page 49) are most commonly used for HTST processing, although tubular heaters may be preferred where space is limited. Fouling is a serious potential problem during HTST pasteurization, but can be minimized by taking rigorous precautions against the incorporation of excess air into the mix. Fouling also restricts the choice of UHT plant to direct steam injection equipment (see page 61) or, less commonly, scraped surface heat exchangers.

CONTROL POINT: HEAT TREATMENT CCP 1

See Chapter 2, page 52.

(e) Homogenization/emulsification

A reduction in the size of fat globules is required during ice cream manufacture to prevent churning and to improve whipping properties and air incorporation by allowing proteins to adsorb onto the surface of fat globules. This process is usually accomplished by homogenization (see page 55). Homogenization is necessary in large plant producing ice cream of high fat content and/or a high overrun, but emulsification is an alternative in small-scale operations where ice cream is of a relatively low fat content and/or overrun. Emulsification involves the application of high speed centrifugal pumps to force the liquid through a mesh screen and thus to disrupt fat globules. The globules, however, are both larger and more variable in size than those of homogenized mix.

Batch pasteurization plant often incorporates either a homogenizer or an emulsifier which operates concurrently with pasteurization. In other cases the homogenizer is situated downstream of the heat treatment. This produces no technological difficulties, but the difficulties of sanitizing homogenizers means a high potential risk of product recontamination. This risk may be obviated by combining the homogenizer with the heat treatment equipment so that the mix is homogenized just before, or during, the highest temperature stage.

Ice cream manufacture does not require that fat globules be reduced to the smallest possible size and over-homogenization is a fault leading to clumping or clustering of fat globules and phase reversal ('buttering') during freezing. Relatively low homogenization pressures are employed, vegetable fats requiring *ca.* 12 MPa and dairy fats *ca.*16 MPa in single stage equipment. Two-stage homogenization is sometimes used at pressures of *ca.* 15 MPa in the first stage and *ca.* 4 MPa in the second, for both types of fat. Pressures used are also lower with ice cream of higher fat content and when the process is operated at higher temperatures.

(f) Cooling and ageing

Following heat treatment the mix must be cooled as rapidly as possible to 4°C, the maximum time permitted in the UK regulations being 1.5 h. The mix is held at 4°C to age, a process which involves the hydration of

CONTROL POINT: HOMOGENIZATION AND EMULSIFICATION CCP 1 OR 2

Control

Ensure break-up of fat globules to optimal size.

Prevent microbial contamination from homogenizer if placed downstream of heat treatment.

Monitoring

Monitor homogenizer/emulsifier operation.

Implement special cleaning and sanitization schedule for homogenizer. Inspect before each production run.

Verification

Quality and structure of end-product.

Examination of plant records.

Microbiological examination of swabs taken from homogenizer.

---

milk proteins, the crystallization of fats and the absorption of water by any added hydrocolloids. Ageing is substantially completed within 24 h and longer periods should be avoided to prevent spoilage by psychrotrophic micro-organisms. Cooling the mix to −1 to 2°C in a scraped surface heat exchanger permits the use of shorter ageing periods.

---

* The basic physical process of freezing involves rapid heat extraction to bring about the change of state of an ice cream solution from liquid to solid. Three related types of heat must be extracted:

1. Specific heat: the thermal capacity (ability to retain heat), which is a characteristic of each ice cream mix and thus variable.
2. Sensible heat: the amount of heat exchanged for a given change in temperature (water requires 1 BTU/lb of water for each degree of temperature change).
3. Latent heat: the amount of heat exchanged during the change of state (1 lb of water loses 144 BTU in changing from a liquid to a solid).

(Jaspersen, W.S. 1989. *Food Technology International Europe*, 85–8).

CONTROL POINT: COOLING AND AGEING CCP 2

Control

Cooling to ageing temperature within stipulated period.

Ensuring mix held for correct period of time.

Monitoring

Monitor mix temperature throughout cooling and ageing. Equipment should be fitted with thermographs.

Institute formal system to ensure mix held for correct period.

Verification

Quality of end-product.

Number of psychrotrophic micro-organisms in end-product.

Inspection of plant records.

(g) Freezing

In traditional ice cream manufacture freezing is a two-stage process. In the first stage the temperature is reduced under agitation, air being incorporated to give an aerated product. The second stage, which is much slower, involves no incorporation of air and takes place under quiescent conditions in a hardening room or tunnel. The process is not complete and even at very low temperatures some water remains unfrozen.

The crystallization stage is of major importance with respect to ice cream quality since the texture is largely determined by the size of the ice crystals. Fast freezing rates are desirable to ensure that crystals are too small to be detectable in the mouth.

Incorporation of air during freezing leads to an increase in volume of the mix, the overrun. Overrun may be calculated either by volume or weight (Figure 9.3) and is also an important quality determinant, a high

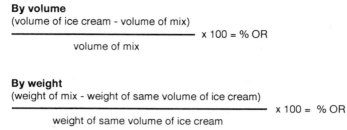

Figure 9.3 Calculation of overrun (OR).

overrun ice cream having less flavour, a drier appearance and a less stiff texture. Ice cream is sold by volume in most countries and it is economically desirable to have an overrun as high as possible without adversely affecting the character of the ice cream.

Two main types of freezer are available for commercial use, the batch and the continuous type. Batch freezers, which may be vertical or horizontal in configuration are suitable primarily for small-scale operations, while the continuous horizontal type forms the basis of very large-scale manufacture. The properties of ice cream produced by the different types differ. Freezing is more rapid in the continuous horizontal type, up to 50% of the water being frozen in a few seconds. This produces a very large number of small ice crystals and consequently a smooth texture. At the same time the means by which air is incorporated also differs. In the batch freezer air is simply 'rolled' into the mix at atmospheric pressure, while in the continuous freezer air is metered into the mix under pressure and subsequently expands producing a large number of small air cells. The overrun obtainable with batch freezers varies from 50 to 100% according to design, but overrun in excess of 130% is readily obtainable with large-scale continuous freezers.

Vertical batch freezers were the earliest commercial ice cream freezers and are the simplest design. The equipment consists of a vertical cylinder or barrel cooled by refrigerated coils or by immersion in a refrigerant. Rotating blades and a beater are fixed within the cylinder to scrape frozen ice cream from the walls. The frozen product and air are incorporated into the unfrozen mix, the process continuing until the whole of the mix is in the frozen state. A mix of 30–32% initial total solids content is required, the overrun varying from *ca.* 25 to 50%.

Horizontal batch freezers are essentially vertical freezers with the cylinder turned on its side. Such equipment is more convenient in use and has largely replaced vertical batch freezers. Refrigeration is now usually achieved by direct expansion systems using halocarbon refrigerants. In most cases the maximum overrun is 50–80%, but injection of air under pressure increases overrun to 100%. The freezer is usually operated at −10 to −20°C, this temperature permitting sufficiently fast freezing to minimize ice crystal size, but also allowing sufficient time for air incorporation.

Continuous horizontal freezers are similar in design to batch horizontal freezers, but the cylinder is continuously fed a mixture of mix and air by means of a pump, or a series of pumps, the frozen product similarly being continuously removed. A dasher is fitted which serves to beat in air, scrape the walls free of frozen product and to move the frozen mix towards the outlet. Modern equipment is invariably fitted with automatic overrun control and, in many cases, the entire operation is controlled by microprocessors. Continuous freezers have a number of advantages and disadvantages independent of the very high throughput obtainable (Table 9.6).

Continuous freezers are operated under pressure and the overrun is only fully realised as the air within the ice cream expands as the product leaves the freezer. Cylinder pressures of 3.5–5.5 atm are used for overruns of up to 130% at drawing temperatures of $ca.$ −7°C, but higher pressure is required when producing ice cream of higher overrun, or when a lower drawing temperature is used.

The temperature of the mix entering the freezer is very important, 0 to −1°C optimizing performance. The entry temperature of the mix should be constant to facilitate control of overrun and freezing rate.

Low temperature freezing involves delivery of the ice cream at −9 to −10°C. The equipment used has a primary cylinder similar to that of a conventional freezer and a secondary cylinder fitted with an eccentrically mounted dasher fitted with short blades. The mix is subject to a rigorous treatment in the freezer which results in changes to the structure and a reduction in ice crystal size from 45–55 to 18–22µm. The ice cream has a very smooth texture and is highly resistant to adverse handling.

The consistency of ice cream leaving the freezer is primarily a function of formulation, although temperature also has some effect. Dasher design is also important, a displacement dasher operated at high speed,

Table 9.6 Advantages and disadvantages of continuous freezers

Advantages

1. The ice cream is of smoother texture due to the small ice crystals formed
2. Rapid freezing favours the formation of small lactose crystals and problems due to 'sandiness' are minimized
3. Less stabilizer required as crystals are formed in the freezer rather than in the hardening room and because a mix of lower viscosity is required
4. Ageing time is reduced as a lower viscosity mix is required and because incorporation of air is less dependent on the character of the mix
5. Less flavour is required because the smaller ice crystals melt more rapidly in the mouth making the flavour more pronounced
6. The product is of more consistent character with less variation between packs
7. Handling is minimized with less potential for contamination
8. Continuous production facilitates manufacture of speciality ice creams such as centre moulds and combinations of colours and flavours

Disadvantages

1. High initial capital cost
2. Ice cream may shrink in volume after hardening (problem may be minimized by modifying mix formulation)
3. Close control is required to prevent excessive overrun
4. Machine parts operate at fine tolerance and are prone to damage if mishandled
5. Special training is required for operatives and service personnel

for example, causing a shearing action which produces a stiff and dry ice cream. Such an ice cream is described as a 'warm-eating' product and melts slowly in the mouth. Ice cream of this type may be less refreshing than other products and the flavour is less readily released, but it is preferred by *ca.* 25% of consumers. Ice cream of stiff consistency is generally undesirable from the engineering viewpoint, having poor flow and filling properties and requiring more energy in handling. The influence of dasher design and operation on the properties of the finished ice cream is well understood and different types of dasher are used in production of different products.

Extrusion technology has been widely applied to ice cream production in recent years, usually for the production of novelties. The ice cream leaves the freezer through an extrusion nozzle which determines its shape and it is a prerequisite that the product has sufficient viscosity and stiffness to maintain its shape after delivery. Nozzles extruding different ice creams may be placed within the main nozzle to permit production of intricate designs. Alternatively layered products can be made by passage under consecutive extrusion nozzles. Ice cream may be extruded horizontally or vertically at a temperature of −6 to −7°C, a hot wire being used to cut the product into portions. The temperature of the ice cream is of considerable importance since at too high temperature the shape is not maintained, while at too low cutting is impaired. After portioning the ice cream is carried to a hardening tunnel on individual stainless steel plates. After hardening the ice cream is released by a pneumatic hammer and is transferred either direct to packaging or receives coatings, decorations, etc.

Flavouring, colouring and finely chopped fruit and nuts may be added to the ice cream directly before freezing. Larger pieces of fruit, nuts, etc., must be added as the ice cream leaves the freezer, accompanying flavouring and colouring having been added earlier.

With the exception of extruded product, ice cream leaving the freezer is packed either directly into the final packaging, or into moulds to impart the desired shape. Precooling of the packaging or moulds is desirable to prevent melting and refreezing of the outer edge of the ice cream. Similarly the ice cream must be transferred to the hardening process as quickly as possible to prevent melting and the formation of large crystals during refreezing.

During hardening the temperature of the ice cream is further reduced to *ca.* −18°C. A hardening room maintained at −20 to −25°C may be used but continuous hardening tunnels are more common in large-scale operations and have the advantage of faster hardening and less formation of large ice crystals. Hardening tunnels are available in a number of configurations and typically operate at air temperatures of −30 to −35°C and an air velocity of *ca.* 180 m/min. Hardening is completed within 2–5 h compared with 10–12 h in a hardening room.

In recent years contact plate freezers have gained popularity, especially for rectangular packs. Equipment of this type is very efficient, hardening being completed in 2 h. Ice cream stick bars may be hardened in moulds using the same equipment as water ice lollipops (see page 419).

(b) Finishing and packaging

After leaving the hardening process the ice cream may be finished by enrobing with chocolate or another candy product, by addition of chopped nuts, etc., or by combining the ice cream with water ice. Where necessary the ice cream is then packaged.

CONTROL POINT: FREEZING CCP 2

Control

Correct operation of freezer.

Ensure immediate transfer to hardening room, etc.

Ensure hardening room, etc., operating at correct temperature.

Ensure hardening period adequate.

Monitoring

Monitor suitability of freezer set-up (e.g. dasher type) for mix being frozen.

Monitor operation of freezer. Recording equipment should be fitted where appropriate.

Determine overrun of frozen but unhardened mix (continuous freezers may be fitted with automatic overrun control).

Transfer to hardening room to be supervised by experienced personnel (small-scale production using manual transfer).

Monitor temperature of hardening equipment on a continuous basis.

Establish formal system for ensuring correct period in hardening room.

Verification

Quality of end-product.

Check calibration of automatic overrun control equipment.

Examination of plant records.

A wide range of packaging material is used for ice cream. Small multi-portion retail packs were commonly packaged in waxed, aluminium foil laminated, or plastic-coated (low density polyethylene) cardboard, but while still used, this material is being replaced by plastic containers (high density polyethylene or polystyrene) which usually incorporate a tamper-proof lid. Plastic containers are used for most large, multi-portion retail packs and are replacing tinned steel containers for bulk commercial packs. Individual hand held and stick held bars are packaged in moisture-proof coated or foil-lined paper. Packaging can have a significant effect on the storage properties of ice cream, although temperature is of greatest importance. Closed plastic packages are of superior performance during retail storage, aluminium foil laminate being the most effective of the cardboard based materials.

(i) Storage and distribution

Ice cream should be stored at constant temperatures since fluctuations lead to migration and accumulation of water and the formation of large crystals on refreezing. A temperature of -20 to $-25°C$ is used for long-term storage, but higher temperatures of -13 to $-18°C$ are acceptable during transport and short-term display.

Although much ice cream is retailed in its final packaging, a significant quantity is portioned from bulk packs at point of sale. This process is the weak link from the hygiene viewpoint since many outlets are mobile vans, kiosks, etc., which have only limited facilities for hand washing and utensil sterilization.

9.2.2 Special types of ice cream

(a) Soft-scoop

Soft-scoop ice cream is made using the same manufacturing technology as standard ice cream, but is formulated to permit serving direct from deep frozen storage. Ice cream of this type must, therefore, have a low

---

BOX 9.5 The good, the bad and the ugly

Problems of low hygiene standards in mobile ice cream outlets are exacerbated by the fact that many operatives work on a part-time basis, often from domestic premises, and thus lack the necessary knowledge of personal hygiene, product handling and sanitation routines. In some cases, the regulatory authorities have difficulty enforcing hygiene regulations because of the unofficial, and unregistered nature of 'cowboy' operations.

freezing point so that the texture is relatively softer at any given temperature. Freezing point may be lowered by including glycerol, glucose, or high-fructose corn syrup. A higher level of stabilizer and emulsifier may be required.

(b) High-fat (premium and super-premium) ice cream

Excessive viscosity of high-fat ice cream may lead to problems associated with cooling and obtaining the desired overrun. Dairy fats are normally emulsified in either fresh, sweetened skim, or whole, milk before the addition of other ingredients. Flavour should be derived from the dairy fat, not milk solids-non-fat (MSNF) and for this reason no additional source of MSNF, such as condensed skim milk or skim milk powder, should be added. The MSNF content is reduced as the fat content increases, a typical value being 7% in a 20% fat product. This may result in a crumbly body, a common defect in high-fat ice cream, although this can be avoided by using high levels of sucrose or glucose as sweeteners. A high sweetener level is usual to meet consumer requirements and has the additional advantage, especially where dextrose is used, of improving meltdown.

Homogenization of the mix is essential to prevent churning during freezing. Low homogenization pressures must be used, typical values being 5 MPa at 18% fat and 3.5 MPa at 20% fat. The drawing temperature is raised with increasing fat content to overcome problems of excessive stiffness and a high serving temperature may also be necessary.

(c) High-milk solids-non-fat ice cream

Ice cream containing 3–4% additional MSNF are popular on the west coast of the US and have a characteristic 'chewy' texture. Casein and delactosed skim milk powder are used as ingredients to provide MSNF while avoiding problems of 'sandiness' due to high lactose content. Dextrose is used to lower the freezing point and facilitate serving, while a low homogenization pressure is necessary to prevent 'curdiness'. Freezers need a high refrigeration capacity due to the high viscosity of the mix and a high overrun is required to prevent 'sogginess'.

(d) 'Italian-style' ice cream

'Italian-style' ice cream is popular in the US. No product definition or standard exists, however, and the nature of the product may vary considerably. The traditional 'Italian-style' product is served semi-hard and is less sweet and less acid than other ice cream, being flavoured by natural extracts. The ice cream has a high milk fat and total solids content and a very low overrun of 0 to 10%. This imparts a distinctive body and texture and flavour release.

(e) Diabetic and dietetic ice cream

A product resembling ice cream, but suitable for consumption by diabetics may be made from a mix containing 4 to 6% fat, sweetened with sugar alcohols such as glucitol or mannitol, the non-caloric sweetener sucaryl or, less commonly, saccharin.

In addition to frozen dairy desserts made with fat replacers and intense sweeteners, low-cholesterol and cholesterol-reduced ice cream has been developed. Various other approaches have been taken to improve the dietary status of ice cream. An ice cream fortified with bran has been marketed in the UK, while in Japan 'therapeutic' ice cream containing *Bifidobacterium* is available.

(f) Soft-serve ice cream

Soft-serve ice cream differs from other types in being consumed directly from the freezer and, usually, without hardening. The product is thus usually manufactured at the point of sale which may be a specialist outlet or cafe, a non-food outlet such as a filling station, or a mobile outlet or kiosk. Soft-serve ice cream may be manufactured from a conventional mix produced on the premises, from a UHT processed, aseptically packaged mix, or from a spray-dried powder mix. In any case the fat content is usually 5–6%, MSNF 11.5–14% and the sugar content 12–15%. Powder mixes may be formulated for reconstitution in either hot or cold water. Hot water mixes are preferable with respect to hygiene, but cold water mixes are often considered more convenient.

Soft-serve freezers are of horizontal configuration, mix being fed from a refrigerated hopper into a small diameter cylinder. Mix may either flow by gravity into the cylinder or be pumped. In the latter case air is incorporated and there is some control of overrun. The cylinder may be fitted with either conventional beater blades or a solid screw-type beater without blades which is usually constructed from a hard-wearing plastic. The screw-type beater clears the cylinder by a small margin (the rubber), the space being filled by a thin layer of ice cream. The presence of this layer reduces heat transfer and increases refrigeration costs, but equipment of this type is more flexible in operation. Ice cream is drawn from the freezer at *ca.* −6°C with an overrun of 30–50%.

Soft-serve ice cream freezers and ancillary equipment should be dismantled, cleaned and sanitized on a daily basis. It must be recognized that maintenance of the necessary hygiene standards can be more difficult in an environment primarily concerned with retailing than in one wholly concerned with manufacturing. Particular difficulties may be encountered in outlets which are predominantly non-food, such as filling stations, and those with inherently limited facilities such as kiosks. The self-pasteurizing soft-serve freezer offers at least a partial solution. Such equipment is designed to heat mix and machine surfaces in contact with mix to 65°C for 30 min before cooling to 5°C. It is usual to 'pasteurize' daily, usually at the end of the working day, and restrict full cleaning and sanitization to a weekly basis. It must be emphasized that self-pasteurizing freezers are not intended to process unpasteurized mix.

(g) End-product testing

In large-scale ice cream manufacturing, end-product testing serves a number of functions. Chemical analysis is used to ensure compositional standards are met and is complemented by determination of physical characteristics. Organoleptic analysis is also important and is used not only for determination of perceived quality, but also for assessment of the performance of functional ingredients such as stabilizers. Microbiological analysis is necessary for assuring good manufacturing hygiene.

End-product testing is not practical in the case of soft-serve outlets, or for very small manufacturers relying entirely on premixed ingredients. In such cases responsibility for quality assurance lies with the manufacturer of the premix.

9.2.3 Low fat frozen desserts

As noted above low fat desserts are a complex group of products which currently lack legal definitions. The manufacturing technology is based on ice cream making, the products being frozen in a continuous freezer.

---

* The body of some types of low fat dessert resembles a firm, springy gel unlike that of ice cream and having similarities to gelified milk. Simple examination of the melting properties suggests that the structure of the product is intermediate between ice cream and an aerated mousse. A weak structure is retained and water remains strongly bound even after many hours at ambient temperature. It would appear that the primary structure is that of a loose hydrocolloid gel, which is stiffened by the formation of a matrix of ice crystals during freezing.

The very low fat content of most desserts means that the structure of the product differs from that of ice cream and it is necessary to use a high level of stabilizers and thickeners to bind water and to provide body.

A complex mixture of stabilizers and thickeners appears to be necessary, a widely used combination being; modified starch, carob gum, guar gum, carrageenan and xanthan gum. Some types also contain calcium sulphate, presumably as firming agent.

A more recent development has been the use of partial, or total fat replacers. The extent to which the fat replacer contributes to body, texture and stability varies according to type. A high level of stabilizer appears to be necessary, however, and it is unlikely that the chemical nature of fat replacers is compatible with their interacting with other ingredients in the same manner as milk fat.

9.2.4 Sherbets and ices

In many countries, including the UK, no legal standards or definitions exist for sherbets and ices. The two products are generally differentiated by the inclusion of milk solids in sherbets but not in ices. This distinction, however, is increasingly blurred and some 'ices' contain very small amounts of milk solids such as whey protein.

Sherbets and ices have a higher fruit acid content (not less than 0.35%) than ice cream and a relatively tart flavour. There is little, or no, overrun and the texture tends to be coarse. Sherbets and ices are perceived to be more refreshing than ice cream and to have a greater cooling characteristic.

In the UK, large numbers of stick-held ices (lollipops) are produced. The mix contains water, sugar, acid, flavours, colours and stabilizers. The mix is frozen in moulds on a continuous basis, the stick being automatically inserted part way through freezing. Extrusion is, however, possible providing a suitable stabilizer system is used.

The sugar content of the mix depends on the desired organoleptic properties but is usually in the range 17–20%. Citric acid is the most commonly used acidulant, although tartaric, lactic, malic, ascorbic and phosphoric acid are all acceptable. The acid content rises to balance increasing sugar content, but is usually in the range 0.3–0.4%.

Stabilizers must be compatible with acids and produce a 'short' body and rapid flavour release. Gelatin, pectin and xanthan gum are all commonly used and each produces the desired body structure. Flavour release is more rapid with pectin than gelatin and xanthan gum. A satisfactory product may also be made using carboxymethyl cellulose or guar gum, but in each case flavour release tends to be rather slow.

In the past considerable use has been made, especially in cheaper products, of artificial colours and flavouring, but there is now an increasing tendency to use natural or nature-identical colouring and natural flavour extracts. Fruit juice may be added in varying amounts and there is now a large market for frozen juice bars in which the organoleptic character is largely derived from fruit juice.

Ices are usually hard frozen, but may be dispensed direct from the freezer as slush ice. Such a product, Slush puppy^{TM}, may be regarded as the marketing equivalent of soft-serve ice cream and is perceived as being particularly refreshing. Individual portions of ices are also hard frozen in a sealed plastic tube from which they are sucked as a mixture of ice and water. This product is almost entirely consumed by children.

9.3 CHEMISTRY

9.3.1 Nutritional status of ice cream

Ice cream is not of major nutritional significance among the general population. Possible exceptions are amongst children and old persons for whom ice cream can be an important source of dietary energy.

The protein, sugar and fat content obviously vary widely according to the formulation. In general levels of water soluble vitamins and minerals are similar to, or slightly higher than, those of milk. In ice cream made from dairy fat, levels of fat soluble vitamins are two to three times higher than those of full cream milk.

9.3.2 Physico-chemical nature of ice cream

(a) Effect of pasteurization

Although the prime role of pasteurization is to ensure the microbiological safety of the product, the heating also affects the physico-chemical structure of the mix. The emulsifier is melted and heat-activated stabilizers brought into colloidal solution. The whey proteins present in the SNF are partially denatured and uncoil exposing the lipophilic portion of the molecule. As a consequence, the whey proteins begin to act as emulsifiers while, at the same time, the water binding capacity is increased. Denaturation also increases the number of available binding sites for protein : hydrocolloid interactions and thus enhances the action of stabilizers such as carrageenan. Pasteurization is generally beneficial to ice cream quality, but excessive heat treatment leads to unacceptable organoleptic deterioration.

(b) Role of stabilizers and emulsifiers

It has been commonly stated that stabilizers interfere with ice crystallization and limit the growth of large crystals. Until recently this has not been studied systematically, although qualitative support is offered by the observation that the use of stabilizers results in an ice cream of smoother body. At a given rate of heat removal, the coarseness of a population of crystals depends on the relative rates of nucleation and growth and it has been postulated that stabilizers act either by increasing nucleation or reducing the rate of growth. Investigations into each of these possibilities showed that while stabilizers have no significant effect on rate of nucleation, the growth of ice, in a model system, is drastically retarded and the morphology altered by the presence of a stabilizer gel network.

The effect on ice growth rate and morphology has been explained qualitatively in terms of mechanical interference with the growth of ice. It is probable that the gel fibres cause the ice-interface to develop a radius of curvature of similar size to the mesh size of the network and thus to depress the freezing point. The gel fibres themselves are placed under stress and may rupture, but the significance of this is unclear. The effect of stabilizers is greater than expected and this may be due to the presence of sucrose as an additional solute.

The effect of stabilizers cannot be explained wholly in terms of the effect on ice growth. The three-dimensional network formed binds, and restricts movement, of water in the frozen product and thus minimizes the potential for formation of large crystals during temperature fluctuations. The viscosity of the mix is also increased as a result of interactions between gum-type stabilizers and partially denatured proteins.

Emulsifiers act as surface active agents and reduce the energy required to maintain the integrity of the fat globules. In an emulsifier such as glyceryl monostearate, the lipophilic portion of the molecule, the fatty acid chain enters the fat phase, while the hydrophilic portion, the

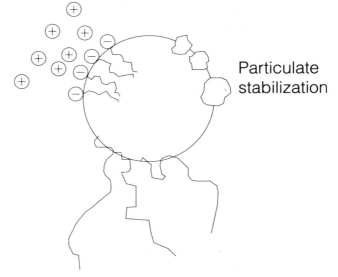

Figure 9.4 Mechanisms of emulsifier activity. Redrawn with permission from Fillery-Travis, A. *et al.* 1990, *Food Science and Technology Today*, 4, 89–93. Copyright 1990, Institute of Food Science and Technology.

glycerol portion enters the water phase (Figure 9.4). Added emulsifiers are more effective than milk proteins due to their smaller molecular size and greater mobility.

Paradoxically an additional beneficial function of emulsifiers is to *destabilize* the fat globule. This permits a degree of clustering and prevents too rapid melting.

(c) Effect of ageing

During ageing a number of important changes occur. Skim milk powder and stabilizers become fully hydrated and the emulsifier-induced desorption of proteins continues. Substantial crystallization of fat also occurs. Triacylglycerols of the highest melting point are the first to

---

* Destabilization is caused by the desorption of proteins, mainly casein, from the outer layer of the fat globules. At homogenization temperatures the surface activity of emulsifiers is low and no desorption occurs. When the temperature is lowered, however, the surface activity becomes sufficiently high to repel the outer protein layer from the fat globule. Desorption is a relatively slow process which continues during ageing.

crystallize and are situated closest to the surface of the fat globules. The crystallization process continues successively with triacylglycerols of progressively lower melting points, creating multiple-shelled fat globules with a core of liquid fat. The quantity of fat remaining liquid depends on the specific fat involved, but a balance between liquid and crystallized fat is essential for good quality. The poor quality associated with ice cream made from unsaturated fats results from the relatively low degree of fat crystallization.

(d) The freezing process and structure of the final ice cream

The final structure of ice cream is determined during the freezing and aeration of the mix. Ice cream is a complex physico-chemical system consisting of air cells dispersed in a continuous liquid phase in which ice crystals are embedded (Figure 9.5). The liquid phase also contains solidified fat particles, insoluble salts, milk proteins, lactose crystals,

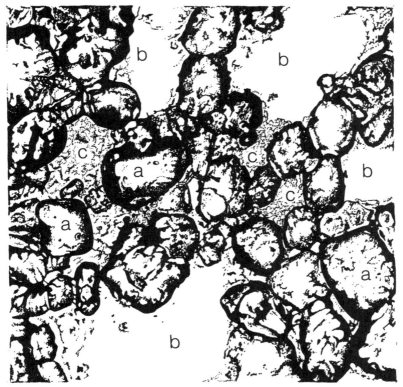

Figure 9.5 Internal structure of ice cream. (a) Ice crystal 45–55 μm diameter. (b) Ice crystals 120–170 μm diameter. (c) Unfrozen material.

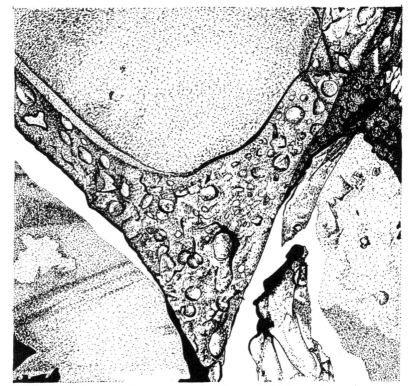

Figure 9.6 Wall separating ice crystals. The wall thickness is *ca.* 2 µm.

stabilizers of colloidal dimensions as well as sugars and salts in true solution.

Over the years many values have been quoted for the size of air cells, values for the mean diameter varying from 30 to 150 µm. These figures largely refer to ice cream made with batch freezers, and with continuous freezers, air cell diameters range from 5 to 300 µm, with a mean value of *ca.* 60 µm. Smaller cells are coated with a layer of fat crystals. Large air cells produce an ice cream of a snowy, flaky texture, while small cells produce a smooth texture. A mean air cell diameter of 60–100 µm is satisfactory for most types of ice cream.

The interface between air cell and the mix is a relatively thin, continuous layer of unstructured material coated with a layer of discrete fat globules which project into the interior of the air cells. Air cells and ice crystals (Figure 9.6) are separated by lamellae of variable thickness which contain both individual fat globules and casein micelles.

The amount of water existing as ice in ice cream varies with temperature, but is *ca.* 50% at an extrusion temperature of −5°C, *ca.* 95% at a hardening room temperature of −30°C and *ca.* 70% at a storage temperature of −11°C. The results of investigations concerning the size of ice crystals have been variable, but the overall mean size would appear to be 35–40 µm, with crystals being some 5–8 µm apart.

Fat globules in ice cream mix are of two types, large globules ranging in size from 0.5 to 4 µm, but with most less than 2 µm diameter, the overall mean being *ca.* 1 µm. In addition very fine globules are also present which range in diameter from 0.04 to 3 µm with a mean diameter of 0.5 µm. Destabilization of the fat emulsion, initiated by the action of emulsifiers, continues as a result of the combined effects of freezing and mechanical agitation during the freezing process. The fat globule membrane is ruptured and the liquid fat released to appear at the mix/air interface. Whipping properties are reduced and some coalescence occurs. A higher level of fat destabilization occurs when at low temperatures and when a high proportion of liquid fat is present (Figure 9.7).

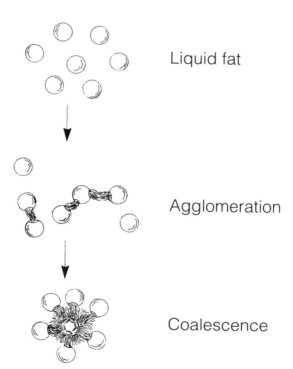

Figure 9.7 Destabilization of fat globules during freezing.

9.3.3 Chemical analysis of ice cream

General analytical methods for ice cream are similar to those for liquid milk (Chapter 2, page 85). Total solids content is an important parameter in ice cream analysis and may be determined by drying to constant weight. Solids-non-fat content can be determined by formol titration, but protein content is often a preferred parameter provided that gelatin and eggs are absent from the recipe. Protein may be determined by micro-kjeldahl analysis or by instrumental methods. Determination of sugar content can cause difficulties, but polarimetry is most widely used. The Gerber method may be modified for determination of fat in ice cream and instrumental methods may also be used. The Rose–Gottleib method, however, may give more reproducible results.

Determination of overrun is important as verification of correct operation of ice cream freezers even where continuous equipment with automatic control is in use. No really satisfactory means exists of determining overrun after hardening, but a simple method suitable for use during production involves determining the weight of a volume of mix and of the same volume of ice cream leaving the freezer.

Organoleptic assessment of ice cream is important not only in determining suitability of taste, but also mouth-feel and body. Simple *ad hoc* methods have been devised for measurement of body firmness but it is doubtful that these offer any significant advantage over organoleptic assessments. Melting characteristics of ice cream are important and are usually assessed visually, or by measuring liquid released over a given time period using a measuring cylinder.

9.4 MICROBIOLOGY

Micro-organisms are unable to grow in ice cream stored at correct temperatures, although many survive for extended periods. Microbiological considerations, therefore, primarily involve the elimination of vegetative pathogens by pasteurization and the prevention of recontamination at all stages up to point of sale, the microbiological status of ingredients with particular reference to thermoduric organisms and preformed toxins and the prevention of microbial growth before freezing.

9.4.1 Ice cream and foodborne disease

To some extent ice cream retains a reputation as a high-risk food. This is unjustified for commercially produced ice cream in developed countries where the safety record over many years has been very good. In the

UK, for example, the last confirmed outbreak of food poisoning associated with ice cream occurred in 1955. This record may be attributed to the use of high quality ingredients and the strict control of pasteurization of the mix and of hygiene during subsequent operations.

Complacency should, however, be avoided and the isolation of *Listeria monocytogenes* from ice cream in the US demonstrates the continuing vulnerability of the ice cream making process and suggests that a re-evaluation of procedures with respect to this organism is necessary. Special attention should be paid to the efficiency of less-rigorous pasteurization procedures and to the control of *Listeria* in the environment. A further potential pathogen, *Aeromonas caviae*, was isolated from 4.7% of ice cream samples examined in Wales, but the public health significance of this finding is not clear.

9.4.2 Bacteriological standards

In many, but not all, countries legislative standards are applied to ice cream. These are usually based on 'total' viable count, coliforms and the absence of pathogenic micro-organisms and vary in severity (Table 9.7). An exception is the UK where, in the absence of standards, the methylene blue dye reduction test is applied by health authorities as a guide to hygiene standards.

9.4.3 Microbiological analysis

Methods used for liquid milk (Chapter 2, pages 97–8) are generally suitable for ice cream. In the case of coliforms, the International Dairy Federation specifies both a routine method involving plating onto violet red-bile agar (incubation 22 h at 30°C), and a reference method involving a most probable number estimation using brilliant green-lactose-bile broth (incubation 48 h at 22°C). Where liquid media are

---

* Ice cream was the vehicle of infection in an outbreak of typhoid fever in Aberystwyth, Wales during 1947, which involved *ca.* 210 cases and four deaths. This outbreak was unusual in that the ice cream was contaminated by the maker, a urinary carrier of *Salmonella typhi*. This person had suffered typhoid fever some years earlier, but had been cleared as a food handler on the basis of negative-stool tests.

  Ice cream was also the vehicle of infection in a single case of *Shigella flexneri* food poisoning. This case was unusual in that the ice cream was infected by a monkey in the pets' corner of a department store, which touched the ice cream during the course of consumption by a child. Pets' corners are no longer a feature of department stores in the UK, but there are unsubstantiated allegations of food poisoning due to contamination of food by monkeys kept by street photographers in Spain.

Table 9.7 Microbiological standards for ice cream

International Dairy Federation	total count 10^5/g
	coliforms 10^2/g
	absence of vegetative pathogens
EEC	
(whole product pasteurized)	total count 10^5/g
	coliforms 10^2/g
Post-pasteurization additions)	total count 2×10^5/g
	coliforms 2×10^2/g
UK	none (see text)
US	total count 5×10^4–10^5/g
	(varies by State)
Australia	total count 5×10^4/g
	coliforms <0.1/g
	absence of pathogens
France	total count 3×10^4/g
	absence of pathogens
Japan	total count 5×10^4/g
(3% milk fat ice cream)	total count 1×10^4/g

used, false-positives can be a problem at low dilutions due to carry-over of sucrose, or other fermentable carbohydrate, from the ice cream.

The methylene blue dye reduction test is technically simple and involves measuring the period of time taken to decolourize methylene blue during incubation at 37°C after preliminary incubation at 20°C for 17 h.

The methylene blue reduction test is recognized as being imprecise and its value in assessing the microbiological quality of ice cream has been questioned on many occasions. In contrast, it has been been argued that despite conceptual shortcomings the worth of the test has been proven in practice and that its application has been of major value in raising the hygienic standards pertaining in ice cream manufacture. Despite this plate count methods are generally considered to be more suitable where hygiene standards are generally high and it is likely that such methods will replace the methylene blue reduction test.

---

* Application of the methylene blue test involves grading according to the time to decolourization (> 4.5 h, grade 1; 2.5–4.0 h, grade 2; 0.5–2.0 h, grade 3; 0 h h, grade 4). Performance is assessed on the basis of a year of sampling when it is expected that 50% of samples are grade 1, 80% grade 1 and 2, not more than 20% grade 3 and none grade 4.

Microbiology

Examination of ice cream for specific pathogens is, in most cases, considered neither necessary nor desirable and the effort involved would be better employed implementing safety assurance procedures. National regulations stipulating 'absence of pathogens' are subject to wide variations in interpretation and it is inconceivable that any sane microbiologist would wish to test for all organisms which could, conceivably, be present in ice cream.

In practice, *Salmonella* and *Listeria monocytogenes* are of greatest concern and may be tested for by standard cultural techniques or by use of rapid methods. Routine examination for these organisms is not, however, necessary.

Additions made to ice cream after pasteurization are potential sources of hazard and, under some circumstances, microbiological examination may be considered appropriate.

EXERCISE 9.1.

Manufacturers of branded ice cream must be willing, and able, to invest heavily in advertising to protect the market share of the brands from the inroads made by the own-label lines of multiple retailers. Smaller manufacturers are often financially unable to support the advertising required and must seek alternative strategies, including diversification into niche markets. You are Marketing Director of an ice cream manufacturer seeking niche markets. Decide the most suitable types of ice cream (in terms of fat content and overrun) for sale in the following markets. Draw up an outline list of flavour combinations, colours and any added-value ingredients.

1. Night clubs catering for energetic, fashion-conscious, 21–25 year olds.
2. Barbecues, held predominantly by relatively affluent, 25–49 year old homeowners.
3. Fitness suites and clubs catering for a clientele ranging from the more dedicated aerobics enthusiasts to serious weight lifters and martial arts practitioners.
4. Henley Regatta.

EXERCISE 9.2.

A wide range of low fat dairy desserts are available leading to considerable confusion amongst consumers and, for this reason, the UK Milk Marketing Board has launched a special advice service in an attempt to clarify the situation. One of the most confusing areas is that of fat content since consumers are attracted by descriptors such as 'less than half the fat of dairy ice cream' and 'almost no fat' yet, despite nutritional labelling, find comparisons between different products (and dairy ice cream) difficult to make. Design a classification scheme for low fat dairy desserts (*cf.* low fat spreads, Chapter 6, page 226, yoghurts, Chapter 8, page 353) deciding the maximum permitted fat content in each category. Should fat content be the sole criterion in classifying foods of this type which are purchased for their perceived 'healthful' qualities?

EXERCISE 9.3.

You are employed as a technologist by a regional ice cream maker supplying a number of seaside resorts. In recent years sales have been falling due to competition from national brands and the declining popularity of the resorts supplied. Your company have been approached by a medium-sized, but aggressively expanding, multiple retailer, who wishes to develop a range of own-label goods including ice cream. The offer is financially attractive, but will require your company to undertake most of the development work and some capital expenditure may be required. An element of financial risk thus exists since the introduction of own-label lines by the retailer is to be on an experimental basis. There is considered to be no difficulty with standard, high overrun, bulk packed ice cream, but there is also a requirement for a premium and a super-premium range. The retailer has stipulated that these should be 'as close as possible to own-label market leaders in quality', but at a 'significantly lower cost'. Before undertaking expensive development work you are asked to consider the feasibility of this requirement and to discuss the advantages and disadvantages of possible approaches. What alternative strategies could your company develop to offset the decline in their traditional market?

EXERCISE 9.4.

What are the major microbiological hazards associated with the following ingredients added after pasteurization? Discuss possible means of minimizing these hazards, indicating those ingredients where microbiological examination is considered necessary and the minimum standards of acceptability required.

1. Chopped mixed nuts.
2. Sugar-preserved fruit.
3. Pour-over fruit sauce (pH value 4.7), containing gelatin as a thickening agent.
4. Desiccated coconut.
5. Chocolate flakes.

BIBLIOGRAPHY

GENERAL

Analysis

Harrigan, W.F. and McCance, M.E. (1976) *Laboratory Methods in Food and Dairy Microbiology*, 2nd edn. Academic Press, London.

Richardson, G.H. (ed.) (1985) *Standard Methods for the Examination of Dairy Products*. American Public Health Association, Washington, DC.

Chemistry

Walstra, P. and Jenness, R. (1984) *Dairy Chemistry and Physics*, Wiley, New York.

Wong, M.B., Jenness, R., Keeney, M. and Marth, E.H. (eds) (1988) *Fundamentals of Dairy Chemistry*. Van Nostrand Reinhold, New York.

Microbiology

Robinson, R.K. (ed.) (1990) *Dairy Microbiology*, vols 1 and 2, 2nd edn. Elsevier Applied Sciences, London.

Varnam, A.H. and Evans, M.G. (1991) *Foodborne Pathogens: An Illustrated Text*. Wolfe Publishing, London.

Milk products and processing

Earle, R.G. (1983) *Unit Operations in Food Processing*, 2nd edn. Pergamon Press, Oxford.

Early, R. (ed.) (1991) *The Technology of Dairy Products*. Blackie and Son, Glasgow.

Kessler, H.G. (1981) *Food Engineering and Dairy Technology*.

Verlag A. Kessler, Freising. Robinson, R.K. (ed.) (1986) *Modern Dairy Technology*, vols 1 and 2. Elsevier Applied Sciences, London.

Nutrition

Renner, E. (1983) *Milk and Dairy Products in Human Nutrition*. Volkswirtschaftlicher Verlag, Munich.

Renner, E. (ed.) (1989) *Micronutrients in Milk and Milk-based Products*. Elsevier Applied Sciences, London.

Quality assurance

ICMSF (1986) *Micro-organisms in Foods, 2. Sampling for Microbiological Analysis: Principles and Specific Applications*. Blackwell Scientific Publications, Oxford.

ICMSF (1988) *Micro-organisms in Foods, 4. Application of the Hazard Analysis Critical Control Point (HACCP) System to Ensure Microbiological Safety and Quality*. Blackwell Scientific Publications, Oxford.

NRC (1985) *An Evaluation of the Role of Microbiological Criteria for Foods and Food Ingredients*. National Academy Press, Washington, DC.

Sutherland, J.P., Varnam, A.H. and Evans, M.G. (1986) *Colour Atlas of Food Quality Control*. Wolfe Publishing Ltd, London.

CHAPTER 1. INTRODUCTION

Schmidt, G.H., van Vleck, L.D. and Hutjens, M.F. (1988) *Principles of Dairy Science,*, 2nd edn. Prentice Hall, Englewood Cliffs, NJ.

Slater, K. (1991) *Principles of Dairy Farming*. Farming Press, Norwich.

CHAPTER 2. LIQUID MILK AND MILK PRODUCTS

Burton, H. (1988) *Ultra-High-Temperature Processing of Milk and Milk Products*. Elsevier Applied Sciences, London.

Chin, J. (1982) Raw milk, a continuing vehicle for the transmission of infectious disease agents in the United States. *Journal of Infectious Diseases*, **46**, 440–1.

Kosikowski, F.V. and Mistry V.V. (1990) Microfiltration, ultrafiltration and centrifugation separation and sterilization processes for improving milk and cheese quality. *Journal of Dairy Science*, **73**, 1411–9.

Renner, E. and Abd El-Salam, M.H. (1991) *Application of Ultra Filtration in the Dairy Industry*. Elsevier Applied Sciences, London.

CHAPTER 3. CONCENTRATED AND DRIED MILK PRODUCTS

Anon (1991) A new concept in powder hygiene. *Dairy Industries International*, **56**(4), 26–7.

Alvarez de Felipe, A.I., Melcon, B. and Zapico, J. (1991) Structural Changes in sweetened condensed milk during storage: an electron microscopy study. *Journal of Dairy Research*, **58**, 337–44.

Augustin, M.A. (1991) Developing non-fat milk powders with specific functional properties. *CSIRO Food Research Quarterly*, **51**, 16–22.

Baldwin, A.J. and Ackland, J.D. (1991) Effect of preheat temperature and storage on the properties of whole milk powder. Changes in physical and chemical properties. *Netherlands Milk and Dairy Journal*, **45**, 169–81.

Baldwin, A.J., Cooper, H.R. and Palmer, K.C. (1991) Effect of preheat temperature and storage on the properties of whole milk powder. Changes in sensory properties. *Netherlands Milk and Dairy Journal*, **45**, 97–116.

de Koning, P.J., de Wit, J.N. and Driessen, F.M. (1992) Process conditions affecting age-thickening and gelation of sterilized canned evaporated milk. *Netherlands Milk and Dairy Journal*, **46**, 3–18.

Evans, A.A. (1989) Misconceptions in evaporator and dryer theory. *Australian Journal of Dairy Technology*, **44**, 97–100.

Mettler, A.E. (1992) Pathogen control in the manufacture of spray dried milk powders. *Journal of the Society of Dairy Technology*, **45**, 1–2.

Nieuwenhuijse, J.A., Sjollema, A., van Boekel, A.J.S. *et al.* (1991) The heat stability of concentrated skim milk. *Netherlands Milk and Dairy Journal*, **45**, 193–224.

Singh, L.H. and Creamer, L.K. (1991) Denaturation, aggregation and heat stability of milk protein during the manufacture of skim milk powder. *Journal of Dairy Research*, **58**, 269–83.

CHAPTER 4. DAIRY PROTEIN PRODUCTS

Mulvihill, D.M. (1991) Trends in the production and utilisation of dairy protein products: production. *CSIRO Food Research Quarterly*, **51**, 145–57.

Pearce, R.J., Dunkerley, J.A., Marshall, S.C. *et al.* (1991) New dairy science and technology leads to novel milk protein products. *CSIRO Food Research Quarterly*, **51**, 145–57.

CHAPTER 5. CREAM AND CREAM-BASED PRODUCTS

Rajah, K.K. and Burgess, K.J. (1991) *Milk Fat: Production, Technology and Utilization*. Society for Dairy Technology, Huntingdon.

Rothwell, J. (1989) *Cream Processing Manual*. Society for Dairy Technology, Huntingdon.

CHAPTER 6. BUTTER, MARGARINE AND SPREADS

Charteris, W.P. and Keogh, M.P. (1991) Table spreads, trends and the European market. *Journal of the Society of Dairy Technology*, **44**, 3–8.

IDF (1991) Utilisation of milk fat. *Bulletin of the International Dairy Federation*, No. 260.

Moran, D.J.P. (1990) The development of yellow spreads. *Dairy Industries International*, **55(5)**, 41–4.

Rajah, K.K. and Burgess, K.J. (1991) *Milk Fat: Production, Technology and Utilization*. Society for Dairy Technology, Huntingdon.

CHAPTER 7. CHEESE

Davies, F.L. and Law, B.A. (eds) (1984) *Advances in the Microbiology and Biochemistry of Cheese and Fermented Milks*. Elsevier Applied Sciences, London.

Eck, A. (ed.) (1987) *Cheesemaking: Science and Technology, (Le*

Fromage). Lavoisier Publishing, New York.

Fox, P.F. (ed.) (1992) *Cheese: Chemistry, Physics and Microbiology*, 2nd edn. Elsevier Applied Sciences, London.

International Dairy Federation (1990) Use of enzymes in cheesemaking. *Bulletin of the International Dairy Federation*, No. 267.

Scott, R. (1986) *Cheesemaking Practice*, 2nd edn. Elsevier Applied Sciences, London.

CHAPTER 8. FERMENTED MILKS

Davies, F.L. and Law, B.A. (eds) (1984) *Advances in the Microbiology of Cheese and Fermented Milks*. Elsevier Applied Sciences, London.

Kurman, J.A., Rasic, J.Lj. and Kroger, M. (1992) *Encyclopedia of Fermented Fresh Milk Products*. Chapman & Hall Ltd, London.

Nakazawa, Y. and Hosono, A. (1992) *Function of Fermented Milk. Challenges for the Health Sciences*. Elsevier Applied Sciences, London.

Robinson, R.K. (1988) Cultures for yogurt–their selection and use. *Dairy Industries International*, **53(7)**, 15–9.

Tamime, A.Y. and Robinson, R.K. (1985) *Yoghurt, Science and Technology*. Pergamon Press, Oxford.

CHAPTER 9. ICE CREAM AND RELATED PRODUCTS

Arbuckle, W.S. (1986) *Ice Cream*, 6th edn. AVI, New York.

Rothwell, J. (1988) *Ice Cream Manufacture*. J. Rothwell, Reading.

Index

Absidia 383
Acetaldehyde
  as flavour compound 326, 349, 365, 372
  formation 349
Acidophilus milk 364–5
Acidulants 419
Acinetobacter 92
Actinomyces 35
Aeromonas 39
Aeromonas caviae 427
Aerosol packing 203–4, 207, 389
Agglutinins, starter failure 285
Alcaligenes 41, 92, 95
Alcaligenes tolerans 94, 154
Alkaline phosphatase, pasteurization index 26, 52, 83, 88, 215
Alnarp process, butter making 230–1
Alpma systems, cheese making 313
Alternaria 388
'*Alteromonas putrefaciens*' 92
Antibiotic residues 29–30, 31, 289, 355
  starter failure 285
  tests for 86–7
Antioxidants 135, 147, 206, 248, 392
APV novel cheese making systems 313, 314
Aqueous phase structuring agents 253, 269
Arthrobacter 311
Aspergillus 153, 271, 306, 337, 338, 383
Aspergillus flavus 30
Aspergillus parasiticus 30
Attrition drying 172, 173
Aureobacterium liquefaciens 338
Auto-sterilization
  canned evaporated milk 154

Bacillus
  in cheese 365, 380
  in concentrated milk 151, 153
  in cream 219, 220
  in milk 41, 93, 95
  in milk power 154
  in whey powder 180
Bacillus cereus
  in cream 218
  food poisoning
    cream 216
    milk powder 154
  mastitis 35
  in milk 58, 91, 93, 95
Bacillus circulans 93
Bacillus coagulans 95, 152
Bacillus licheniformis 95
Bacillus polymyxa 271
Bacillus stearothermophilus 67, 86, 95, 110, 152, 219
Bacillus subtilus 68, 95, 152, 306
Bacteriocins, as preservatives 281
Bacteriophage
  detection 341
  host relationships 284–5
  precautions against 282–4, 290–2
  resistance to 280, 285, 292
  starter failure 283, 284–5, 290, 358
Bacteriophage-insensitive mutants 285, 292, 296, 358
Bactofugation 91, 288, 298, 301
Bel paesa cheese 304
Berridge cold renneting principle 313
  see also Rennet, cheese manufacture
Bifidobacterium
  in bifid-amended products 71, 360, 417

as starter cultures 348, 369
selective media for 384
therapeutic properties 350, 354, 360, 365
Bifidobacterium bifidum 359, 369
Bifidobacterium breve 369
Bifidobacterium longum 359
Biogenic amines, in cheese 304, 336–7
Bitty cream 93
Bleu d'Auvergne cheese 290, 305
Block milk 121
  see also Sweetened condensed milk
Boudon cheese 310
Brainerd diarrhoea 38
Brevibacterium 311
Brevibacterium linens
  and accelerated cheese ripening 307
  amino acid metabolism 329
  in blue vein cheese 300, 305
  cheese spoilage 338
  methanethiol production 327
  in smear-ripened cheese 306, 311
  in soft cheese 311, 312, 330
Brie cheese 310–11, 330, 334, 335
Brined cheese 307–8
Broken texture of cream 199
Brucella 39
  in cheese 336
Brucella abortus 38
Brucella suis 38
Bulk condensed milk 114–16
Bulking agents 208, 395
Butter
  added value ingredients 241
  chemical analysis 266–7
  control points 229, 230, 232, 239, 242
  cream processing for 227–30
  fat in 225, 244, 264–5
  handling 233–4, 238
  ingredient use 392
  manufacture 227–44
  microbiological analysis 271
  microbiological stability 268–9
  nutritional properties 257
  packaging 238, 239–41
  recombined 244
  spoilage 270–1
  taints 236, 264
  whipped 243
  food poisoning 270
Butter fat, *see* Milk fat
Buttermilk 227, 228, 236, 237, 317, 384
  cholesterol-lowering effect 317–18
  microbial growth in 237
  stabilization of recombined products 122
  see also Cultured buttermilk
Buttermilk powder, ingredient use 253, 393

Cacciocavallo cheese 203
Caerphilly cheese 290, 295, 208, 303
Calcium lactate, crystals in cheese 304, 339
Camembert cheese 281, 283, 310–11, 330
Campylobacter
  control by heat 286, 354
  food poisoning, milk 44, 88, 89
  mastitis 35
  source in milk 36, 38
  survival in dairy products 40, 381
Campylobacter jejuni 89
Candida 338, 339, 340, 383
Candida cylindraceae 256
Candida kefyr 348, 366
Candida lipolyticum 218, 271
Canned evaporated milk
  manufacture 116–18
  spoilage 152
  sterilization 118
Carcinogens, suspect in packing material 240
Casein 8–11, 328, 362, 376, 422
  alpha 9–11
  beta 9–11, 327, 375–6
  gamma 9
  kappa 9–10, 142, 143, 144, 170, 375–6
  hydrolysis by proteinases 322–3
  interaction with whey proteins 374, 375
Casein micelles 8, 10–11, 13, 145, 176
  age-thickening of concentrated milk 144–5
  aggregation 84, 140
  binding of plasmin 327

complex formation with whey proteins 139–40
destabilization by chymosin 322–3
effect of concentration 141
effect of heat 12, 80, 374
feathering of cream 213–14
heat stability of concentrated milk 142–4
in ice cream 424
sedimentation 85
see also Milk protein; Protein; Whey protein
Casein products
  caseins
    acid 161, 170, 171
    commercial classification 161
    fractionation 175
    functional properties 161
    ingredient uses 178–9, 319
    manufacture 170–2
    rennet casein 161, 170, 171
  caseinates 166, 172–4
    functional properties 174–5
    ingredient use 135, 178–9, 210, 253
    manufacture 172–4
  casein-derived peptides 177
  casein, whey co-precipitates 175–6
  microbiology 180
  nutritional properties 170–80
Caseobacter 304, 305
Centriwhey system, cheese making 314
Chabris cheese 310
Cheddar cheese
  added value ingredients 315
  a_w level 332
  food poisoning 337, 339
  low-fat 290, 316, 317
  manufacture 283, 290–1, 293, 294–6, 330, 331
  ripening 304, 305–6, 326, 327, 328
  texture 325
  yield 286
Cheese
  accelerated ripening 306–7
  acid-set 307–9
  acidity and curd texture 295–6, 328
  added value 315–16
  bitter taste 283, 288, 315, 316, 317, 328–9

blowing 287, 338–9
brining 298, 300, 308
cheddaring 295
chemical analysis 331
classification 276
control points 289, 294, 297, 299, 308
curd formation 321–4
defects in 296, 304, 331, 337, 339, 340
end-product testing 320–1
environment for micro-organisms 332
eye formation 279, 302, 304, 330
fat content 286, 287, 321
flavour compounds 326–7, 328, 329–30, 331
food poisoning 332–6
manufacture 277, 290–303, 307–11
microbiological analysis 341–3
moisture content 276, 290
non-conventional manufacture 313–15
nutritional properties 321
nutritionally modified 316–18
package 337–8, 339
ripening 278, 300, 303–7, 311–12, 326–31
smoking 315
spoilage 337–40
Cheese analogue 319–20
Cheese base 318, 320
Cheese milk 286–9, 307, 309–10
  acidified 314–15
  composition 286
  concentration of 288, 313–14, 307, 317, 320
  endospores in 339
  heat treatment 286–8, 298
  preservatives in 298, 302
  recombined 286, 315
  source of enzymes in cheese 303, 312
Cheese spread 318
Cheshire cheese 295, 303, 305–6
Chlorogenic acid, feathering of cream 214
Cholesterol 13, 14, 18, 22, 321
  reduction 20–1, 317–18
Chymosin 9, 292, 325, 327, 329
  mode of action 322–3
Citrobacter 91, 92, 152
Cladosporium 271, 338
Cleaning-in-place 53, 110, 111

Clostridia, lactate fermenting 287–8, 339
Clostridium 41, 94, 151, 152, 319
Clostridium botulinum 319
　food poisoning
　　cheese 335–6
　　yoghurt 381–2
Clostridum butyricum 339
Clostridium perfringens 35, 151
　food poisoning, milk 151
Clostridium sporogenes 339
Clostridium tyrobutyricum 339
Coffee whitener, *see* Cream substitutes
Coliforms 98, 151, 153, 155, 385
　detection in ice cream 427–8
　taints in milk powder 153
Colouring 292, 398, 420
Concentrated milk 113–14
Concentrated milk products 103, 104, 113, 114
　age thickening 144–5
　chemical analysis 148
　control points 115, 117, 118, 120
　end-product testing 122
　heat stability 141–4
　ingredient use 73, 392–3
　microbiological analysis 154
　nutritional properties 138
　spoilage 151–2
Concentration
　methods 104
　see also Evaporation; Freeze concentration; Reverse osmosis; Ultrafiltration
　physico-chemical changes during 140–1
Concentration factor 140–1
Continuous buttermakers 234–5
Copper, as pro-oxidant 19, 83, 227, 231, 265
Co-precipitation, caseins and whey products 175–6
Co-randomization, fats 246, 259–60
Corynebacterium bovis 35
Corynebacterium pyogenes 35
Coryneform bacteria 35, 36, 94
Cottage cheese 281, 285, 309, 310, 315, 340
　and healthful diet 316
Coxiella burnetii 40, 41, 48

Cream
　added value 208
　coffee 130, 198, 201
　　feathering 201, 213–14
　clotted 204–5, 219, 220
　chemical analysis 215–16
　control points 186, 195, 196, 199
　double 200–1, 212
　dried 205–6
　end-product testing 208–9
　fat content 184, 198, 200, 202, 204
　flavour and aroma 211–13
　frozen 206–7
　half 200
　handling 194
　ingredient use 206, 209–11, 300, 392
　low fat 184
　manufacture 185–99
　microbiological analysis 220–2
　milk for 184–5
　nature of 183
　nutritional properties 211
　prewhipped 203–4
　single 200
　spoilage 217–20
　taints 197
　viscosity 199–200, 213
　whipping 201–2, 209–10
　see also Cultured cream
Cream cakes and desserts 209–19, 220
　food poisoning 209, 210, 217
Cream liqueurs 210–11
Cream substitutes 207–8
Cryptococcus 271, 340
Cryptococcus neoformans 35
Cryptosporidium 38
Cryptosporidium parvum 44
Cultured buttermilk 365–6, 382
　see also Buttermilk
Cultured cream 207, 365–6, 366
　see also Cream
Custards 388

Dairy industry 3–6
Dairy products
　adulteration 86, 87, 266–7, 379
　bifid amended 71, 360, 417
　cholesterol reduced 71, 225, 317–18, 369, 393, 417

Debaryomyces 311, 383
Dehydrogenation, milk fat 261
Demi-sel cheese 310
Derby cheese 295, 315
Diacetyl 230, 265, 371, 373
  detection 267
  production by starter cultures 278, 282, 349
Diafiltration 118, 167, 176, 314
Double Gloucester cheese 290, 295
Drying, principles 122–4
Dunlop cheese, 295
Dutch-type cheese 279, 298, 331

Edam cheese 298, 313, 314, 325, 328, 339
Electrodialysis 164
Emmental cheese 301, 303, 304, 325, 327, 339
Emulsifiers
  in cream liqueurs 210
  in cream substitutes 207
  in ice cream 397–8, 420, 421–7
  in margarine and spreads 241, 247, 252, 253
  mechanism of 421–2
  in proceessed cheese 318
Emulsions
  changes during processing and storage 14–15, 234–5, 255
  instability 198, 210
  phase reversal 247
  structure of butter, margarine and spreads 252, 263
Endiothia parasitica 293
Endospore-forming bacteria 58, 36, 93, 151, 218
  detection 99, 154
Endospores
  in cheese milk 287–8, 298, 301, 339
  of *Coxiella burnetii* 41, 48
  dormancy 219
  enumeration 97
  in fermented milks 354
  inactivation 46, 58, 94, 198, 219
  in processed cheese 319
  of thermophilic bacteria 69, 95
  survival of heat treatment 45, 93, 95, 197, 218, 219

Enterobacter 91, 92, 151
Enterobacter agglomerans 340
Enterobacter sakazakii, meningitis, milk powder 151
Enterobacteriaceae
  in butter 271
  in cheese 338–9
  in concentrated milk products 152, 153
  in cream cakes 220
  effect of storage temperature 92
  in milk powder 151
  milk spoilage 41
Enterococcus 41, 94, 153, 154, 180, 21
Enterococcus faecalis 152
Enterococcus faecium 271, 336
Environmental chemicals, in milk 32–3
Enzyme-modified cheese 306, 320
Escherichia coli 34, 40, 272
  and cheese flavour 339
  detection in cheese 342
  food poisoning
    milk 44
    soft cheese 334–5
  growth in cheese 312, 332
  as index organism 92, 153, 272, 335
  mastitis 35
Evaporation 104, 133, 320
  principles 104–5
Evaporators
  falling film design 105–7, 163
  multiple effect 108–9, 110
  recompression 109–10, 163
Extended heat treatment, cream 197, 218–19

Fat
  analysis 267–8
  adulteration 266–7
  in butter 229–30, 243, 265, 263–5
  crystallization and crystal morphology 17–18, 256, 257–8
  in ice cream 387–9, 390, 399, 407, 421–2, 423
    function of 391
    sources of 392–3
  in margarine and spreads
    crystallization 249–52
    importance of 244–5

sources of 241, 243, 244–5, 253
structure 263–4
medium-chain triacylglycerol 318
oxidation 19, 83, 146, 147, 212–13, 265
zero-calorie 225
see also Marine oils; Milk fat; Vegetable fat
Fat content
determination 85, 86, 148, 266, 426
novel methodology 86, 194, 215, 331
Fat replacers
cheese products 417
cream substitute 208
low fat frozen desserts 389, 393, 419
spreads 225, 254, 269
Fatty acids
carbon number 16
in cheese 326, 329–30
in cream 191, 212, 215
dietary importance 21, 22, 225, 226
effect on fat crystals 260
fat deterioration 19, 26, 83 84
in fermented milks 373
monounsaturated 16, 225, 245
$n-3$ 226
polyunsaturated 21, 225, 245, 246, 253, 317
Fermented milks
chemical analysis 379
classification 347
end-product testing 370
environment for micro-organisms 380–1
flavour compounds 372–4
ingredient use 364
microbial slime in 378–9
microbiological analysis 383–5
nutritional properties 370–2
spoilage 382–3
therapeutic properties 359–60, 364–5, 368–9
see also Yoghurt and other individual types
Fermented milk analogues 369
Feta cheese 308
Filmjolk 368
Filtermat dryer, see Spray drying
Flavobacterium 92, 95

Flavoured milk, see Milk, added-value
Fluidized bed dryers 128–9, 169, 172
see also Spray drying
Foam spray drying 129, 147
see also Spray drying
Formate production, yoghurt fermentation 358
Fractionation, fats 104, 113
Freeze drying, yoghurt 364
Fromage frais 310

Gamma ray sterilization 328
Gammelost cheese 301
Gels
acid 322–4, 375–7
acid:rennet 324
rennet 322–5
Genetic modification, micro-organisms 280, 281, 293
Geotrichum 271
Geotrichum candidum 218, 304, 311, 368, 373
Glucono-delta-lactone 293, 376
Glycolysis, cheese ripening 327, 330–1
Gorgonzola cheese 300, 304–5
Gouda cheese
a_w level 332
manufacture 288, 292, 298, 313
ripening 327
spoilage 338, 339

Hafnia 92, 152
Half and half, see Cream, half
Halloumi cheese, 136
Heat stable enzymes, detection 87
see also Lipases; Proteinases; Psychrotropic bacteria
Heart disease and dairy products 21–2, 78, 225–6, 317–18
Homogenization
of anhydrous milk fat 255
of cheese milk 286, 300
of cream 196, 200, 202, 207, 366
control of feathering 201
control of broken body defect 199
effect on viscosity 213
prevention of creaming 198
of cream liqueurs 210
of ice cream mix 407, 416

of milk 53, 55, 58, 63
    effect on fat digestibility 78
    pre-concentration 114, 115, 117
    of spreads 254
    principle 55
    of soft cheese curd 307, 310
Hutin–Stenne system, cheese making 313
Hydrogen peroxide, as sterilizing agent 6, 68, 197
Hydrogenation, of fats 245, 253, 260–1

Ice cream
    added value 398–9, 413
    bacteriological standards 427, 428
    chemical analysis 426
    consumer preferences 390–1
    control points 400, 406, 407, 408, 409, 414
    end-product testing, 418, 426
    food poisoning 398, 426–7
    hygiene, retail outlets 415, 418
    manufacture 391–415
    microbiological analysis 427
    nutritional properties 420
    off-flavours 181, 395
    overrun 399, 407, 409–10, 411, 426
    packaging 414–15
    special types of
        dietetic 417
        high fat 416
        high solids-non-fat 416
        Italian-style 416–17
        soft-scoop 415–16
        soft-serve 417–18
    structure 420–1, 423–5
        vehicle of *Salmonella typhi* 427
        vehicle of *Shigella dysenteriae* 427
Ice cream freezers 399, 410–13, 417–18
Ice cream mix
    ageing 408, 422–3
    blending 404–5
    freezing 408, 409–13
    heat treatment 404, 405–6, 417, 418
Ice cream products, nature of 387–90
Imitation cream, *see* Cream substitutes
Impastata cheese 309
In-bottle sterilization 68–9, 76, 77, 78
    see also Milk, in-bottle sterilized

In-container sterilization, cream 198, 219
Infant food 127, 130, 164, 177, 364
    food poisoning 150
    microbiological quality 149, 180
Insolubility index, milk powder 145, 146, 147
Interesterification, fats 246, 261–2
Iodine number (value), fats 267–8
Ion exchange 164–5, 168–9, 170, 171

Jones system, bulk starter preparation 283

Kefir 348, 366–7, 373
Kefir grain 348, 366–8
Kluyvera 340
Kluyveromyces lactis 165, 178, 293
Kluyveromyces marxianus
    var. *lactis* 348, 368
Kluyveromyces marxianus
    var. *marxianus* 348, 366, 368
Kopanisti cheese 329, 339
Koumiss 348, 368

Lactate metabolism, in cheese ripening 304, 305, 312
Lactic acid 93, 164, 339
    acid casein manufacture 161
    as acidulant 419
    cream ripening 231, 232
    as flavour compound 277, 323
    inhibition of *Campylobacter* 40, 381
    inhibition of micro-organisms 269, 280, 332, 380, 381
    metabolism by *Propionibacterium* 304
    production by starter cultures 276, 296–7, 302, 341
Lactic acid bacteria
    as starter cultures 276–8, 348
    non-starter strains in cheese 296, 304
Lactic cheese 307, 309
Lactobacillus
    in cheese 296, 304, 305, 339
    cream liqueur spoilage 210
    genetic modification 280
    lactose metabolism 279
    in milk 41, 92, 93, 94
    starter species 278, 281
    in yoghurt 349, 384

Lactobacillus acidophilus 350, 359, 364–5, 366, 368, 369
Lactobacillus brevis 338
Lactobacillus buchneri 336
Lactobacillus casei 306, 359, 366
  ssp. *casei* 369
Lactobacillus delbrueckii
  ssp. *bulgaricus* 356, 371, 382
    as cheese starter 278, 301, 302, 308
    as fermented milk starter 357–8, 365, 368, 369
    lactose metabolism 279
    nodulation of yoghurt 359, 377
    stimulation by formate 374
  ssp. *lactis* 278, 279, 301
Lactobacillus fermentum 336
Lactobacillus helveticus 278, 301, 302, 329, 336
  ssp. *jugurtii* 358, 359
Lactobacillus kefir 366, 367
Lactobacillus kefiranofaciens 366, 367
Lactococcus
  acetaldehyde formation 369
  bacteriophage 282, 284
  genetic modification 280
  lac⁻ strains as starters 307
  milk spoilage 92
  starter species 278, 281, 310
Lactococcus lactis 207, 231, 301, 356
  ssp. *cremoris*
    as butter starter 230
    as fermented milk starter 365, 366
    plasmid-borne properties 280
    slime forming strains 379
  ssp. *lactis*
    as butter starter 230
    cheese spoilage 340
    as cheese starter 278, 283, 290, 302, 307, 310
    fast acid producing strains 296
    as fermented milk starter 365, 366, 368
    milk spoilage 93
    nisin producing strains 281
    plasmid-borne properties 280
    slime forming strains 379
  ssp. *lactis* var. *diacetylactis*
    antagonist production 281
    as butter starter 230
    cheese spoilage 283
    as cheese starter 278, 298, 300, 310
    diacetyl production 279, 280, 375
    plasmid-borne properties 280
Lactones
  flavour of dairy products
    butter 265, 266
    cream 211, 212
    milk 14, 77
    yoghurt 373
Lactoperoxidase system 26, 34, 47, 286, 287, 288
Lactose
  in cheese 321, 328
  crystallization defect
    in milk powder 147
    in ice cream 393, 394, 395, 400, 412
    in sweetened condensed milk 119
  effect of heating 79, 85
  enzymatic hydrolysis 71
  extraction from whey 164
  importance as milk constituent 22–3
  intolerance 23, 71, 72, 321, 369, 371
  removal by diafiltration 314
  seeding with 119, 133
  structure of ice cream 423
Lancashire cheese 298, 334
Langfil 379
Lecithin, as emulsifier 247, 397
Lecithinization, whole milk powder 135
Leuconostoc 278, 279, 281, 349, 366, 373
Leuconostoc mesenteroides
  ssp. *cremoris* 230, 231, 278, 310, 365
Lewis system, bulk starter preparation 283
Limburger cheese 299, 304, 311, 329
Lipases 19–20, 25, 82–3, 197, 256
  in butter 271
  in cream 191, 196, 206, 207, 212, 222
  in freshly drawn milk 26
  spoilage of UHT milk 84
  in Swiss-type cheese making 301, 330
  see also Psychrotrophic bacteria
Lipolysis 19, 83, 212
  cheese ripening 300, 327, 329–30, 337
Lipolytic micro-organisms, enumeration 271
Listeria monocytogenes 34, 35–6, 39
  behaviour in cheese 332, 333

control by antagonists 281
food poisoning
  cheese 335
  cream 216
  milk 44, 89
growth in butter and spreads 269, 270
in ice cream 427, 429
resistance to pasteurization 48, 89–90, 287, 405
survival in yoghurt 381
Low fat frozen desserts 364, 389, 390, 391, 395–6, 418–19
Lysinoalanine formation 12, 78–9, 173
Lysogenic cultures, *see* Bacteriophage, host relationships

Maillard reaction 84–5, 118, 140, 166, 173, 199
  cooked milk flavour 80
  nutrient losses 78, 138, 139
  Margarine 224–5, 247
  chemical analysis 267–8
  control points 248, 249, 250
  end-product testing 252
  ingredients 245–8
  manufacture 248–52
  microbial stability 268–9, 271
  microbiological analysis 271
  nutritional properties 257
Marine oils 226, 245, 253
  see also Fats
Mastitis 4, 26, 35, 214
Methanethiol, flavour of cheese 326–7
Methylene blue test 98, 220, 428–9
Metschnikowia 383
Micelia 383
Microbacterium 94
Microbacterium lacticum 94
Micrococcus
  butter spoilage 270, 271
  cheese ripening 304, 305, 306, 312
  cheese spoilage 338
  methanethiol production 327
  in milk 35, 36, 94
  sweetened condensed milk spoilage 153
Microfiltration 176
Micromonospora 152
Micro-organisms
  enumeration in milk 96–8

non-starter in cheese ripening 303, 304, 327–8, 329, 330–1
Milk
  added value 72–3
  anti-microbial systems in 33–4, 47
  biosynthesis 6–8
  chemical analysis 85–8
  composition 1, 2, 8–27
  flavour 27–8
  food-borne disease 5–6, 35–6, 38–40, 44, 88–91
  gelified 74–5
  in-bottle sterilized 45, 70, 68–9, 95–6
  see also In-bottle sterilization
  lactose-hydrolyzed 71–2
  microbiological analysis 96–100
  nutritionally modified 43, 69–71
  nutritional properties 13, 21–2, 76–80
  pasteurized
    control points 48, 52, 56, 57
    end-product testing 58
    microbiological standards 98
    packacing 53, 54–7
    processing 46–57, 58, 59, 94
    spoilage 91–4
    see also Pasteurization
  production 1–4
  taints in 27–8, 56
  Ultra-heat-treated 19–20, 58–68, 91
    age-thickening 84
    control points 64, 65, 67
    definition 45
    flavour 80–2
    packaging 64–6
    processing 60–4
    sedimentation 85
    spoilage 94–5
    sterility testing 98–9
    see also Ultra-heat-treatment
Milk analogues (substitutes) 43, 75–6
Milk fat
  dietary modification 18–19, 243
  fatty acid composition 13, 14, 15–16, 18–19, 24, 259
  importance as milk constituent 13–18
  lipid composition 14, 15
  nutritional properties 16, 21–2
  triacylglycerols of 15–18, 213, 229–30, 258, 261–2, 422–3

see also Fat; Vegetable fat
Milk fat globules 14–15, 78, 141, 207, 228, 309, 369
  buttermaking 234–5, 262–3
  in cheese curd 326
  in cream separation 184, 185, 186, 187, 189
  effect of homogenization 53, 213, 300, 374
  in ice cream 124, 126
  types of 263
  in whipping of cream 214–15
  in yoghurt coagulum 377
Milk fat globule membrane
  and fat oxidation 83, 213
  flavour of cream 211
  production of sulphydryl groups 135
  rupture in processing 255, 425
  stability of fat globules 14
  whipping of cream 201
Milk fat products 256–7
  anhydrous milk fat 121, 135, 244, 255, 315, 392
  butter oil 210, 255
  dried cream 205–6
  fractionated milk fat 226, 256, 392–3
  functional properties 257
  ghee 255
  hydrolyzed milk fat 256
  ingredient use 121, 135, 257, 392–3
Milk ice 388–9
Milk powder
  change during storage 147
  chemical analysis 148
  end-product testing 136
  food poisoning 149–51
  high fat 135
  ingredient use 73, 134, 137, 246, 318, 393–4
    fermented milks 353, 373, 379–80
    functional properties 136–7
    recombined products 122, 244, 315
    special types 132–3
  manufacture 130–6
  microbiological analysis 155–6
  micro-organisms in 153–4
  nutritional properties 138–9
  packaging 135–6
  physical and chemical properties 145–6
  skim 130–4, 176
  heat treatment of milk for 132–3
  high heat 122, 132, 147
  high protein 134
  lipase-reduced 132–3
  low heat 132
  low protein 134
  low calcium 132
  medium heat 132, 393, 384
  whole milk 134–5
Milk proteins 12–13, 78–9, 423
  as emulsifiers 13
  importance as constituents of milk 2, 8–11
  nutritional properties 13
  see also Casein; Proteins; Whey proteins
Milk protein concentrate 176
Minerals 23–5, 79–80, 314
Monilia 383
Monterey cheese 303, 304
Moulds
  cheese ripening 286, 300, 304–5, 311, 312
  control by preservatives 269, 380
  detection 221, 383
  source of milk clotting enzymes 293
  spoilage
    butter 271
    cheese 332, 338, 340
    clotted cream 220
    fermented milks 383
    sugar syrup 394
    sweetened condensed milk 120, 153, 155
Mousse 74, 204, 208, 389
Mozzarella cheese 302, 309, 316, 321
Mucor 301, 311, 383
Mucor miehi 293, 333
Mucor pusillus 293
Mutagens, in heated milk 83
Mycobacterium bovis 35, 38, 39
Mycobacterium tuberculosis 35, 38, 39, 48
Mycotoxins 30–1, 300, 328, 338

Nisin 118, 205, 219, 281, 319
NIZO systems
  cheese making 313
  cream ripening 231
Nocardia 35

Parev ice 388
Parmesan cheese 290, 301, 303, 395–6, 339
Pasta filata cheese 302–3
Pasteurization
  cream 196, 197, 295, 227, 229
  definition 44–5
  effect on nutrients 76, 77, 78
  high protein dryer feed 134
  ice cream mix 404–5, 496, 418, 426, 427
  margarine ingredients 249
  milk 46, 47–51, 73
  technology 47–51, 406, 407
Pearson's square 192–3, 402–4
Pediococcus 278, 304, 331, 339
Penicillium
  cheese ripening 301, 306
  spoilage of dairy products 153, 218, 271, 337–8, 383
Penicillium camembertii 311, 312
Penicillium roquefortii 300, 304, 305, 328, 330
Peptides
  bitter in cheese 292, 293, 317, 328–9, 339
  taste quality 328
Peroxide value, of fats 267
pH value
  anhydrous milk fat manufacture 255
  of cheese 332, 340
  in cheese making 294, 296, 297, 298, 325, 341
  in cheese ripening 305, 312, 327–8, 336
  in cream ripening 231
  deterimination of 331
  effect on micro-organisms 268, 312, 332, 333, 380
  of fermented milks 359, 372
  and gel formation 323, 324, 375–6
  lowering by Maillard degradation 85, 140
  milk protein product manufacture 168–9, 170, 174, 176
  and preservative action 162, 163, 269
  and stability of UHT milk 48, 87
  of whey 159, 161
  yoghurt manufacture 361, 362, 380

Phage carrier state, *see* Bacteriophage host relationships
Phenylalanine$_{105}$-Methionine$_{106}$ bond 9, 170, 322–3
Phenylketonuria 177
Phoma 338
Pichia 340, 383
Plasmin 25–6, 84, 327, 328, 329, 376
  see also Proteinases; Proteolysis
Poliovirus 44
Port du salut cheese 299
Preservatives
  cheese 298, 310, 319
  fermented milks 380
  margarine and spreads 248, 253, 269
  whey 162, 163
Processed cheese 318, 319
Proline, flavour of Swiss-type cheese 329
Propionibacterium 302, 329, 330
Propionibacterium freudenreichii 304, 305
Propionic acid, flavour of cheese 304
Protein
  determination 86, 148, 331, 426
  hydration 375, 407–8
  in milk analogues 75
  nodulation in yoghurt 377
  sources
    in ice cream 393–4
    margarine and spreads 252, 246
  supplementation of yoghurt 352–3, 370
Proteinases 25–6, 41, 47, 352, 376, 383
  age thickening 84, 95, 148
  cheese ripening 306, 327, 307
  inhibitors, in soya beans 75
  of starter cultures 278, 279, 280
  see also Psychrotrophic bacteria
Proteolysis
  biogenic amine formation 336
  cheese
    ripening 327–9, 330
    spoilage 337, 338
  cheese milk quality 286
Provolone cheese 302, 315
Pseudomonadaceae 397
Pseudomonas
  spoilage of dairy products 41, 92, 152, 218, 220, 340

synthesis of heat-stable enzymes 95, 271
Pseudomonas fluorescens 306, 340
Pseudomonas fragi 92
Pseudomonas putida 340
Psychrobacter 92
Psychrotrophic bacteria
  enumeration 87, 96, 97, 98
  growth in ice cream mix 408
  heat stable enzymes of 19–20, 41, 47, 84, 148, 212
    see also Proteinases, Lipases
  in milking equipment 37
  spoilage of dairy products 41, 92, 93, 218, 271
Pullaria 383

Quarg cheese 288, 309, 314, 340, 363
Queso blanco cheese 308, 336

Radio-isotopes, in milk 31–2
Recombined concentrated milk 121–2, 147
Red Leicester cheese 292
Reichert-Meissel number, fats 267
Rennet
  active components 170, 292, 322
    see also Chymosin
  bitter peptide formation 317
  casein manufacture 170
  cheese manufacture 292, 293, 294, 309, 310, 311
    see also Berridge cold renneting principle
  cheese ripening 293, 303, 304, 327
  high-lipase 301, 330
  in ice cream improver 397
  in modified cheese 288
  substitutes 293
Rennet coagulation time 322, 323
Reverse osmosis 104, 111–13, 133, 163, 288, 353
Rhizopus 301, 383
Rhodococcus equi 21
Rhodotorula 271, 383
Ricotta cheese 302
Ripening cultures 312, 317
Roller drying 122, 124, 169, 172, 173
Romadour cheese 327, 329, 330
Romano cheese 327, 329, 330
Roquefort cheese 299, 300, 304–5

Saccharomyces 366, 383
Saccharomyces cerevisiae 366, 373
St Nectare cheese, microbial succession in 311
Salmonella 34, 36, 38, 39, 40, 73
  behaviour at low pH values 381
  detection 155, 156, 429
  food poisoning
    cheese 333, 334
    milk 88, 89
    milk powder 149–50
  inactivation by heat 286, 287, 354
  mastitis 35
  in raw milk 44, 46
  survival in cheese 284, 333
Salmonella agona, in raw milk 40
Salmonella dublin 36
Salmonella ealing, in milk powder 150
Salmonella enteritidis, in home made ice cream 398
Salmonella typhi, in ice cream
Salmonella typhimurium 40, 333
Sanitizers, starter failure 285
Scandinavian ropy milk 368, 379
Scania system, cream manufacture 192
Scraped surface heat exchangers 243, 249, 320, 408
Separation of milk 185–6, 191–2
Separators 189–91, 227, 309
Serratia 91, 92
Sherbets 388, 389, 419–20
Shigella flexneri, in ice cream 427
Small, round, structured viruses, food poisoning 217
Solid fat index, margarine blends 246
Sorbets 389
Sporobolomyces 340
Spray drying 124–30
  caseinates 172, 173
  cream 205
  dryer design 125, 128–9, 130
  Filtermat dryer 129–30, 206
  milk 130–5
  yoghurt 364
    see also Drying, principles; Fluidized bed dryers; Foam spray drying
Spreadable butter 241, 243
Spreads 241, 244
  chemical analysis 268

fats in 225, 226, 255
growth of pathogens in 269, 270
ingredients 253–4
manufacture 254–5
microbial stability 254–5, 268–9
microbiological analysis 271
nutritional properties 257
spoilage 271
Stabilizers
  in cheese 310, 318–19, 320
  in cream 198, 203, 209
  in fermented milks 354–5, 359, 362, 365, 372, 376
  in ice cream products 396–7, 412, 416, 419–20, 421, 423
  in low fat frozen desserts 389, 419
  in milk drinks 72, 73, 74, 75
  nutritional properties 372
Standardization 115, 116, 192–4, 290, 352–3
Staphylococcus aureus 35, 273, 385
  detection 155, 341–2
  food poisoning
    chocolate-flavoured milk 91
    cheese 333–4
    cream-based products 217
    milk powder 150–1
    whipped butter 270
  growth and enterotoxin production 217, 270, 284, 333
Starter cultures
  antagonist production 280–1
  butter 230
  cheese 276–7, 290–1, 299
  cheesebase 320
  citrate metabolism 279, 280
  classification 281–2
  defined strains 282, 290–2
  deterioration of yoghurt 383–3
  direct-in-vat inoculation 283–4, 340, 358
  effect of a_w level 356, 381
  failure 283, 284–6
  fermented milks 307–8, 348–51, 364, 365, 366–8, 369
  flavour defects in cheese 329, 339 –40
  genetics 279–80

lactose metabolism 278–9, 280, 296–9, 301–2
mesophilic 281, 283
  cheese manufacture 290, 298, 299, 300, 307
  microbiological examination 340–1, 383–4
mixed strain 230, 282, 285, 302, 398
multiple strain 296, 358
non-conventional cheese making 313–14
production 282–4
protein metabolism 279, 280, 292
ripening of cheese 303, 307, 326, 327, 328–9
slime in fermented milks 359
stimulation in heated milk 354
therapeutic properties 348, 349–51
thermophilic 281, 282, 300, 307–8, 369
Sterility testing 99, 154–5
Stilton cheese 299, 300, 305, 336, 334
Streptococcus 35, 36, 41, 94, 278, 384
  Group A 34, 44
Streptococcus agalactae 35
'*Streptococcus cremoris*' 311
Streptococcus dysgalactae 35
Streptococcus pyogenes 35
Streptococcus salivarius
  slime production 378
  ssp. *thermophilus* 281, 382
    acetaldehyde production 349
    as cheese starter 278, 279, 301, 302
    effect of a_w level 356
    genetic modification 280
    lactose metabolism 279
    vitamin synthesis 371
    as yoghurt starter 357–8, 359, 365, 384
Streptococcus uberis 35
Streptococcus zooepidemicus 44, 336
Sulphydryl groups 12, 25, 135, 146
  as antioxidants 146·7, 212, 227
  cooked milk flavour 80–1, 82
Sweetened condensed milk
  ingredient use 166, 392–3
  manufacture 119–21
  microbiological analysis 155
  spoilage 152–3

sugar ratio 119
see also Block milk
Sweeteners 72, 208, 356, 389, 391, 394–5
  as ice cream ingredient 394–5, 416
Swiss-type cheese
  biogenic amines in 336
  food poisoning 334
  manufacture 301–2
  ripening 303, 304, 327, 339

Thermoactinomyces 152
Thermoduric bacteria 36, 73, 219, 426
  detection 155
Thickeners 72, 419
Torulaspora 383
Torulopsis 152, 271, 311, 340
Trichosporon 383
Tricothecium roseum 311
Turbidity test 45, 69, 88, 100

Ultrafiltration 104, 111–13, 313–14
Ultra-heat-treatment
  aerated desserts 389
  cream 198, 219
  effect on nutrients 76, 77, 78
  evaporated milk 118
  ice cream mix 396, 495, 406, 417
  kinetics 62–3, 65
  milk 58–9
    direct process 61, 62–3
    indirect process 59–61, 62
  verification 67–8
  see also Milk, ultra-heat-treated
Ultraosmosis 165
Ultraviolet sterilization 121, 197, 238, 282, 337, 394

Vacherin Mont d'Or cheese 334, 335
Vacreator 197, 227–8
Vegetable fat
  in cheese analogue 316, 320
  in cream substitutes 207–8
  fatty acid composition 257, 259
  in 'filled' cheese products 317
  flavour reversion 246, 266
  hardening flavour 266
  modification 260–2
  resistance to lipolysis 272

Viili 348, 368, 373, 379
Virulent phage, *see* Bacteriophage host resistance
Vitamins
  in dairy products 21, 26–7, 257, 420
  loss during processing 77–8, 138, 139, 321, 370
  photo-oxidative losses 69, 77
  supplementation of dairy products 69, 71, 245, 247
  synthesis in yoghurt fermentation 371
Votator 249, 250, 252, 254

Water activity (a_w) level
  cheese 332
  concentrated milk products 119
  effect on micro-organisms 149, 332, 336, 356, 381
  loss of protein quality 180
  milk 149
  milk powder 147, 149
  spreads 269
Water ice 388, 389, 419–20
Wensleydale cheese 305
Whey 159–61, 177–8
Whey cheese 309
Whey proteins 11, 12, 13, 146
  age thickening of concentrated milks 144
  complexes with casein micelles 139–40
  cooked milk flavour 80–2
  denaturation 80, 139, 145, 374
  incorporation into cheese curd 288
  interaction with kappa casein 374, 365
  thickening of canned evaporated milk 116–17
  see also Casein; Milk proteins; Proteins
Whey protein products
  concentrated whey 162–3
  ingredient use 176–7, 246, 318, 353, 304, 395
  lactalbumin, commercial 169
  lactose-hydrolyzed whey 165–6, 394, 395
  microbiology of 180–1
  nutritional properties 179–80
  whey protein concentrate 166–7
  whey protein isolate 167–9
  whey protein fractions 169–70

whey protein powder 162, 163–4
  delactosed 164
  demineralized 164–6
Whipping of cream 202–3, 211, 214–15

Xanthomonas campestris 397

Yakult products 368–9
Yarrowia 383
Yarrowia lipolytica 271
Yeasts, 120, 152, 272, 348, 394
  in cheese, 304, 305, 311, 312, 338, 340
  in fermented milks 380–1, 383
  enumeration 155, 220–1, 384
Yersinia 91

Yersinia enterocolitica 40, 44, 269, 270
  food poisoning
    milk 90–1
    milk powder 151
Yersinia frederickensii 91
Yoghurt 347, 351–4, 357, 378
  added-value ingredients 356–7, 361
  concentrated 363
  control points 355, 357, 361, 362
  dried 363
  drinking 361–2
  frozen 363–4, 391
  high pressure preservation 365
  manufacture 351–63
  nutritional properties 370–2
  nodule formation 353, 359, 360, 377
  physico-chemical changes 374–7
  see also Fermented milks